Climate Change and Food Security

CABI Climate Change Series

Climate change is a major environmental challenge to the world today, with significant threats to ecosystems, food security, water resources, and economic stability overall. In order to understand and research ways to alleviate the effects of climate change, scientists need access to information that not only provides an overview of and background to the field, but also keeps them up to date with the latest research findings.

This series addresses many topics relating to climate change, including strategies to develop sustainable systems that minimize impact on climate and/or mitigate the effects of human activity on climate change. Coverage will encompass all areas of environmental and agricultural sciences. Aimed at researchers, upper-level students, and policymakers, titles in the series provide international coverage of topics related to climate change, including both a synthesis of facts and discussions of future research perspectives and possible solutions.

Titles Available

1. Climate Change and Crop Production
   Edited by Matthew P. Reynolds

2. Crop Stress Management and Climate Change
   Edited by José L. Araus and Gustavo A. Slafer

3. Temperature Adaptation in a Changing Climate: Nature at Risk
   Edited by Kenneth Storey and Karen Tanino

4. Plant Genetic Resources and Climate Change
   Edited by Michael Jackson, Brian Ford-Lloyd and Martin Parry

5. Climate Change Impact and Adaptation in Agricultural Systems
   Edited by Jürg Fuhrer and Peter J. Gregory

6. Livestock Production and Climate Change
   Edited by Pradeep K. Malik, Raghavendra Bhatta, Junichi Takahashi, Richard A. Kohn and Cadaba S. Prasad

7. Climate Change and Insect Pests
   Edited by Christer Björkman and Pekka Niemelä

8. Climate Change and Agricultural Water Management in Developing Countries
   Edited by Chu Thai Hoanh, Vladimir Smakhtin and Robyn Johnston

9. Climate Change Challenges and Adaptations at Farm-level: Case Studies from Asia and Africa
   Edited by Naveen Prakash Singh, Cynthia. Bantilan, Kattarkandi Byjesh and Swamikannu Nedumaran

10. Climate Change Impacts on Urban Pests
    Edited by Partho Dhang

11. Climate Change Impacts and Sustainability: Ecosystems of Tanzania
    Edited by Pius.Z. Yanda, Claude G. Mung'ong'o and Edmund B. Mabhuye

12. Climate, Ticks and Disease
Edited by Pat Nuttall

13. Climate Change, Adaptation and Gender: Policy, Practice and Methodological Underpinnings
Mamta Mehar and Narayan Prasad

14. Climate Change and Sustainable Agroecology in Global Drylands
Edited by Adel El-Beltagy, Rattan Lal and Kauser Malik

15. Climate Change and Food Security
Edited by Sukanta Mondal, Pankaj Singh and Ram Lakhan Singh

Climate Change and Food Security

Edited by

Sukanta Mondal

ICAR-National Institute of Animal Nutrition and Physiology, Adugodi, Bangalore

Pankaj Singh

Dr. Rammanohar Lohia Avadh University, Ayodhya

Ram Lakhan Singh

Dr. Rammanohar Lohia Avadh University, Ayodhya

CABI

CABI is a trading name of CAB International

CABI
Nosworthy Way
Wallingford
Oxfordshire OX10 8DE
UK

Tel: +44 (0)1491 832111
E-mail: info@cabi.org
Website: www.cabi.org

CABI
200 Portland Street
Boston
MA 02114
USA

T: +1 (617)682-9015
E-mail: cabi-nao@cabi.org

A catalogue record for this book is available from the British Library, London, UK.

ISBN-13: 9781800625013 (hardback)
9781800625020 (ePDF)
9781800625037 (ePub)

DOI: 10.1079/9781800625037.0000

Commissioning Editor: Ward Cooper
Editorial Assistant: Helen Elliott
Production Editor: Rosie Hayden

Typeset by Exeter Premedia Services Pvt Ltd, Chennai, India
Printed and bound in the UK by CPI Group (UK) Ltd, Croydon, CR0 4YY

Contents

Contributors

Abhijith P.: Department of Studies in Biotechnology, University of Mysore, Mysuru-570006, Karnataka, India; Email: abhijithpadukana@gmail.com

Aditya K.A.: Department of Studies in Biotechnology, University of Mysore, Mysuru-570006, Karnataka, India; Email: adithyamysore21@gmail.com

Basade, Yasmeen: Indian Council of Agricultural Research, Fisheries Science Division, KAB II, Pusa-110012, New Delhi, India; Email: yasmeen.basade@icar.gov.in

Biswas, Bikash Kanti: Principal Scientist, DREF, W.B. University of Animal & Fishery Sciences, 37, K. B. Sarani, Kolkata – 700037, India; Email: bkbiswasmicro@rediffmail.com

Biswas, Olipriya: Assistant Professor, Department of Fishery Engineering, W.B.U.A.F.S., Kolkata & Former Food Safety Officer under Govt. of W.B., India; Email: olipriya.online16@gmail.com

Bose, Saronik: Department of Studies in Biotechnology, University of Mysore, Mysuru-570006, Karnataka, India; Email: saronikbosebiotech@gmail.com

Chakraborty, Amrita: Department of Biotechnology, Techno India University, West Bengal, EM-4, Salt Lake, Sector- V, Kolkata- 700091, West Bengal, India; Email: amritachakrabortyy@gmail.com

Chatterjee, Sirshendu: Department of Biotechnology, Techno India University, West Bengal, EM-4, Salt Lake, Sector- V, Kolkata- 700091, West Bengal, India; Email: sirshendu.chatterjee@gmail.com

Chhatwal, Himani: National Institute of Plant Genome Research (NIPGR), New Delhi, India; Email: himanichhatwal@nipgr.ac.in

Chirag, Radadiya: Department Genetics and Plant Breeding, B. A. College of Agriculture, Anand Agricultural University, Anand, India; Email: radadiyachirag111@gmail.com

Das, Sudip Kumar: Assistant Director of Research, W.B. University of Animal & Fishery Sciences, 37, K. B. Sarani, Kolkata–700037, India; Email: sudipvet@gmail.com

Dobaria, Jalpa: B.A. College of Agriculture, Anand Agricultural University, Anand, India; Email: jalpadobaria@aau.in

Geetha N.: Department of Studies in Biotechnology, University of Mysore, Mysuru-570006, Karnataka, India; Email: geetha@appbot.uni-mysore.ac.in

Ghosh, Pranabesh: School of Bio-Science, Seacom Skills University, Santiniketan, Bolpur, Birbhum-731236, West Bengal, India; Email: doctorpranabeshghosh@gmail.com

Kanaujia, Pawan Kumar: Department of Biotechnology, Allied Health Sciences, Mahayogi Gorakhnath University Gorakhpur, Uttar Pradesh, India; Email: pkkanaujia@gmail.com

Konar, Abhishek: School of Bio-Science, Seacom Skills University, Santiniketan, Bolpur, Birbhum-731236, West Bengal, India; Email: abhishek.konar60@gmail.com

Mohanty, Bimal Prasanna: Indian Council of Agricultural Research, Fisheries Science Division, KAB II, Pusa-110012, New Delhi, India; Email: bimal.mohanty@icar.gov.in

Mondal, Sukanta: ICAR-National Institute of Animal Nutrition and Physiology, Adugodi, Bangalore, India; Email: sukanta781@gmail.com

Muralidhar Y.: Dept of Vety Pharmacology & Toxicology, CVSc, Tirupati, Andhra Pradesh, India; Email: murali4357@gmail.com

Ojha, Anupama: Department of Medical Biochemistry, Allied Health Sciences, Mahayogi Gorakhnath University Gorakhpur, Uttar Pradesh, India; Email: anupama.ojha0@gmail.com

Pandey, Ashutosh: National Institute of Plant Genome Research (NIPGR), New Delhi, India; Email: ashutosh@nipgr.ac.in

Pandey, Yash: Division of Plant Breeding and Genetic Resource Conservation, CSIR—Central Institute of Medicinal and Aromatic Plants, Lucknow, India; Email: yeshjio129@gmail.com

Parihar, Akarsh: Department Genetics and Plant Breeding, B. A. College of Agriculture, Anand Agricultural University, Anand, India; Email: akarsh@aau.in

Patil, Satish S.: Department of Environmental Science, Dr. Babasaheb Ambedkar Marathwada University, Aurangabad-431004 (M.S.) India; Email: sushshrey@rediffmail.com

Ramya P.: Dept of Vety Public Health & Epidemiology, CVSc, Tirupati, Andhra Pradesh, India; Email: puttururamya@gmail.com

Singh, Pankaj: Department of Biotechnology, Dr. Rammanohar Lohia Avadh University, Ayodhya, UP, India; Email: singhpankaj0984@rediffmail.com

Singh, Ram Lakhan: Department of Biochemistry, Dr. Rammanohar Lohia Avadh University, Ayodhya, UP, India; Email: drrlsingh@rediffmail.com

Singh, Samar: National Institute of Plant Genome Research (NIPGR), New Delhi, India; Email: samarsingh@nipgr.ac.in

Singh, Saumya: Department of Biotechnology, Allied Health Sciences, Mahayogi Gorakhnath University Gorakhpur, Uttar Pradesh, India; Email: saumyasingh130502@gmail.com

Singh, Shreya: Department of Biotechnology, Allied Health Sciences, Mahayogi Gorakhnath University Gorakhpur, Uttar Pradesh, India; Email: shreyasinghsrinet7@gmail.com

Sravanthi M.: State Level Diagnostic Laboratory, SVVU, Tirupati, Andhra Pradesh, India; Email: sravanthi.mannem@gmail.com

Sudheer K.: Dept of Livestock Products Technology, CVSc, Tirupati, Andhra Pradesh, India; Email: drsudhirkorukonda@gmail.com

Swain, Himanshu Sekhar: ICAR- Central Institute of Freshwater Aquaculture, Bhubaneswar-751002, Odisha, India; Email: himanshu.swain@icar.gov.in

Yadav, Mukesh Pratap: University of Maryland School of Medicine Baltimore USA, Research Specialist lab lead, Department of Diagnostic Radiology and Nuclear Medicine, Frank C Bressler Building, 10th Floor room no-10-038, 655 West Baltimore Street, Baltimore, Maryland, USA 21201; Email: myadav@som.umaryland.edu

Preface

Our environment is currently facing immense pressure due to the burgeoning global population, limited agricultural land and water resources, low-yielding plant varieties, and the escalating demand for nutritious, safe, and affordable food. The gradual yet relentless threats posed by anthropogenic activities have further exacerbated the situation. Over the past two to three decades, we have witnessed significant alterations in our environment, which have had profound repercussions on our ecosystem and biodiversity. Climate change, in particular, has emerged as a paramount concern, given its potential to inflict severe and possibly irreversible damage on our planet's geological, biological, environmental, and ecological systems. These alterations also herald the advent of both abiotic and biotic hazards, including those that jeopardize human health. Notable changes include the production of greenhouse gases, ozone depletion, extreme weather events, biodiversity loss, and stress on food production systems. Moreover, climate change influences microbial biodiversity in soil, air, and water, while also contributing to an increase in the populations of pests and other pathogenic vectors. Food security is intrinsically linked to climate change, as it impacts livestock, crops, forestry, aquaculture, and fisheries in multifaceted ways, potentially leading to economic repercussions, diminished livelihoods, trade disruptions, and adverse health outcomes.

The aim of our book entitled *Climate Change and Food Security* is to provide invaluable literature for graduate, postgraduate, and research students engaged in the study of environmental science, plant biology, genomic science, and biotechnology on a global scale. Climate change affects the microbial population in the macroenvironment (soil, air, and water) and the populations of pests and other vectors, thereby contributing to the occurrence and severity of biotic diseases attributable to microorganisms such as fungi, bacteria, viruses, and insects. Rising temperatures, in particular, have both direct and indirect effects on animal production. Heat stress, resulting from animals' inability to dissipate environmental heat, can have detrimental impacts on health, growth, and reproduction. Furthermore, changes in the nutritional environment, such as the availability and quality of livestock feeds, pastures, and forage crops, can have indirect effects. These impacts are expected to be most pronounced in temperate regions. Climate change presents significant challenges to global food safety. Long-term alterations in temperature, humidity, rainfall patterns, and the frequency of extreme weather events are already influencing farming practices, crop production, and the nutritional quality of food crops. Elevated temperatures, fluctuating water availability, deteriorating soil quality, extreme weather events, rising sea levels, and ocean acidification, among other factors, impact food production, processing, storage, and distribution, ultimately threatening our entire food system. The sensitivity of pathogens, potentially toxin-producing microorganisms, and other pests to climatic factors suggests that climate change could affect the prevalence and intensity of certain foodborne diseases. Additionally, changing conditions may favour the establishment of invasive alien species harmful to plant and animal health. The warming of surface seawater and increased nutrient input lead to the proliferation of toxin-producing algae, causing outbreaks of seafood contamination.

Climate Change and Food Security is an outstanding collection of large-scale, high-throughput technologies, and efficient tools and methodologies to elucidate the many aspects of environmental sciences, plant and animal sciences, molecular breeding as well as genomics to accelerate the crop and livestock improvement and productivity. This book holds particular appeal for scientists, researchers, students, educators, and professionals in the fields of environmental sciences, agriculture and biotechnology.

Key features:

- Ensures a strong grasp of climate change for sustainable food security and safety through well-illustrated discussion.
- Covers updated information on food security through sustainable livestock and agricultural production in the changing climate.
- Describes food regulation, policies and social issues of food safety.

Our book *Climate Change and Food Security* consists of 12 chapters that focus on the impact of climate change on crop production as well as livestock production.

Chapter 1, "Introduction to Impact of Climate Change on Food Security and Safety" covers climate change due to abiotic and biotic factors, climate change hotspot indicators, effect of climate change on food safety, noncommunicable diseases, approaches for new sustainable food resources such as sustainable agriculture, organic farming, aquatic food system, cell based food production, regulations for food safety and security; and challenges in food safety and security.

Chapter 2, "Climate Change and Sustainable Food Security", elucidates the adverse effect of climate change, implications of climate change on food systems, agriculture, direct influence of atmospheric CO_2 and tropospheric O_3 on agriculture, direct influence of temperature and water on agriculture, livestock and pastoral system, fisheries and aquaculture systems, sustainable food security, impacts on food availability, access of food, food utilization, food stability, and overall food productivity.

Chapter 3,"Sustainable Agriculture, Food Safety and Climate Change", focuses on sustainable agriculture, integrated farming systems, methodology to organize farming systems, recommended approaches for integrated farming systems, biological and ecological processes, management practices to maximize nutrient cycling and nutrient-use efficiency, integrated pest management and foodborne diseases and malnutrition.

Chapter 4, "Improving Sustainable Agriculture and Food Security", describes climate change impacts on soil, forest, land, water resources, and crop yield, sustainable agriculture and its key principles, the main characteristics of and governmental initiatives on climate smart agriculture (CSA), policy interventions, factors that contribute to food insecurity in the US, the Indian food system and factors that contribute to food insecurity in India, improvement in next-generation plants by root structure, new transpiration modules, transpiration at elevated vapour pressure deficit, G-protein regulation, stomatal response against climate change, and microRNAs in plant drought tolerance.

Chapter 5, "Climate Neutral Agriculture", discusses the global challenges in sustainable agriculture, greenhouse gas emissions, soil degradation, water scarcity, loss of biodiversity, chemical toxicity in soil, principles for pursuing carbon-neutral agriculture, sustainable soil management, agroecological approaches, renewable energy integration into agriculture, climate smart crop breeding, sustainable livestock practices, social consequences of food access and nutritional challenges, climate change and agricultural adaptation, soil carbon and bioenergy, renewable energy production including solar, wind, ocean, nuclear, geothermal, hydrogen energy, determining growth in food demand for estimating land area, carbon neutrality and potential offsets as well as challenges and opportunities.

Chapter 6, "Hotspots Associated with Climate Change and Food Security Risks", focuses on the major hotspots of the world which face an immediate threat from climate change: the impact of climate change on food security; effects of climate change on agriculture and livestock production; the impact on forest and environmental services, fisheries, food market; measures that can be taken to ensure food security; adaptive measures for increasing crop production; managing genetic resources; forest protection and renewal; diversified and integrated fishing systems; increasing investments in research and development; and government laws and policies to reduce the risks.

Chapter 7, "Food Security Through Sustainable Livestock Production", presents the challenges of climate change and global warming with regard to common fodders cultivated for intensive dairy animal farming, competition between man and animal for grains, animal health problems, lack of adaptive animal breeds in adverse weather, and mitigation strategies for heat stress in dairy cattle in a hot and humid climate.

Chapter 8, "Aquatic Food Security: Challenges and Opportunities", deals with climate change and aquatic food security, the causes and elements of climate change, the effects of climate change, mitigations, adaptation and challenges to adaptation mechanisms.

Chapter 9, "Climate Change Impact on Human Health", covers the consequences and impact of climate change on human health, mitigating strategies to protect human health from varying climatic conditions, and action for resilience to indirect impacts.

Chapter 10, "Climate Change Impact on Food Safety", highlights sources of contamination of food, monitoring of microbial contamination in the food chain, detection of microorganisms in food samples, immunological methods, good hygienic and manufacturing practices, food spoilage, food preservation, food poisoning, impact of climate change on pathogen prevalence, prevention of foodborne infections, impacts of climate change on chemical hazards such as mycotoxins, pesticides, and heavy metals, risk of contamination of food by chemical hazards, and future challenges to food safety and security.

Chapter 11, "Policies and Social Issues of Food Safety", describes food policies, action frameworks for a food safety policy, considerations for framing or revising a food safety policy, critical need for effective food safety policies, statutory bodies for forming and implementing food safety policies, tools for addressing food safety hazards, hazard analysis critical control point, existing food safety policies, social issues of food safety, approaches for resolving social issues of food safety, and the way forward.

Chapter 12, "Future of Food Safety: the Way Forward", focuses on climate change and food security, risks to food security, quantifying the impacts on food security, uncertainties and limitations, food security, food systems and the link to climate, food security and the risks of a changing climate, the impact of climate variability on food production, and global warming and future climate.

It is a pleasure to publicly acknowledge our enormous debt to contributors who assisted materially in the preparation of this book. We are grateful to our families who cheerfully tolerated and supported many hours of absence to finish the book project. Thanks are also due to Lucy Pritchard, Rosie Hayden, Ward Cooper, Helen Elliott and the publishing team at CABI for their extra care and patience in publishing the book.

Sukanta Mondal
Pankaj Singh
Ram Lakhan Singh

1 Introduction to Impact of Climate Change on Food Security and Safety

Yash Pandey[1], Ram Lakhan Singh[2], Sukanta Mondal[3] and Pankaj Singh[4*]

[1]Division of Plant Breeding and Genetic Resource Conservation, CSIR—Central Institute of Medicinal and Aromatic Plants, Lucknow, India; [2]Department of Biochemistry, Dr. Rammanohar Lohia Avadh University, Ayodhya, India; [3]ICAR-National Institute of Animal Nutrition and Physiology, Adugodi, Bangalore, India; [4]Department of Biotechnology, Dr. Rammanohar Lohia Avadh University, Ayodhya, India

Abstract

Climate change is an abnormal variation in the average weather patterns that have both direct and indirect impacts on agricultural and livestock production systems. Climate change will increasingly cause adverse impacts on agricultural production through a decline in water supplies, increased heat, drought, and insect outbreaks, increased prevalence of pests and plant diseases. Scientists are executing various approaches to develop a sustainable food security system through development of new food sources based on plants and animals, organic farming, cell-based food production, aquatic food systems, development of integrated systems of plant and animal production practices. Although researchers are developing new food resources but facing many challenges to food safety and security like new and emerging bacteria, toxins, and antimicrobial resistance, changes in consumer preferences and habits, food contamination, food production and imported foods supply, and so on, hopefully scientists will resolve the hurdles in coming decades.

Keywords: Climate change, Food security, Sustainable food, Organic farming, Livestock

1.1 Introduction

Climate change is the significant variation of average weather conditions that affects the physical environment as well as all aspects of both natural and human systems (Abbass *et al.*, 2022). It is directly contributing to humanitarian emergencies in scale, frequency, and intensity, such as tropical storms and hurricanes, heat waves, wildfires, floods, and onset of various diseases. A recent report shows that about 3.6 billion people are currently highly susceptible to climate change, and it is expected that between 2030 and 2050 approximately 250,000 deaths per year will occur from malnutrition, malaria, diarrhea, and heat stress due to climate change alone (Lugten and Hariharan, 2022). Climate change is one of the biggest challenges in the world that is negatively impacting human health and agriculture in recent years (Caminade *et al.*, 2019). Climate change may be at local, regional, and global level, and it may be due to many natural factors, such as volcanic activity, change in reflectivity or absorption of the sun's energy, changes in the

*Corresponding author: singhpankaj0984@rediffmail.com

DOI: 10.1079/9781800625037.0001

earth's rotation, variation in solar activity, high amounts of heat-trapping greenhouse gases and many anthropogenic activities, such as deforestation, burning of coal, oil and gas fossil fuels, high use of chemical fertilizers and pesticides and transportation. The major five zones of climate change include tropical, arid, temperate, cold, and polar, as well as 30 different climate sub-groups equatorial, tropical, temperate, and polar regions (Yadav *et al.*, 2023). Climate change is affecting our planet's environment from one pole to another, including rises in sea level, increases in frequency or intensity of extreme weather events, changes in precipitation, shrinking of glaciers and sea ice, unpredictable weather, increases in flood and storms, and a rise in global temperature by about 1.8°F (1°C) from 1901 to 2020. Increased temperatures, drought, diseases, loss of biodiversity, and water stress are creating challenges for the farmers to put food on our tables. These conditions have raised a global alarm for food security and safety. Our food production, storage, and supply depend on favorable climate and weather conditions, although farmers are adapting some agricultural techniques and technologies to reduce the effect of climate change. Climate variability has had a substantial effect on agricultural production and the locations where different crops can be grown. Climate change has changed the season length and is projected to increase further expansion of the warm season, which will lead to increase in warm crops in areas where they were not previously grown.

As a result, climate change impacts food security in four dimensions directly and indirectly: availability, access, utilization, and stability (Porter *et al.*, 2014). Availability of aquatic foods and aquatic species distribution will vary through changes in habitats (Barange and Perry, 2009). Cheung *et al.* (2010) reported that marine fish availability in the tropical belt and coastal regions has been predicted to decrease substantially. The Fourth Assessment Report of the IPCC (2007) indicated that 200 million to 600 million more people could suffer from hunger by 2080 due to change in climate because they totally or partly depend on agriculture, fisheries, aquaculture, and related industries (Fischer *et al.*, 2005; Yohe *et al.*, 2007; HLPE, 2014). Climate change also has increased the incidence and prevalence of foodborne diseases. It has been found that mycotoxins and pesticide residues have been detected in samples due to climate change effects in Europe (Miraglia *et al.*, 2009). Food security is one of the major concerns associated with climate change that affects crops, livestock, forestry, fisheries, and aquaculture, which can cause adverse health impacts (Parry *et al.*, 2009; Chakrabarty, 2016; Raj *et al.*, 2022).

1.2 Climate Change

Climate change refers to long-term shifts in global or regional climate patterns, often attributed to human activities. Undoubtedly, climate change has had a significant impact on agricultural production worldwide. Its effects on soil fertility, microbial activity and diversity, carbon sequestration, plant growth, and productivity have far-reaching implications for global food security.

1.2.1 Abiotic factors

The performance of food crops concerning yield and quality is predominantly influenced by their genotype, environmental factors, and how they are cultivated. When crops are exposed to unfavorable environmental conditions, known as abiotic stresses, their productivity can be severely reduced, resulting in a loss of yield ranging from 50% to 70% (Boyer, 1982). Abiotic stresses are a common occurrence and may include cold, heat, drought, flooding, salinity, nutrient deficiency, and high and low light intensities, which may also include ultraviolet radiation.

1.2.2 Biotic factors

Biotic factors are living components of an ecosystem, and they play a crucial role in shaping the environment. The impact of climate change is jeopardizing food security on a global scale. This threat is twofold: it directly affects crop physiology and productivity, and it also indirectly affects plant-associated microbiota such as plant pathogens (Singh *et al.*, 2023). A comprehensive survey conducted worldwide on the major food crops revealed that pests, pathogens, and weeds are the primary culprits behind the average yield losses, which range from 17.2% in potatoes to as much as 30.0% in rice (Savary *et al.*, 2019). The continuous coevolution of plants and their pest and pathogens has been extensively studied. This coevolution entails enhanced defense and protection mechanisms to prevent or mitigate injuries on the plant side, and shifting infectivity and aggressiveness (or growth rate and reproduction) on the pathogen, or pesticide (Rausher, 2001; Zhan *et al.*, 2014).

1.3 Climate Change Hotspot Indicators

In the realm of scientific discourse, climate change hotspots may be examined through two distinct lenses: vulnerability and climate response. The vulnerability approach identifies hotspots as regions where the environmental or economic effects of climate change are particularly pronounced. Conversely, the climate response perspective defines hotspots as areas that display a heightened sensitivity to global changes (Giorgi, 2006).

1.4 Effect of Climate Change on Food Safety

Risks associated with food safety can be impacted by climate change and fluctuation at several points in the food chain, from production to consumption, posing significant challenges to global nutrition, agrodiversity, food productivity, and the health of livestock. This multifaceted impact demands a comprehensive understanding and proactive measures to address the evolving risks associated with climate change.

1.4.1 Nutritional value of foods

The nutritional value of foods is intricately linked to climate patterns and conditions. Changes in temperature, precipitation, and atmospheric composition influence crop yields and alter the nutritional content of various foods, such as the expected rise in the atmospheric concentration of carbon dioxide to 550 parts per million by the year 2050 (Ciais *et al.*, 2014). Research has shown that higher levels of CO_2 in atmosphere may have an adverse effect on crop nutrition, as it can lead to a decrease in mineral concentration within the seeds. In fact, studies have reported a significant reduction (averaging around 8%) in various minerals found in both edible and foliar tissues of C_3 crops under elevated CO_2 conditions. Furthermore, concentrations of protein, zinc, and iron have been found to decrease by 3–17% (Loladze, 2014; Myers *et al.*, 2014). Additionally, extreme weather events, including droughts and floods, may compromise the integrity of the soil and hinder the absorption of nutrients by plants. Plant growth and yield can be affected by insufficiencies of micronutrients, which can limit the biosynthesis or expression of crucial mechanisms of energy capture and/or metabolism. This, in turn, can have adverse effects on the nutritional quality of crop plants (Hajiboland, 2012). Climate change affects the phytochemical constituents of plants which have antioxidant and pharmacological properties (Gupta *et al.*, 2015, Singh *et al.*, 2023).

1.4.2 Loss of agro diversity

Loss of agro diversity is a significant concern associated with climate change. Agrobiodiversity, a word used to quantify biodiversity in agricultural environments, does not have a commonly agreed definition due to its complexities. Various academics and institutes offer different views (Matthies *et al.*, 2023). The impact of climate conditions on crop suitability can have far-reaching consequences, including the potential loss of agricultural biodiversity. Such losses adversely affect ecosystem resilience while also limiting the availability of diverse and nutritious foods, leading to nutritional deficiencies in communities that rely on these crops.

1.4.3 Food productivity

Food productivity is a critical aspect of global food security, and climate change poses a substantial threat to agricultural yields. Changes in the environment due to climate change have had an impact on food security, both directly and indirectly. The direct impact is seen through alterations in the conditions that affect agriculture, such as the productivity and patterns of crops. On the other hand, the indirect impact is seen through disruptions in the growth and distribution of incomes, which results in a change in the demand for agricultural products (Brown and Funk, 2008; Vermeulen *et al.*, 2012). Crop yields are more negatively affected across most of the tropical and subtropical regions as compared to the temperate regions, with crop yields in the former being generally constrained to crop varieties, sowing time, soil degradation, nutrient availability, heat stress, and water availability. By adapting agriculture to the changing climatic conditions, it is possible to reduce the effects of climate change.

1.4.4 Climate-sensitive livestock diseases

Livestock, crucial for global food systems, is susceptible to climate-sensitive diseases that can impact food safety and security. The temperature of the earth's surface has been gradually increasing since the 1800s, primarily due to the growing amount of gases like carbon dioxide and methane that trap heat in the atmosphere. This has resulted in various significant effects, including alterations in rainfall, prolonged droughts, and damage to ecosystems (Martin *et al.*, 2008). These alterations significantly affect infectious disease transmission patterns. Based on the information available, there are two broad categories of impact, which are commonly known as "direct" and "indirect." When factors such as temperature, precipitation intensity, flooding, and humidity increase, the metabolic processes, reproductive rates, and/or population densities of vectors and pathogens also increase. This leads to an increased risk of disease due to greater contact between the vector, pathogen, and host (Bett *et al.*, 2017). Indirect effects are changes in disease transmission patterns caused by climate-induced disruptions. Prolonged droughts in pastoral areas can lead to more frequent and longer animal movements, exposing them to new disease agents (Lindahl and Grace, 2015).

1.5 Effect of Climate Change on Human Health

The impact of climate change on human health is extensive and diverse, ranging from food and water insecurity to the transmission of vector-borne diseases, malnutrition, heat-related illnesses, respiratory issues, and exposure to external environmental factors. Furthermore, the increasing prevalence of non-communicable diseases is a growing concern. It is clear that the consequences of climate change on human health are significant and cannot be ignored.

1.5.1 Food, water, and vectorborne diseases

The biological systems of the planet could be impacted by climate change. This impact could extend to the proliferation and survival of pathogens as well as their transmission routes. Foodborne diseases arise from ingesting food containing toxic or infectious agents. Climate change is linked to the rise in microbial contamination, including bacteria, viruses, and pathogens, in food and water due to changes in climatic conditions like temperature and humidity. In 2010, more than 2 billion illnesses were documented, with bacteria, viruses, and parasitic protozoa contributing to 31 million disability-adjusted life years (DALYs) (Springmann *et al.*, 2016). Approximately 29% of these diseases were confirmed to spread through contaminated food. Waterborne diseases stem from pathogenic microorganisms or their toxins transmitted through water, encompassing drinking water and food exposed to polluted water. As per the World Health Organization (WHO), waterborne diseases account for about 3.6% of the global DALY burden, leading to approximately 1.5 million annual fatalities. The increased occurrence of severe weather events, such as heavy rain, flash floods, and droughts, has had a significant impact on the presence and sustainability of infectious agents in our surroundings. Predicted changes to the hydrological cycle as

Table 1.1. Some water and foodborne microbial pathogens.

Organisms	Disease	Climatic condition
Giardia duodenalis	Giardiasis	While moisture and relatively humid
Entamoeba histolytica	Amoebiasis	Humid environment and moderate temperature
Vibrio cholerae	Cholera	Monsoons can influence the concentration of water
Campylobacter spp.	Campylobacteriosis	Relatively high temperature
Dracunculus medinensis	Dracunculiasis	Temperatures between 25 and 30°C are best for larval development in freshwater
Norwalk-like Viruses	Viral gastroenteritis	Survival was measured under various temperatures (4 to 40°C) and relative humidity of 50% and 70%

a result of climate change, such as increased flooding and droughts, are expected to increase the occurrence and spread of waterborne illnesses. Prolonged hot weather can also intensify the temperature of water sources, raising the frequency of the appearance of cyanobacterial blooms, which contain harmful toxins (Thomas *et al.*, 2006; Costello *et al.*, 2009; Paerl and Huisman, 2009). Several foodborne and waterborne pathogenic bacteria that are a significant cause for concern in the context of climate change are listed in Table 1.1.

Also, the effects of climate change on the biology of vectors and transmission cycles will amplify the impact of vectorborne diseases on global human and animal health. This influence can occur directly or indirectly through socioeconomic mechanisms that impact the vulnerability of humans and animals to these diseases as well as our ability to prevent and manage those (Campbell-Lendrum *et al.*, 2015). The identification and assignment of climate change effects on malaria, much like the previous discourse on predicting climate change impacts, have been subjects of controversy. It is widely acknowledged that the existing geographic spread of malaria is heavily influenced by historical and ongoing initiatives to control both vectors and transmission, in addition to climate factors (Reiter *et al.*, 2004).

1.5.2 Health issues due to unhealthy diets, malnutrition and *Koishshikore*

The impact of climate change on human health extends to nutrition and diet-related concerns. At a global scale, climate change has adverse effects on key health determinants, such as agriculture and water access. These impacts disproportionately affect populations that may bear minimal responsibility for the issue, thereby intensifying challenges related to social and economic inequality among countries (Watts *et al.*, 2019). Around 45.4 million children below the age of 5 went through wasting due to acute food shortages and/or infections in the year 2020. This state is linked to a weakened immune system, a higher probability of catching infectious diseases, and death (Victora *et al.*, 2021). Hidden hunger can occur even in the presence of an energy-rich diet. An obese child may suffer from micronutrient deficiencies, as modern diets are high in calories but lacking in nutrients (Vishwakarma *et al.*, 2013; Singh *et al.*, 2014). The impact of climate change on the agroecosystem has the potential to affect food production and nutritional security. Changes in the quantity, quality, accessibility, and affordability of food from agricultural, fishery, and livestock sectors can worsen nutrient deficiencies and chronic malnutrition, thus increasing vulnerability among the most food-insecure population groups (Swinburn *et al.*, 2019; Watts *et al.*, 2019). Projections suggest that under high-emission climate scenarios by the year 2100, climate change is anticipated to lead to a decrease in crop yields, with estimated declines of 20% to 45% for maize, 5% to 50% for wheat, 20% to 30% for rice, and 30% to 60% for soybeans. Increased levels of carbon dioxide (CO_2) have been noted to lower protein concentrations in wheat, barley, rice, and potato crops (Uleberg *et al.*, 2014; Gitz *et al.*, 2024).

In specific regions, such as the Sundarbans delta in Bangladesh, climate change is associated with a unique health concern known as *Koishshikore* or "heat syndrome." This condition, characterized by symptoms like weakness, dizziness, and dehydration, is attributed to the rising temperatures in the region. The increasing frequency and intensity of heat waves associated with climate change amplify the risks of heat-related illnesses, including heatstroke and heat exhaustion.

1.5.3 Heat-related illness

Extended exposure to elevated temperatures can lead to heat-related ailments, encompassing conditions such as heat cramps, heat syncope, heat exhaustion, heatstroke, and, in severe cases, fatality. Among these, heat exhaustion stands out as the most prevalent heat-related illness (Lugo-Amador *et al.*, 2004). Individuals may experience signs and symptoms such as extreme thirst, profuse sweating, feebleness, pallor, unease, nervousness, lightheadedness, exhaustion, fainting, sickness or retching, and a headache. Anticipated changes in global climate patterns indicate that all areas of the world will experience increased temperatures in the coming years. The average global temperature is projected to rise within the range of 1.8–4.0°C by the year 2100. Specifically, in densely populated tropical regions where maximum temperatures already surpass 40°C, there could be an increase of 1–3°C by 2020 and a more substantial rise of 3–5°C by 2080 (IPCC, 2007). The "heat island effect" is expected to contribute an additional 4–5°C (Oke, 1973) of heat exposure in numerous urban areas, heightening the risk of heat-related issues like heatstroke within cities.

1.5.4 Respiratory illness

Climate change has a substantial impact on respiratory health. The prevalence of various infectious diseases, such as malaria, tickborne illnesses, and those transmitted through food and water, is closely tied to the local environment. Climate shifts have the potential to influence respiratory infections by directly or indirectly impacting vectors and host immune responses (Dobson, 2009). Some respiratory infections are already experiencing effects from changing weather patterns. The prevalence and geographic distribution of seasonal conditions may be impacted by climate change, but the effects are often unpredictable. In the United States, pneumonia and influenza are among the top ten leading causes of death (Johnson *et al.*, 2014). In an investigation spanning 3 years and involving 11,614 cases, the incidence of invasive pneumococcal disease exhibited a seasonal pattern inversely linked to ambient temperature (Dowell *et al.*, 2003). Within tropical and subtropical regions of Asia and Africa, pneumonia demonstrates elevated prevalence and mortality rates during the rainy season, underscoring the correlation between pneumonia patterns and temperature and precipitation (Paynter *et al.*, 2010). Notably, childhood pneumonia exhibited a notable association with the variable termed "temperature change between two neighboring days" (TCN), measured as the mean temperature difference from one day to the next. This health effect persisted for a duration of up to 3 weeks (Xu *et al.*, 2014). The diurnal temperature range (DTR), representing the maximum daily temperature minus the minimum daily temperature, serves as an additional metric for assessing the impacts of global climate change.

The activity of viruses can be influenced by changes in temperature, precipitation, relative humidity, and air pollution, which may also contribute to the severity and scale of epidemics. For instance, in Los Santos, Panama, a respiratory disease outbreak caused by Hantavirus was reported in late 1999 and early 2000. Temperature change has a significant impact on the spread of influenza; a study conducted by Towers and colleagues showed that following warm winters, there tends to be an early onset of severe influenza incidence in the following season (Towers *et al.*, 2013).

1.5.5 External exposome promoted respiratory disorders

Recent research into the genetic origins of diseases has illuminated the fact that non-genetic influences play a far more substantial role than previously believed. These non-genetic factors can contribute up to 90% of the risk of

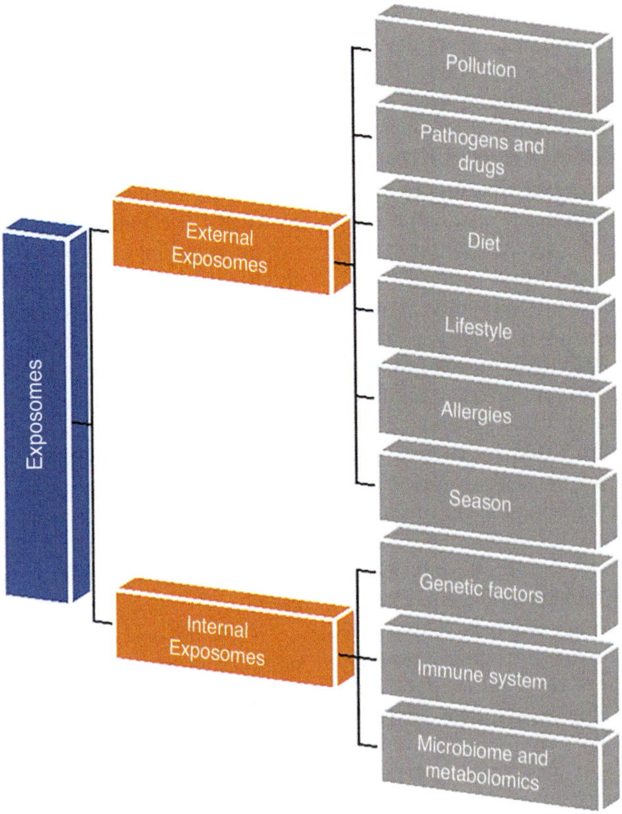

Fig. 1.1. External and internal exposome parameters which are related to respiratory disorders.

chronic illnesses. As a result, exposome science has emerged as a new field of study that aims to understand all the various exposures that individuals encounter from the moment of conception onward (Wild, 2005). A more specific proposed definition of the exposome is the cumulative effects of environmental exposure and the associated biological response (Miller and Jones, 2014). Some internal and external exposers of the exposome are shown in Fig. 1.1.

External exposome refers to the cumulative environmental exposures an individual experiences throughout their life. Climate changes can affect the composition of the exposome, influencing respiratory health and promoting respiratory disorders. Increased exposure to air pollutants and allergens due to climate-related events can contribute to the development and aggravation of respiratory conditions. There is now enough proof to suggest that various air pollutants present indoors and outdoors are not just worsening but also instigating an increase in allergic diseases. Exposure to fine particles that are less than 2.5 mm in diameter (PM2.5), O_3 or NO_2 consistently has been linked to heightened asthma exacerbations, greater medication usage, and aggravation of atopic dermatitis (Ahn, 2014; Guarnieri and Balmes, 2014).

1.5.6 Non-communicable diseases

The impact of climate change also extends to non-communicable diseases (NCDs). NCDs are typically long-term conditions and can include both acute illnesses that frequently occur due to exposure to climate conditions and acute exacerbations or complications of chronic illnesses. The prevalent non-communicable diseases (NCDs), including heart disease, stroke, cancer, type 2 diabetes, and respiratory diseases, contribute to 60% of the total annual global deaths (58 million) and constitute 46% of the global burden of disease measured in DALYs (WHO and Public Health Agency of Canada, 2005). The risk of cardiovascular disease (CVD) may be amplified by climate change, which can occur through three main exposure pathways: directly through air pollution and through extreme temperatures, and indirectly through changes to dietary options (Friel *et al.*, 2011). The contamination of aflatoxin may result in an increase in liver cancer as the climate warms and weather patterns become more erratic. Cereals and pulses are contaminated by these fungal metabolites, and dry, hot conditions during crop growth can increase the risk of contamination, while the harvesting and storage of mature crops are susceptible to contamination during wet conditions (Cotty and Jaime-Garcia, 2007).

A comprehensive approach is required to address the health impacts of climate change. This involves measures to enhance food and water security, developing adaptive strategies to combat vectorborne diseases, promoting sustainable agricultural practices, and implementing policies to reduce greenhouse gas emissions. Further, public health interventions must prioritize raising awareness, improving healthcare infrastructure, and strengthening community resilience to climate-related health risks. To summarize, the complex relationship between climate change and human health highlights the importance of tackling this global challenge urgently. Climate change affects health in various ways, from food and water safety to the spread of vectorborne diseases, and the consequences are interconnected. Therefore, implementing comprehensive strategies to mitigate these risks is crucial to safeguard the well-being of current and future generations.

1.6 Approaches for New Sustainable Food Resources

Sustainable food security (SFS) is a food system that delivers nutritious food availability for all in such a way that the economic, social, and environmental bases are not compromised (Vågsholm *et al.*, 2020). The development of a sustainable food system is the main goal of the United Nations' Sustainable Development Goals (SDGs). SDGs were adopted in 2015 and were focused on all required major transformations in agriculture and food systems in order to fulfill food requirement and improve nutrition by 2030. The aim of the SDGs development is not only limited to local levels but also aimed at the development of SDGs at national, regional, and global levels. However, development of the global food system is complex and has systemic challenges that require highly productive and environmentally sustainable crops, more disease-resistant plants and livestock to deliver healthy and nutritious diets to all. Food security is associated with sustainability in which agricultural practices have a high impact on the interrelationships between food security and sustainability (Lin *et al.*, 2020; Guiné *et al.*, 2021).

Helland and Sorbo (2014) reported that sustainable food security will need to include four major parameters:

1. Sufficient food production and availability of food
2. Access for all to purchase food
3. Nutritious food having high energy, proteins, and micronutrients as well as safety
4. Stability of food system

1.6.1 Sustainable agriculture

Sustainable agriculture is an integrated system of plant and animal production practices that fulfill human food and fiber needs, develop a greener environmental system for quality of life for farmers and society. The primary goals to develop the sustainable agriculture include more profitable farming, moderate consumption of non-renewable resources, protection and maintenance of soil quality with moderate use of synthetic fertilizers and pesticides, and the minimization of the adverse impact on safety, wildlife, water quality, and environmental resources. Food and Agriculture Organization of the United Nations (FAO) statement indicates that worldwide food production will increase in upcoming years (FAO, 2011).

Sustainable agriculture approaches include regular crop rotation which provides nitrogen to N-fixing plants, assists in pest control, reduces soil depletion, mitigates farming risks, and increases soil biota for nitrogen fixation; use of advanced irrigation technologies; use of green manure; avoidance of regular plowing in conventional farming; and employing integrated pest management (IPM), integrated weed management (IWM) and permaculture technique (Muhie, 2022). Other important approaches for sustainable agriculture include polyculture, integration of livestock and crops, biodynamic and organic farming, using biofuels and zero-emissions transport, conservation tillage, sustainable agriculture intercropping, and crop residue management (CRM). The selection of suitable sustainable agriculture practices depends on the crop type and farm specifics. However, holistic management of entire systems and landscapes, where all aspects are interconnected and equally valued, remains the ideal method.

1.6.2 Organic farming

As defined by the United States Department of Agriculture (USDA) study group, "organic farming is a form of farming which avoids the use of synthetic chemicals and utilizes biological fertilizers, such as manures, vermin-compost, green leaves, crop rotation, biofertilizers, animal manure, and organic waste, which promote the crop yield and nutrition in products. Organic farming gives us beneficial, energy-rich products, and it is less harmful impact on the ecosystem and improves the ecological balance" (Kumar *et al.*, 2022). There are two types of organic farming: pure organic farming and integrated organic farming. In pure organic farming, we use natural ingredients such as manure, biopesticides, and green leaves; in integrated farming, we use the amalgamate approach (Rana, 2011). The major principle of organic farming includes the principle of health, the principle of ecology, the principle of fairness, and the principle of care (Sankar and Reddy, 2022).

Organic farming requires no use of expensive fertilizers, leading to a good return on investment. With the use of cheaper, local inputs and high demand, organic products are more nutritional, tasty, and good for health and are environmentally friendly. Use of high amounts of chemical fertilizer reduces soil fertility and also causes water and soil pollution, which then enters into the food chain, leading to acute and chronic diseases. In India, the government has promoted organic farming by implementing different programs like Farming Paramparagat Krishi Vikas Yojana, Rashtriya Krishi Vikas Yojana (RKVY), One District One Product, National Mission on Oilseed and Oil Palm, and the Organic Farming Action Program. The main challenges in organic farming include a shortage of biomass, inequality of stock, time-consuming process, and the absence of an adequate architect.

1.6.3 Aquatic food system

Aquatic foods provide food and nutritional security to more than 800 million people in the world, but they are very sensitive to climate change, poverty, conflict, and humanitarian emergencies (Cheung *et al.*, 2010; Tigchelaar *et al.*, 2021). They include animals (finfish, shellfish, aquatic feed like single-cell organisms, synthetic substitutes like animal cell-based alternative food proteins) and plants (aquatic plants as well as algae) grown in or harvested from water. Aquatic foods have a high content of vitamins, minerals, and healthy fats and are beneficial to people living in developing countries. Almost three-quarters of the surface of the earth is covered with water. Marine biotechnology plays an important role in the production of new food-producing animal species. For example, genetically modified salmon can be developed by the use of recombinant DNA technology to grow 5–10 times faster than wild-type salmon. Our research on aquatic food systems will establish inclusive, efficient, sustainable, nutritious, and healthy food to achieve the SDGs by 2030 (Lala-Pritchard and Johnstone, 2022).

1.6.4 Cell-based food production

Cellular production of food products such as proteins, fats, carbohydrates, and tissues through a combination of biotechnology, tissue engineering, molecular biology, and synthetic biology is a new method to overcome traditional agriculture dependency. Cell-cultivated foods based on yeast, microalgae, filamentous fungi, animal cells, and plant cells have several advantages, such as high nutritional characteristics, including vitamins, minerals, pigments, fiber, and other nutrients, as well as therapeutic value for humans. An early report on consumption of yeast as a "single-cell protein," in massive quantities, was reported in Germany during World Wars I (*Saccharomyces cerevisiae*) and II (*Candida utilis*) in place of imported protein (Joseph and Bachhawat, 2014). Recently, Nyyssölä *et al.* (2022) reported that commercial products like Marmite®, Cenovis®, and Vegemite®, based on *S. cerevisiae* yeast biomass, are available in the market for utilization as food products. It also found that microalgae supplements in the form of capsules and powder are currently found in the market as an ingredient for many food products such as baked goods, drinks, and candies (Wells *et al.*, 2017). Stoffel *et al.* (2021) reported the use of fungal biomass in chocolate cookies, replacing wheat flour with *Pleurotus albidus* mycoprotein flour. Animal cell-based food products include applications to alternative meat products like whole-cut *in vitro* steak and pork (Kang *et al.*, 2021; Handral *et al.*, 2022).

1.7 Regulations for Food Safety and Security

A food safety and security policy is essential for food safety management to ensure the food is free from foodborne pathogens and toxicants, uphold hygiene standards, and ensure availability of safe and nutritious food. The food safety and security policies also contribute to improved livelihoods and consumer welfare and support local economic growth in competitive edge businesses. The Food and Agriculture Organization (FAO) also recommends developing good practices at all stages of food production to maintain food safety and security (Ndondo, 2023). FAO strategic framework for the transformation of agri-food systems is based on 4Bs namely, better production, better nutrition, better environment, and a better life (FAO, 2015). In all countries, food safety and security is governed by strict laws and regulations to ensure food safety and quality. The term "food law" covers the legislation that regulates food production, quality, and safety and controls food trade across the entire food chain. Food safety policies differ from country to country due to various factors such as differences in food safety hazards, food testing and interpretation of food safety standards, level of food safety capacity, knowledge and responsibilities of regulatory agencies, technical skills, historical outbreaks, and tradition.

1.7.1 International organizations for food law and regulations

1.7.1.1 World Health Organization (WHO) and Food and Agriculture Organization (FAO)

FAO and WHO provide scientific advice on various aspects of food quality, safety, and nutrition covered by Codex Alimentarius Commission. They provide independent scientific expert advice to the Codex Alimentarius Commission to ensure safe, good food for everyone, everywhere.

1.7.1.2 Food and Drug Administration (FDA)

The Food and Drug Administration (FDA) is an agency of the United States Department of Health and Human Services, and it is responsible for protecting and promoting public health through regulation and supervision of food safety of nutritional supplements, production of pharmaceutical drugs, vaccines, biopharmaceuticals, medical devices, veterinary products, and cosmetics. It approves food products and ingredients before they are used by consumers in the food, which includes food additives, food colors, etc.

1.7.1.3 International Organization for Standardization (ISO)

ISO is a standard dealing with food safety developed by the International Organization for Standardization. It is the most effective food safety system established for food hazard analysis and has 22,000 families of standards.

1.7.1.4 *Codex Alimentarius Commission (CAC)*

CAC was established in 1963 by FAO and WHO to develop food standards, guidelines, and codes of practice to maintain consumer's health and fair food trade and to increase coordination among all food standards at international governmental and non-governmental organizations.

1.7.1.5 *Sanitary and Phytosanitary Measures (SPS)*

SPS governs how governments can apply food safety and animal and plant health measures.

1.7.1.6 *FAOLEX*

FAOLEX is one of the world's largest electronic collections of legislative database on food, agriculture, and renewable natural resources.

1.7.1.7 *International Plant Protection Convention (IPPC)*

IPPC was established in 1952 to protect cultivated and wild plants.

1.7.1.8 *World Organization for Animal Health (OIE)*

OIE is an intergovernmental organization recognized by the World Trade Organization (WTO) and had a total of 178 Member Countries in 2013. This organization is responsible for improving animal health worldwide.

1.7.1.9 *Organization for Economic Co-operation and Development (OECD) for Agriculture and Fisheries*

The mission of OECD is to promote policies that will improve the economy by introducing new plant varieties for high production and develop new aquatic food resource for well-being of people around the world.

1.7.1.10 *International Trade Centre's Standards Map*

Standards Map provides comprehensive and transparent information on food quality and safety to strengthen the producers and exporters to participate in more sustainable production and trade.

1.7.2 National organizations for food law and regulations

1.7.2.1 *Food Safety and Standards Authority of India (FSSAI)*

The main agency for food safety regulation in India is the Food Safety and Standards Authority of India (FSSAI). It is an autonomous body that works under the Ministry of Health and Family Welfare, Government of India. It was established under Food Safety and Standards, 2006, for setting standards for food, regulating their manufacture, storage, distribution, sale, and import. The headquarters of FSSAI is in New Delhi and also has six regional offices in Delhi, Guwahati, Mumbai, Kolkata, Cochin, and Chennai.

1.7.2.2 *AGMARK*

AGMARK is a certification mark used for agricultural products in India, and standards are approved by the Directorate of Marketing and Inspection, falling under the Department of Agriculture, Cooperation and Farmers Welfare under the Ministry of Agricultural and Farmers Welfare, Government of India. The term AGMARK is made up of "AG," referring to agriculture, and "mark" for a certification mark. The main objective of this organization is to provide unadulterated food products to consumers with high nutritious quality. The head office of AGMARK is in Faridabad (Haryana) and has one Central AGMARK Laboratory (CAL) in Nagpur and 11 Regional AGMARK Laboratories (RALs) in Mumbai, New Delhi, Chennai, Kolkata, Kanpur, Kochi, Guntur, Amritsar, Jaipur, Rajkot, and Bhopal. AGMARK covers quality guidelines for 224 different commodities that include pulses, cereals, essential oils, vegetable oils, fruits, and vegetables.

1.7.2.3 *Food Products Order (FPO)*

FPO mark is used for food products ensuring that the product is fit for consumption and has been effective since 1955. The list of food products that requires FPO mark includes syrups, synthetic beverages, vinegar, pickles, squashes, barley water, beverages containing fruit juice, jellies, jams, tomato products, preserved and crystallized fruits, chutneys, and frozen vegetables and fruits. FPO is controlled by the FSSAI.

1.7.2.4 Federation of Indian Chambers of Commerce and Industry (FICCI)

FICCI is an allied body of Indian Food Trade and Industry and was established in 1927. FICCI is the largest and oldest food business-controlling organization in India. FICCI provides a platform for networking between different sectors of Indian industry and policymakers and the international business community. The headquarters of FICCI are in New Delhi.

1.8 Challenges to Food Safety and Security

Currently, we are facing enormous critical global concern regarding food safety, as we are aware that different types of food contaminants such as common microbes, toxins, poisonous mushrooms, pesticides, pollutants, and other substances can cause various illnesses, including mild to chronic gastroenteritis, as well as severe diseases such as hepatitis A or cholera. Natural disasters and extreme events are also impacting our agricultural productivity in several ways such as destruction of agricultural infrastructure, loss of animals and crops, environmental contamination with pathogens, and negative health outcomes (Wu *et al.*, 2017; Andrade *et al.*, 2018; MacLachlan *et al.*, 2018).

The main challenges in food safety include the emergence of new bacteria; contamination of food through unidentified toxins at any point during production, processing, and distribution; development of antimicrobial resistance; unsafe food storage and labeling; preparation of food by untrained or unwell employees; changes in consumer preferences; lack of food testing laboratories that can diagnose foodborne illness and food allergies; improper sanitation; and unsafe food holding. These challenges are of major concern in developing and undeveloped countries and creating pressure on food producers to ensure food safety. It is estimated that at the end of the year 2050, food production will need to increase by more than 50% as compared to current food production (Garcia *et al.*, 2020).

In India, inefficient food supply chains, storage, transport facilities, financing, population, poverty, unsafe practices, illiteracy, gender bias, and unemployment are big issues related to food safety. The main threats related to food security include an exponential increase in world population and food demand, high food prices, loss of biodiversity, shrinkage of agricultural field, lack of high-yield plant species, water scarcity, underdeveloped technological approach toward agriculture, lack of education, corruption, and increase in food losses and food waste (Tilman *et al.*, 2011). The objective of food safety and security is to provide nutritious food to increasing populations while reducing foodborne illnesses or health issues related to food consumption. A multitude of approaches can be used to improve food safety and security, like combating against climate change; developing sustainable agriculture, new food resource and regulatory frameworks; reducing food waste and food loss; improving infrastructure; engagement of the private sector in local, regional, and global partnerships; promoting fair trading practices; reducing the yield gap and addressing the indirect causes of food insecurity.

1.9 Conclusions

Food security is a measure of people's access to sufficient, safe, and nutritious food that fulfills their nutritional requirements for an active and healthy life. Climate change is causing more frequent and extreme weather conditions like heat waves, droughts, and floods, which has caused irreversible damage to food security. Global warming due to climate change progressively weakens soil health and the ecosystem, causes biodiversity loss, leads to an increase in diseases, and it reduces aquatic as well as land animal biomass and plants' productivity, leading to malnutrition and micronutrient deficiencies in many regions.

Food security scholars have recently started to integrate the impact of climate change on sustainable food security. As a result, knowledge gaps between the impacts of climate change and food systems are reducing day by day. The main strategy of food security is to develop a "sustainable, nature-positive, more resilient, environmentally friendly and healthier affordable food system" which includes developing new food resources like plant-based alternatives, organic farming, aquatic food system, integrated system of plant and animal production practices, sea animals, seaweeds and cell-based food production.

References

Abbass, K., Qasim, M.Z., Song, H., Murshed, M., Mahmood, H. *et al.* (2022) A review of the global climate change impacts, adaptation, and sustainable mitigation measures. *Environmental Science and Pollution Research International* 29(28), 42539–42559. DOI: 10.1007/s11356-022-19718-6.

Ahn, K. (2014) The role of air pollutants in atopic dermatitis. *The Journal of Allergy and Clinical Immunology* 134(5), 993–999. DOI: 10.1016/j.jaci.2014.09.023.

Andrade, L., O'Dwyer, J., O'Neill, E. and Hynds, P. (2018) Surface water flooding, groundwater contamination, and enteric disease in developed countries: a scoping review of connections and consequences. *Environmental Pollution (Barking, Essex)* 236, 540–549. DOI: 10.1016/j.envpol.2018.01.104.

Barange, M. and Perry, R.I. (2009) Physical and ecological impacts of climate change relevant to marine and inland capture fisheries and aquaculture. In: Cochrane, K., De, C., Soto, D. and Bahri, T. (eds) *Climate Change Implications for Fisheries and Aquaculture, Overview of Current Scientific Knowledge. FAO Fisheries and Aquaculture Technical Paper No.530*. FAO, Rome, Italy, p. 212.

Bett, B., Kiunga, P., Gachohi, J., Sindato, C., Mbotha, D. *et al.* (2017) Effects of climate change on the occurrence and distribution of livestock diseases. *Preventive Veterinary Medicine* 137(Pt B), 119–129. DOI: 10.1016/j.prevetmed.2016.11.019.

Boyer, J.S. (1982) Plant productivity and environment. *Science* 218(4571), 443–448. DOI: 10.1126/science.218.4571.443.

Brown, M.E. and Funk, C.C. (2008) Food security under climate change. *Science* 319(5863), 580–581. DOI: 10.1126/science.1154102.

Caminade, C., McIntyre, K.M. and Jones, A.E. (2019) Impact of recent and future climate change on vector-borne diseases. *Annals of the New York Academy of Sciences* 1436(1), 157–173. DOI: 10.1111/nyas.13950.

Campbell-Lendrum, D., Manga, L., Bagayoko, M. and Sommerfeld, J. (2015) Climate change and vector-borne diseases: what are the implications for public health research and policy? *Philosophical Transactions of the Royal Society of London. Series B, Biological Sciences* 370(1665), 20130552. DOI: 10.1098/rstb.2013.0552.

Chakrabarty, M. (2016) *Climate Change and Food Security in India*. ORF Issue Brief no.157, Observer Research Foundation Delhi, India and the Peace Research Institute, Oslo, Norway.

Cheung, W.W.L., Lam, V.W.Y., Sarmiento, J.L., Kearney, K., Watson, R. *et al.* (2010) Large-scale redistribution of maximum fisheries catch potential in the global ocean under climate change. *Global Change Biology* 16(1), 24–35. DOI: 10.1111/j.1365-2486.2009.01995.x.

Ciais, P., Sabine, C., Bala, G., Bopp, L., Brovkin, V. *et al.* (2014) Carbon and other biogeochemical cycles. In: Stocker, T.F., Qin, D., Plattner, G.K., Tignor, M. and Allen, S.K. (eds) *Climate Change 2013, the Physical Science Basis. Contribution of Working Group I to the Fifth Assessment Report of the Intergovernmental Panel on Climate Change*. Cambridge University Press, Cambridge, UK, and New York, pp. 465–570. DOI: 10.1017/CBO9781107415324.015.

Costello, A., Abbas, M., Allen, A., Ball, S., Bell, S. *et al.* (2009) Managing the health effects of climate change, lancet and University College London Institute for Global Health Commission. *Lancet* 373(9676), 1693–1733. DOI: 10.1017/CBO9781107415324.015.

Cotty, P.J. and Jaime-Garcia, R. (2007) Influences of climate on aflatoxin producing fungi and aflatoxin contamination. *International Journal of Food Microbiology* 119(1–2), 109–115. DOI: 10.1016/j.ijfoodmicro.2007.07.060.

Dobson, A. (2009) Climate variability, global change, immunity, and the dynamics of infectious diseases. *Ecology* 90(4), 920–927. DOI: 10.1890/08-0736.1.

Dowell, S.F., Whitney, C.G., Wright, C., Rose, C.E. and Schuchat, A. (2003) Seasonal patterns of invasive pneumococcal disease. *Emerging Infectious Diseases* 9(5), 573–579. DOI: 10.3201/eid0905.020556.

FAO (2011) *The State of the World's Land and Water Resources for Food and Agriculture (SOLAW)—Managing Systems At Risk*. Food and Agriculture Organization of the United Nations, Rome, Italy.

FAO (2015) *Report of the Conference of FAO*. Food and Agriculture Organization of the United Nations, Rome, Italy.

Fischer, G., Shah, M., N. Tubiello, F. and van Velhuizen, H. (2005) Socio-economic and climate change impacts on agriculture: an integrated assessment, 1990–2080. *Philosophical Transactions of the Royal Society B* 360(1463), 2067–2083. DOI: 10.1098/rstb.2005.1744.

Friel, S., Bowen, K., Campbell-Lendrum, D., Frumkin, H., McMichael, A.J. *et al.* (2011) Climate change, noncommunicable diseases, and development: the relationships and common policy opportunities. *Annual Review of Public Health* 32, 133–147. DOI: 10.1146/annurev-publhealth-071910-140612.

Garcia, S.N., Osburn, B.I. and Jay-Russell, M.T. (2020) One health for food safety, food security, and sustainable food production. *Frontiers in Sustainable Food Systems* 4, 1. DOI: 10.3389/fsufs.2020.00001.

Giorgi, F. (2006) Climate change hot-spots. *Geophysical Research Letters* 33(8), L08707. DOI: 10.1029/2006GL025734.

Gitz, V., Meybeck, A., Lipper, L., Young, C.D. and Braatz, S. (2024) Climate change and food security, risks and responses. *Food and Agriculture Organization of the United Nations (FAO). Report* 10(2). DOI: 10.1029/2006GL025734.

Guarnieri, M. and Balmes, J.R. (2014) Outdoor air pollution and asthma. *The Lancet* 383(9928), 1581–1592. DOI: 10.1016/S0140-6736(14)60617-6.

Guiné, R.P.F., Pato, M.L.J., Costa, C.A.D., Costa, D.V.T.A.D., Silva, P.B.C.D. *et al.* (2021) Food security and sustainability: discussing the four pillars to encompass other dimensions. *Foods* 10(11), 2732. DOI: 10.3390/foods10112732.

Gupta, R., Singh, R.L. and Singh, P. (2015) Quantification of phytochemicals and evaluation of antioxidant potential of ethanolic leaf extract of terminalia bellerica, terminalia chebula and emblica offinalis vis-a-vis triphala. *International Journal of Pharmaceutical Sciences Review and Research* 32(2), 14–22.

Hajiboland, R. (2012) Effect of micronutrient deficiencies on plants stress responses. In: Ahmad, P. and Prasad, M. (eds) *Abiotic Stress Responses in Plants*. Springer, New York, pp. 283–329. DOI: 10.1007/978-1-4614-0634-1_16.

Handral, H., Hua Tay, S., Wan Chan, W. and Choudhury, D. (2022) 3D printing of cultured meat products. *Critical Reviews in Food Science and Nutrition* 62(1), 272–281. DOI: 10.1080/10408398.2020.1815172.

Helland, J. and Sorbo, G.M. (2014) *Food Security and Social Conflict*. CMI Report 2014:1. Christian Michelssen Institute, Bergen, Norway. DOI: 10.1016/B978-0-12-384730-0.00361-X.

HLPE (2014) *Sustainable Fisheries and Aquaculture for Food Security and Nutrition*. A report by the High Level Panel of Experts on Food Security and Nutrition of the Committee on World Food Security, Rome, Italy. DOI: 10.1016/B978-0-12-384730-0.00361-X.

IPCC (2007) The physical science basis. In: Solomon, S., Qin, D., Manning, M., Chen, Z. and Marquis, M. (eds) *Contribution of Working Group I to the Fourth Assessment Report of the Intergovernmental Panel on Climate Change*. Cambridge University Press, Cambridge, UK, and New York, pp. 113–119. DOI: 10.1016/B978-0-12-384730-0.00361-X.

Johnson, N.B., Hayes, L.D., Brown, K., Hoo, E.C. and Ethier, K.A. (2014) CDC national health report, leading causes of morbidity and mortality and associated behavioral risk and protective factors—united states, 2005–2013. *MMWR* Suppl 63(4), 3–27. DOI: 10.1016/B978-0-12-384730-0.00361-X.

Joseph, R. and Bachhawat, A.K. (2014) Yeasts, production and commercial uses. In: Batt, C.A. and Tortorello, M.L. (eds) *Encyclopedia of Food Microbiology*, 2nd edn. Academic Press, Oxford, UK, pp. 823–830. DOI: 10.1016/B978-0-12-384730-0.00361-X.

Kang, D.H., Louis, F., Liu, H., Shimoda, H., Nishiyama, Y. *et al.* (2021) Engineered whole cut meat-like tissue by the assembly of cell fibers using tendon-gel integrated bioprinting. *Nature Communications* 12(1), 5059. DOI: 10.1038/s41467-021-25236-9.

Kumar, R., Rana, N., Kaur, M., Bhowmik, S., Kumar, M. *et al.* (2022) Organic farming status in india: a review. *The Pharma Innovation Journal* 11(12), 2964–2671.

Lala-Pritchard, T. and Johnstone, G. (2022) *2030 Research and Innovation Strategy: Aquatic Foods for Healthy People and Planet*. Strategy: 2020-38, Penang, Malaysia.

Lin, J., Frank, M. and Reid, D. (2020) No home without hormones: how plant hormones control legume nodule organogenesis. *Plant Communications* 1(5), 100104. DOI: 10.1016/j.xplc.2020.100104.

Lindahl, J.F. and Grace, D. (2015) The consequences of human actions on risks for infectious diseases: a review. *Infection Ecology & Epidemiology* 5(1), 30048. DOI: 10.3402/iee.v5.30048.

Loladze, I. (2014) Hidden shift of the ionome of plants exposed to elevated CO_2 depletes minerals at the base of human nutrition. *eLife* 3, e02245. DOI: 10.7554/eLife.02245.

Lugo-Amador, N.M., Rothenhaus, T. and Moyer, P. (2004) Heat-related illness. *Emergency Medicine Clinics of North America* 22(2), 315–327, . DOI: 10.1016/j.emc.2004.01.004.

Lugten, E. and Hariharan, N. (2022) Strengthening health systems for climate adaptation and health security: key considerations for policy and programming. *Health Security* 20(5), 435–439. DOI: 10.1089/hs.2022.0050.

MacLachlan, M., Ramos, S., Hungerford, A. and Edwards, S. (2018) Federal natural disaster assistance programs for livestock producers, 2008–16. *US Department of Agriculture, Economic Research Service*. DOI: 10.22004/ag.econ.276251.

Martin, V., Chevalier, V., Ceccato, P., Anyamba, A., De Simone, L. *et al.* (2008) The impact of climate change on the epidemiology and control of Rift Valley fever. *Revue Scientifique et Technique* 27(2), 413–426.

Matthies, A.E., Fayet, C.M.J., O'Connor, L.M.J. and Verburg, P.H. (2023) Mapping agrobiodiversity in Europe: different indicators, different priority areas. *Ecological Indicators* 154, 110744. DOI: 10.1016/j.ecolind.2023.110744.

Miller, G.W. and Jones, D.P. (2014) The nature of nurture: refining the definition of the exposome. *Toxicological Sciences* 137(1), 1–2. DOI: 10.1093/toxsci/kft251.

Miraglia, M., Marvin, H.J.P., Kleter, G.A., Battilani, P., Brera, C. *et al.* (2009) Climate change and food safety: an emerging issue with special focus on Europe. *Food and Chemical Toxicology* 47(5), 1009–1021. DOI: 10.1016/j.fct.2009.02.005.

Muhie, S.H. (2022) Novel approaches and practices to sustainable agriculture. *Journal of Agriculture and Food Research* 10, 100446. DOI: 10.1016/j.jafr.2022.100446.

Myers, S.S., Zanobetti, A., Kloog, I., Huybers, P., Leakey, A.D. *et al.* (2014) Increasing CO_2 threatens human nutrition. *Nature* 510(7503), 139–142. DOI: 10.1038/nature13179.

Ndondo, J.T.K. (2023) Review of the Food and Agriculture Organisation (FAO) strategic priorities on food safety. In: Ahmad, R.S. (ed.) *Food Safety—New Insights*. IntechOpen. DOI: 10.5772/intechopen.112132.

Nyyssölä, A., Suhonen, A., Ritala, A. and Oksman-Caldentey, K.M. (2022) The role of single cell protein in cellular agriculture. *Current Opinion in Biotechnology* 75, 102686. DOI: 10.1016/j.copbio.2022.102686.

Oke, T.R. (1973) City size and the urban heat island. *Atmospheric Environment (1967)* 7(8), 769–779. DOI: 10.1016/0004-6981(73)90140-6.

Paerl, H.W. and Huisman, J. (2009) Climate change: a catalyst for global expansion of harmful cyanobacterial blooms. *Environmental Microbiology Reports* 1(1), 27–37. DOI: 10.1111/j.1758-2229.2008.00004.x.

Parry, M., Evans, A., Rosegrant, M.W. and Wheeler, T. (2009) *Climate Change and Hunger, Responding to the Challenge*. World Food Programme, Rome, Italy.

Paynter, S., Ware, R.S., Weinstein, P., Williams, G. and Sly, P.D. (2010) Childhood pneumonia: a neglected, climate-sensitive disease? *The Lancet* 376(9755), 1804–1805. DOI: 10.1016/S0140-6736(10)62141-1.

Porter, J.R., Xie, L., Challinor, A.J., Cochrane, K., Howden, S.M. *et al.* (2014) Food security and food production systems. In: Field, C.B., Barros, V.R., Dokken, D.J., Mach, K.J. and Mastran, M.D. (eds) *Climate Change 2014, Impacts, Adaptation, and Vulnerability. Part A, Global and Sectoral Aspects, Contribution of Working Group II to the Fifth Assessment Report of the Intergovernmental Panel on Climate Change*. Cambridge University Press, Cambridge, UK, and New York, pp. 485–533.

Raj, S., Roodbar, S., Brinkley, C. and Wolfe, D.W. (2022) Food security and climate change, Differences in impacts and adaptation strategies for rural communities in the global South and North. *Frontiers in Sustainable Food Systems* 5, 691191. DOI: 10.3389/fsufs.2021.691191.

Rana, S.S. (2011) *Organic Farming*. Department of Agronomy, College of Agriculture, CSK Himachal Pradesh Krishi Vishvavidyalaya, Palampur, India, p. 90.

Rausher, M.D. (2001) Co-evolution and plant resistance to natural enemies. *Nature* 411(6839), 857–864. DOI: 10.1038/35081193.

Reiter, P., Thomas, C.J., Atkinson, P.M., Hay, S.I., Randolph, S.E. *et al.* (2004) Global warming and malaria: a call for accuracy. *The Lancet Infectious Diseases* 4(6), 323–324. DOI: 10.1016/S1473-3099(04)01038-2.

Sankar, A.S. and Reddy, K.V. (2022) Recent trends in organic farming in India: a study. *International Journal of Creative Research Thoughts* 10(7), 860–869.

Savary, S., Willocquet, L., Pethybridge, S.J., Esker, P., McRoberts, N. *et al.* (2019) The global burden of pathogens and pests on major food crops. *Nature Ecology & Evolution* 3(3), 430–439. DOI: 10.1038/s41559-018-0793-y.

Singh, P., Gupta, R., Shukla, S., Gupta, A. and Singh, R.L. (2023) Estimation of phytochemical constituents and analysis of antioxidant activity in different parts of plant. *International Journal of Pharmaceutical Chemistry and Analysis* 10(3), 194–201. DOI: 10.18231/j.ijpca.2023.032.

Singh, P., Rayeen, F., Singh, R., Pathak, N., Singh, R.P. *et al.* (2023) Enhancing agricultural sustainability through microbial-mediated abiotic stress tolerance. *Journal of Ecophysiology and Occupational Health* 23(4), 233–247. DOI: 10.18311/jeoh/2023/34777.

Singh, R.L., Vishwakarma, S.P. and Singh, P. (2014) Vitamins and minerals, roles and plant sources. In: Prakash, D. (ed.) *Phytochemicals of Nutraceutical Importance and Their Role in Human Health*. CAB International, Wallingford, UK, pp. 248–265.

Springmann, M., Mason-D'Croz, D., Robinson, S., Garnett, T., Godfray, H.C.J. *et al.* (2016) Global and regional health effects of future food production under climate change: a modelling study. *The Lancet* 387(10031), 1937–1946. DOI: 10.1016/S0140-6736(15)01156-3.

Stoffel, F., Santana, W.O., Fontana, R.C. and Camassola, M. (2021) Use of pleurotus albidus mycoprotein flour to produce cookies: evaluation of nutritional enrichment and biological activity. *Innovative Food Science & Emerging Technologies* 68, 102642. DOI: 10.1016/j.ifset.2021.102642.

Swinburn, B.A., Kraak, V.I., Allender, S., Atkins, V.J., Baker, P.I. *et al.* (2019) The global syndemic of obesity, undernutrition, and climate change: the Lancet Commission Report. *The Lancet* 393(10173), 791–846. DOI: 10.1016/S0140-6736(18)32822-8.

Thomas, K.M., Charron, D.F., Waltner-Toews, D., Schuster, C., Maarouf, A.R. *et al.* (2006) A role of high impact weather events in waterborne disease outbreaks in Canada, 1975–2001. *International Journal of Environmental Health Research* 16(3), 167–180. DOI: 10.1080/09603120600641326.

Tigchelaar, M., Cheung, W.W.L., Mohammed, E.Y., Phillips, M.J., Payne, H.J. *et al.* (2021) Compound climate risks threaten aquatic food system benefits. *Nature Food* 2(9), 673–682. DOI: 10.1038/s43016-021-00368-9.

Tilman, D., Balzer, C., Hill, J. and Befort, B.L. (2011) Global food demand and the sustainable intensification of agriculture. *Proceedings of the National Academy of Sciences* 108(50), 20260. DOI: 10.1073/pnas.1116437108.

Towers, S., Chowell, G., Hameed, R., Jastrebski, M., Khan, M. *et al.* (2013) Climate change and influenza, the likelihood of early and severe influenza seasons following warmer than average winters. *PLoS Currents* 5. DOI: 10.1371/currents.flu.3679b56a5616dc7c043fb944c6f138.

Uleberg, E., Hanssen-Bauer, I., van Oort, B. and Dalmannsdottir, S. (2014) Impact of climate change on agriculture in northern norway and potential strategies for adaptation. *Climatic Change* 122(1–2), 27–39. DOI: 10.1007/s10584-013-0983-1.

Vågsholm, I., Arzoomand, N.S. and Boqvist, S. (2020) Food security, safety, and sustainability—getting the trade-offs right. *Frontiers in Sustainable Food Systems* 4, 16. DOI: 10.3389/fsufs.2020.00016.

Vermeulen, S.J., Aggarwal, P.K., Ainslie, A., Angelone, C., Campbell, B.M. *et al.* (2012) Options for support to agriculture and food security under climate change. *Environmental Science & Policy* 15(1), 136–144. DOI: 10.1016/j.envsci.2011.09.003.

Victora, C.G., Christian, P., Vidaletti, L.P., Gatica-Domínguez, G., Menon, P. *et al.* (2021) Revisiting maternal and child undernutrition in low-income and middle-income countries: variable progress towards an unfinished agenda. *The Lancet* 397(10282), 1388–1399. DOI: 10.1016/S0140-6736(21)00394-9.

Vishwakarma, S.P., Singh, P., Singh, U. and Singh, R.L. (2013) Antioxidant activities of some tuberous plants leaves. *International Journal of Pharmaceutical Sciences Review and Research* 20(1), 28–33.

Watts, N., Amann, M., Arnell, N., Ayeb-Karlsson, S., Belesova, K. *et al.* (2019) The 2019 report of the Lancet Countdown on health and climate change: ensuring that the health of a child born today is not defined by a changing climate. *The Lancet* 394(10211), 1836–1878. DOI: 10.1016/S0140-6736(19)32596-6.

Wells, M.L., Potin, P., Craigie, J.S., Raven, J.A., Merchant, S.S. *et al.* (2017) Algae as nutritional and functional food sources: revisiting our understanding. *Journal of Applied Phycology* 29(2), 949–982. DOI: 10.1007/s10811-016-0974-5.

Wild, C.P. (2005) Complementing the genome with an "exposome": the outstanding challenge of environmental exposure measurement in molecular epidemiology. *Cancer Epidemiology, Biomarkers & Prevention* 14(8), 1847–1850. DOI: 10.1158/1055-9965.EPI-05-0456.

World Health Organization and Public Health Agency of Canada (2005) *Preventing Chronic Diseases, a Vital Investment*. World Health Organization, Geneva, Switzerland.

Wu, L., Taylor, M.P. and Handley, H.K. (2017) Remobilisation of industrial lead depositions in ash during Australian wildfires. *The Science of the Total Environment* 599–600, 1233–1240. DOI: 10.1016/j.scitotenv.2017.05.044.

Xu, Z., Hu, W. and Tong, S. (2014) Temperature variability and childhood pneumonia: an ecological study. *Environmental Health* 13(1), 51. DOI: 10.1186/1476-069X-13-51.

Yadav, C.P., Hussain, S.S.A., Mullick, R., Rahi, M. and Sharma, A. (2023) Climate zones are a key component of the heterogeneous presentation of malaria and should be added as a malariometric for the planning of malaria elimination. *PLOS Global Public Health* 3(6), e0001878. DOI: 10.1371/journal.pgph.0001878.

Yohe, G.W., Lasco, R.D., Ahmad, Q.K., Arnell, N.W., Cohen, S.J. *et al*. (2007) Perspectives on climate change and sustainability. In: Parry, M.L., Canziani, O.F., Palutikof, J.P., van der, P.J. and Hanson, C.E. (eds) *Climate Change 2007, Impacts, Adaptation and Vulnerability, Contribution of Working Group II to the Fourth Assessment Report of the Intergovernmental Panel on Climate Change*. Cambridge University Press, Cambridge, UK, and New York, pp. 811–841.

Zhan, J., Thrall, P.H. and Burdon, J.J. (2014) Achieving sustainable plant disease management through evolutionary principles. *Trends in Plant Science* 19(9), 570–575. DOI: 10.1016/j.tplants.2014.04.010.

2 Climate Change and Sustainable Food Security

Pawan Kumar Kanaujia[1]*, Anupama Ojha[2], Mukesh Pratap Yadav[3], Shreya Singh[1] and Saumya Singh[1]

[1]Department of Biotechnology, Allied Health Sciences, Mahayogi Gorakhnath University Gorakhpur, Uttar Pradesh, India; [2]Department of Medical Biochemistry, Allied Health Sciences, Mahayogi Gorakhnath University Gorakhpur, Uttar Pradesh, India; [3]Department of Diagnostic Radiology and Nuclear Medicine, University of Maryland School of Medicine, Baltimore, Maryland

Abstract

Climate change refers to significant and long-term changes in the earth's average atmospheric temperature, weather patterns, and other climate parameters. It is primarily driven by human activities, such as burning fossil fuels, deforestation, and industrial processes, which release greenhouse gases (GHGs) like carbon dioxide (CO_2) into the atmosphere, leading to various environmental impacts. Climate change has far-reaching implications on food systems, posing significant challenges to global food security. Climate change has adverse impacts on food systems at different levels, including ocean acidification, water scarcity, drought, soil degradation, pest resistance, disease pressures, crop-yield variability, and loss of biodiversity. Inadequacy of nutrients in foods raises the risk of malnutrition. This challenge requires a multidisciplinary approach involving scientists, farmers, governments, and communities. Sustainable agricultural practices, coupled with international cooperation and policy initiatives, are essential to ensure food security in the context of climate change.

Keywords: Climate Change, Sustainable Food, Nutrients, Food Systems, Biodiversity

2.1 Introduction

Climate changes heighten the existing susceptibility of human populations, leading to food and water scarcity, population movement, and economic setbacks. The procedure of the earth's climate system is straightforward. Solar energy from the sun coming to earth is thrown back into space, and the planet's temperature decreases. When the earth absorbs solar radiation or when gases present in the atmosphere prevent heat released by the earth into space, the planet's temperature increases. Several factors associated with natural and anthropogenic activity generate GHGs that cause rapid climate changes. Climate change has an extreme effect on the environment in which agricultural activities are performed. The effect of certain climate changes is different on each element of the ecosystem; even a slight change in the climate leads to reduced productivity or even the disappearance of several plant and animal species. There is a cascade of risks to agroecosystem and agricultural production due to climate change.

*Corresponding author: pkkanaujia@gmail.com; drpawan.kanaujia@mgug.ac.in

DOI: 10.1079/9781800625037.0002

Climate change is also causing notable concern for the availability of potable water in several regions of the earth. It will influence the precipitation pattern, runoff, and glacier melt, which will in turn affect the hydrodynamics system as well as quality of water, its temperature, and groundwater recharge. Climate change also affects the sea level with a negative and substantial impact on the salinity of seawater in seaside areas, affecting food security.

According to the World Food Summit 1996, food security exists when all people at all times have a physical and economical avenue to abundant, safe, and nutrition-rich food to meet their dietary needs and food choices for an active and healthy life. Four out of eight key risks induced by climate change identified by the Intergovernmental Panel on Climate Change (IPCC) have direct consequences for food security:

1. Loss of rural livelihood and income
2. Loss of marine and coastal ecosystems and livelihoods
3. Loss of terrestrial and inland water ecosystems and livelihoods
4. Food insecurity and breakdown of food systems

Combating hunger, achieving food security, and improving nutritional value are the core of SDGs. Simultaneously, climate change is already affecting agriculture and food certainty and will make it difficult to satiate hunger and avoid malnutrition (Fig. 2.1).

2.2 Adverse Effect of Climate Change

The IPCC confirmed in its reports that the change in weather patterns is true, and its main cause is anthropogenic GHG emission. The effects of global climate change, which were predicted so far by scientists, are now occurring in the atmosphere. The IPCC's sixth assessment report, published in 2021, found that human emissions of heat-trapping gases have already warmed the climate by nearly 2°F (1.1°C) since 1850–1900 (IPCC, 2021, sixth assessment report). The scientific evidence is unambiguous that present climate change is an ultimatum to human welfare and the condition of the earth. Any further delay in collaborative global action will miss the brief, accelerated closing window to assure a livable future (Fig. 2.2).

Climate change impacts our health, ecosystem, and financial system. The quicker assimilation of heat energy and CO_2 by the water bodies, especially the ocean, results in a series of collateral changes in seawater carbonate chemistry, including reductions in pH and carbonate saturation state, as well as

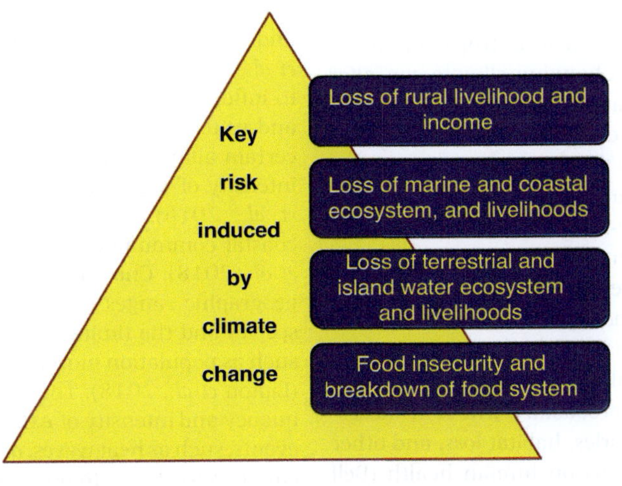

Fig. 2.1. Four key risks induced by climate changes identified by the IPCC have direct consequences for food security.

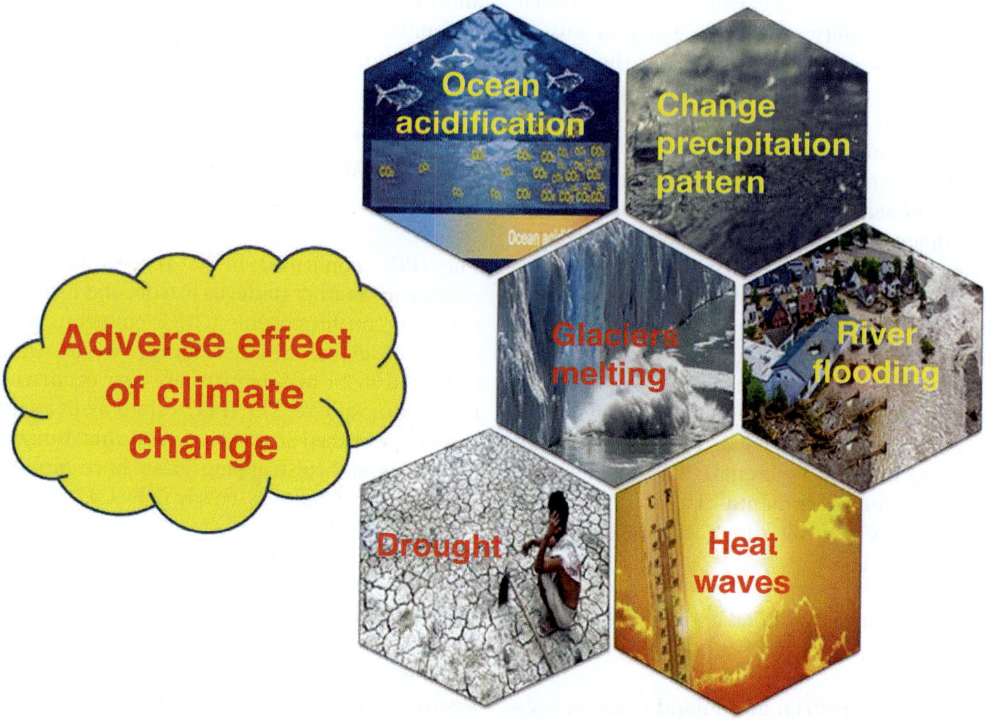

Fig. 2.2. Some of the adverse effect of climate change—ocean acidification, change in precipitation pattern, glacier melting, river flooding, drought, heat waves, and so on.

rises in dissolved CO_2 and bicarbonate ions. This phenomenon is defined as ocean acidification. Acidification of the ocean has the potential to severely affect many marine organisms living in different marine habitats from tropical to high-latitude biosphere. Changing climate can alter the precipitation pattern, namely the intensity and frequency. The elevation in the temperature of oceans increases the rate of evaporation and increases the amount of water in the air. When additional moisture-laden air proceeds over land or comes together into a storm system, it has the potential to produce more intense rainfall, such as torrential rain and typhoons (Wuebbles *et al.*, 2017). The potential risks of massive rainfall include agricultural harm, soil erosion, and a raised risk of flood probability, which in succession can lead to injuries, habitat loss, and other flooding-related effects on human health (Bell *et al.*, 2016). In addition, runoff originating from precipitation can diminish water quality due to the deposition of pollutants on land that are washed away into water bodies (Fig. 2.3).

Warmer temperatures increase the frequency, intensity, and duration of heat waves and can pose a health risk to mankind (Vose *et al.*, 2017). Climate change has the potential to influence human health by deteriorating air and water quality, increasing the expansion of certain ailments, and altering the frequency or intensity of excruciating weather events (Fann *et al.*, 2016). Ascent in sea level threatens coastal communities and ecosystems (Fleming *et al.*, 2018). Changing ecosystems impact the geographic ranges of many plant and animal species and the timing of their lifecycle events, such as population movement and reproduction (Lipton *et al.*, 2018). The abrupt rise in the frequency and intensity of extreme climate change events, such as heat waves, droughts, and floods, can increase losses to property, cause costly disruptions to society, and reduce the affordability of insurance (Ebi *et al.*, 2018).

Fig. 2.3. The adverse effect of climate change.

2.3 Implication of Climate Change on Food System

One of the biggest challenges of the 21st century is countering the adverse effects of climate change on the components of global food security, and mitigating these effects is crucial to meeting the future food demands of an ever-growing population (Table 2.1).

Crop productivity has already been endangered by climate change, particularly in the key food crops (rice, maize, and wheat), which are staple foods in many nations (Lobell *et al.*, 2011; Syed *et al.*, 2022). While the effects of climate change on key food crops have received extensive attention, the effects on fisheries and livestock which are also crucial to food security have received less attention (Herrero *et al.*, 2015; Campbell *et al.*, 2016).

The variables of climate that are changing slowly but unquestionably will adversely affect crop production in the upcoming decades include stresses caused by low and high temperatures, changes in the frequency and intensity of precipitation, and other climate change-induced disasters like drought, salinity, and rising sea levels (Raza *et al.*, 2021; Ullah *et al.*, 2021). As a result, farmers will have to cope with climate change and extreme weather events, which will make farming more challenging (IPCC, 2012). Therefore, before a serious problem arises, strategies for climate mitigation and adaptation must be implemented. Extreme weather and climate change will become major issues that will negatively impact food security and lead to a rise in conditions such as undernourishment and malnutrition (Hughes, 2020). Because of the ensuing vulnerabilities brought on by climate change, the agriculture sector, which includes crops and cattle, is negatively impacted and at risk (Fig. 2.4). There are many different ways that climate variability affects agriculture and food productivity, ranging from regional to global (Wu *et al.*, 2014). Three main elements that contribute to climate change are natural processes, the activity of humans such

Table 2.1. Effects of climate change on global food production.

No.	Climatic components	Livestock	Food and feed crops	Labor and capital	References
1.	Precipitation variation	1. Drinking water shortage 2. Pests and diseases	1. Reduced yield 2. Reduced food and food quality 3. Shifts in production systems	1. Shifts in human health 2. Migration 3. Reduced yield 4. Conflicts among and within a community	Connolly-Boutin and Smit, 2016; Phiiri *et al.*, 2016; Biglari *et al.*, 2019
2.	Temperature	1. Heat stress • Reduction in livestock productivity • Enhanced mortality 2. Pests and diseases • Reduced resistance against pests • Biodiversity loss	1. Reduced yield 2. Reduced food and food quality 3. Shifts in production systems		
3.	CO_2		1. Partial or total stomata closure 2. Shifts in production systems		

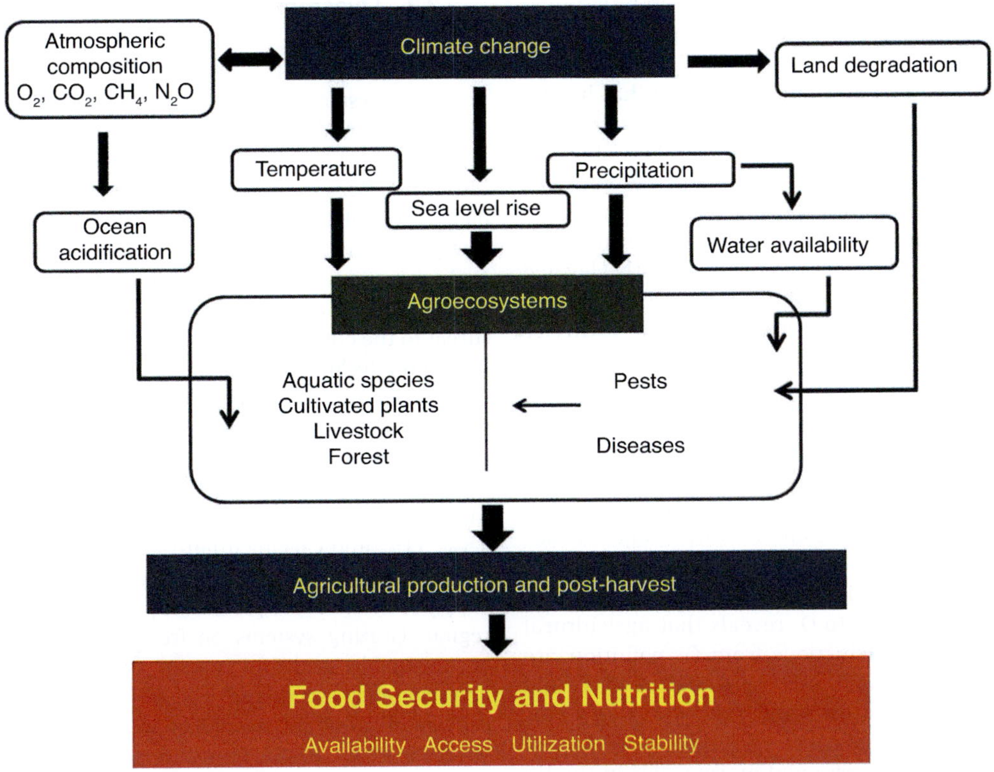

Fig. 2.4. Schematic representation of the cascading effect of climate change on agroecosystems and food security.

as the release of GHG and methane (CH_4) emissions, and changes in land use. Implications for some sectors are discussed below.

2.3.1 Agriculture

Temperature variations, shifts in rainfall patterns, global warming brought on by an increase in GHGs emissions, and soil abiotic stress from various heavy metals affect major food crops (Shah *et al.*, 2020; Raza *et al.*, 2022). These factors ultimately affect biological setups such as crop cycles, insect invasions, pest and disease invasions, and growth periods. The most commonly noted biological shift in reaction to global warming in the northern hemisphere throughout the 20th century was a longer crop life cycle (Livensperger *et al.*, 2019). During the past three decades, wheat and rice have been among the main food crops that have suffered from global warming; nevertheless, yield responses to these effects are still being studied, and satisfied grain yields have been reported in several locations (Raza *et al.*, 2019). Food production (net of food used for biofuels) has to expand by 70% by 2050 to cover the needs of an extra 2.3 billion people (total of 9.6 billion; FAO, 2009; Molotoks *et al.*, 2021). This is assuming mean rising temperature conditions of around 4°C. Uncontrolled and uneven climate change, however, will make it more difficult to meet this century's food requirements.

2.3.1.1 Direct influence of atmospheric CO_2 and tropospheric O_3 on agriculture

Plant primary production is stimulated by increases in CO_2 concentrations (Ainsworth *et al.*, 2020), which raises some species' potential yields. More growth stimulation is applied to plants that have a photosynthetic C_3 pathway, like wheat, rice, soybeans, and temperate grasses, than to plants that having a C_4 route, like maize, sorghum, sugarcane, and tropical grasses. However, the effect of CO_2 fertilization can also lower the animal feed quality (Myers *et al.*, 2014; Augustine *et al.*, 2018; Smith and Myers, 2018). For instance, Myers *et al.* (2014) found that with elevated CO_2 concentrations in the range of 546–586 ppm (−6.3% in wheat grains and −7.5% in rice grains), C_3 crops other than legumes exhibited decreased grain protein contents. Grain from C_4 crops was less affected. In the total organic matter of non-leguminous C_3 plants, higher CO_2 was also observed to increase the total non-structural carbohydrate (primarily starch and sugars) to mineral ratios and reduce comprehensive mineral concentrations (−8%) (Loladze, 2014). Moreover, the synthesis of crop responses to O_3 reveals that agricultural production reductions from O_3 pollution are comparable to those from heat, aridity, and nutritional stress (Mills *et al.*, 2018). For instance, it is calculated that O_3 pollution decreased soybean and maize yields in the US by 5–10% over the previous 30 years based on factual ground-level monitoring data (Ainsworth *et al.*, 2020).

2.3.1.2 Direct influence of temperature and water on agriculture

Forage, crop yields, and quality of feed can all be significantly impacted by temperature and changes in availability of water. Since the low moisture content in soil reduces evaporative cooling by the landscape and raised temperatures increase water loss by the crops, high temperatures are frequently associated with water stress (Lobell *et al.*, 2013). C_4 species are more likely to thrive compared with C_3 species when temperatures are higher and the weather is dryer. A hotter, drier climate is also anticipated to increase the concentration of ergot alkaloids and other potentially hazardous secondary chemicals (e.g. hydrogen cyanide in cassava and forage sorghum) (Bourguignon *et al.*, 2015; Brown *et al.*, 2016; Gleadow *et al.*, 2016).

2.3.2 Livestock and pastoral system

One of the biggest troubles facing the world's livestock systems now is climate change. The health and productivity of animals are impacted by feed and water availability, and also by changes in mean climatic variables and climate variability caused by global warming. The processing, storage, transportation, selling, and consumption of cattle products are all impacted by climate change. Thus, there is a threat to the current livestock system's capacity to sustain livelihoods and supply the increasing demand for animal products. Presently, the farm animal's industry is figuring out factors for both meal safety and delivery.

According to FAOSTAT (2020), livestock merchandise meat, milk, and eggs account for 15%, 31%, and varying quantities, respectively, of the global consumption, in step with capital supply of calories and protein, depending on the region. Grazing systems on frequently flawed or unsuitable land for cropping contribute to approximately 30% and 6% of the sector's delivery of ruminant meat and milk, respectively (Herrero *et al.*, 2013). Farm animals also give some other benefits, including drought, electricity, transportation, nutrient for deficient soils, profit generation, diversification, and financial capital, all of which can enhance the overall resilience and well-being of several groups of the population (CIRAD, 2016). Climate change could have an impact on farm animals' contributions to food security and different sustainable elements, though it is uncertain how much and to what extent. Significant development may have an effect on the food supply chain for cattle, consisting of rising atmospheric carbon dioxide (CO_2). Currently, it is anticipated that weeds, diseases, and pests diminish major crop yield by a combined 25–40% (Flood, 2010).

Elevating temperatures change the spatial dispersal of pests and pathogens and enhance the winter survival of insect pests and herbivory rate (Bebber *et al.*, 2013). Although there is great variance within taxonomic groupings, the distribution of pests and pathogens has shifted toward the poles by an average of 2.7 km yearly since 1960. The spatial mismatch among pests and natural predators could worsen in certain areas due to weather trade, which could harm biological control mechanisms (Selvaraj *et al.*, 2013). Excessive climate conditions can occasionally ruin crop resistance ability and open doorways for pests and weed outbreaks (Table 2.1).

2.3.3 Fisheries and aquaculture systems

The sustainability of capture fisheries and the growth of aquaculture in freshwater and marine ecosystems are facing increasing challenges because of climate change, unpredictability, and extreme weather events. The progression of atmospheric warming and related physical (sea surface temperatures, ocean circulation, waves, storm systems) and chemical (salinity content, oxygen concentration, acidity) changes in the aquatic environment are both responsible for the adverse impacts (IPCC, 2014). Extreme weather events, such as cyclones and deep-sea ocean swells, can have an impact on ecosystems like coral reefs and mangroves that are vital to people's livelihoods and food security (Table 2.1). The aquatic systems are experiencing and will continue to experience changes because of weather trade and carbon absorption, which includes raised water temperatures, expanded thermal stratification, altered salinity and freshwater content, altered oxygen concentrations, and higher acidity of the ocean. Rising temperatures and growing acidification will position coral reef systems, which aid one in four marine species, at more hazard.

Since the majority of aquatic animals have cold blood, the temperature and oxygen availability of their surroundings have a significant impact on their metabolic rates. Temperature variations can have a major impact on fish reproductive cycles, including how quickly fish grow, attain sexual maturity, and spawn. Shortly, lower catch potentials will come from worldwide reductions in the maximum body weight of fish species due to the fall in available oxygen levels brought on by surface water warming. Additionally, animals that are hypoxic-intolerant—like tuna—will suffer a reduction in the area of their habitat, which may result in lower future productivity.

Invasive species from decreased range areas have been discovered to be added into the Mediterranean at a pace of one every 4 weeks in current years. For the duration of the 1980s, the majority of non-local species have been seen emigrating 300 km northward on average, trying to retrace their natural chemical and physical habitat. There can be an intense shortage of trap fisheries production because of the growing stress that weather exchange and human stresses like pollution and overfishing are putting on aquatic resources.

2.4 Sustainable Food Security

The amount and quality of agricultural crop production, fisheries, farm animals, and woodland farming are dependent on a selection of physical and biophysical assets, together with the suitability of the soil, the supply and viability of natural inputs (water, sunlight, CO_2, temperature), and every so often pollination channels. Meal manufacturing is frequently declining due to diseases, viruses, natural calamities, and changes in weather patterns. In certain situations, the availability of labor to perform agricultural activity has a significant impact on each meal's output and food delivery (Tables 2.1 and 2.2). It has been noted that over the past few decades, climate change has reduced the number of digestible calories in a typical diet (Ray *et al.*, 2019). The assumption is that because climate change affects all aspects that determine food security, together with production, availability, stability, and access, it has made household meal instability more troublesome (Mekonnen *et al.*, 2021). All aspects of food production and security are impacted by weather change; however, interactive characterization of these effects is required (Fig. 2.2).

2.4.1 Impacts on food availability

Availability of food refers to both the supply side of the food chain and whether or not people have access to enough food to meet their nutritional requirements. Food production, available technology, stocks, supply chain effectiveness, and trade regulations at the global and regional levels all influence the availability of food. Numerous studies examined the effects

Table 2.2. Possible effects of climate factors on the animal food supply chain.

No.	Supply Chain	Potential impacts of climate change	References
1.	Feed resources	**Productivity** Water-stressed regions may experience: • Increased yields in high latitudes • Increased soil salinity in coastal areas • Shifting pest and disease distribution due to rising temperatures **Nutritional quality** • Warmer temperatures and drier conditions favor C_4 species, increasing toxicity in some plants. • Elevated eCO_2 reduces plant protein and mineral concentration, favoring C_3 plants and woody encroachment. **Variability in feed availability** Global climate variability is predicted to rise, causing a decrease in feed production and unpredictable supply, and extreme events could restrict animal access and disrupt feed production.	Godde *et al*., 2021
2.	Water resources	Rising temperatures and extreme events, including floods and droughts, can: • Increase water requirements • Increase pressure resources • Disrupt glacier flow • Decrease water quality for animal consumption	
3.	Animal health and production	Animal production, welfare, and life expectancy are likely to be negatively impacted by: • Decreased feed availability • Heat stress • Diseases • Mortality from extreme climate events, with warmer temperatures potentially reducing these effects	
4.	Processing, storage, transport, and retailing	Rising sea levels, higher temperatures, and extreme weather events are putting additional stress on built-up capital, increasing animal heat stress, and reducing food quality and shelf life. Increased variability in production and climate events will make trade patterns less regular.	
5.	Livestock products consumption	Climate change threatens: • Livestock availability, quality, and safety • Nutritional appeal • Food availability, leading to increased prices and volatile prices, and potentially impacting diets in high-income countries	

Continued

Table 2.2. Continued

No.	Supply Chain	Potential impacts of climate change	References
6.	Labor	Climate change is expected to negatively impact labor availability and productivity due to: • Heat stress • Increased risk of disease outbreaks • Extreme events • Decreased air quality • Changes in food supply	
7.	Prices	Climate change is expected to increase: • Supply chain costs • Commodity prices • Price volatility, primarily impacting animal product prices through changes in feed availability	

of upcoming climate prediction at different CO_2 concentrations and found that higher CO_2 concentrations were likely to improve crop productivity because they improved photosynthesis and increased water-use efficiency as a result of high carbon fertilization (Lee *et al.*, 2020). However, research has also shown that, in reality, crop productivity would grow at a slower rate than that predicted by crop models.

Therefore, because crop models anticipate different and opposing outcomes, the amount of variation in crop yield will differ on the global, regional, national, and local levels. Food production is impacted by changes in weather patterns unevenly and irregularly, with tropical areas experiencing more severe effects than higher latitudes. Furthermore, climate change is expected to have a particularly severe impact on countries with higher hunger index scores when it comes to livestock and food productivity. The impacts of climate change on food production and availability are found to follow a robust and rational pattern. It is anticipated that uneven climate change could have a terrible impact on areas where undernourishment and meal insecurity are already common.

Regarding the general impact of climate change on the supply of food, there are still several restrictions. First and foremost, models that forecast and evaluate the impact of climate change are limited to large cereal crops, tubers and a selected few roots. Climate change affects some crops, such as vegetables and pulses, which are regarded as locally significant (but minor worldwide) income-driven food crops. These inferences are drawn from the same plant features. Second, there is a dearth of research on the yield and quality of cattle-feeding crops in grasslands, which hinders our ability to comprehend the relationship between livestock and climate change. Thirdly, rather than accounting for weather extremes, which can occasionally have greater detrimental effects on crop, livestock, and wild forest productivities, the majority of crop studies concentrate on the effects of mean climatic changes. Finally, there is a lack of knowledge regarding how climate change affects the measurement of food provision and revenues from forest farming, which further complicates the issue of food security (Table 2.2).

2.4.2 Impacts on access to food

Accessibility to food can be defined as the ability of a household to obtain enough nutrient-rich food to satisfy their dietary requirements (Masipa, 2017). Only when residents have enough money to buy food to sustain healthy nutrition levels can they obtain an adequate quantity of high-quality food (Gartaula *et al.*, 2017; Gupta *et al.*, 2019). The ability, rights, and household income play a major role in determining access to food. Crops, fisheries, cattle, and forest farming are among the food products whose production is impacted and altered by unfavorable local, regional, and global climate changes. In the end, these effects have an impact on biomass output, which encompasses fuel, feed, fiber, and food production. The geography of the food system, which includes farming for crops, livestock, and forests, as well as the profits made by individual farms, are impacted by the changes in land use brought about by climate change (Peña-Lévano *et al.*, 2019). This has an impact on food accessibility in general. Climate change is predicted to have an impact on food access, which will strain people's ability to afford food. Several IPCC assessment reports have demonstrated how climate change will have a detrimental impact on food affordability through projections based on integrated climate, crops, livestock, forest farming, and economic models. These interconnected models forecast prices for food commodities and trade related to agriculture and agroforestry in the future under various social, economic, and climate change scenarios. The anticipated outcomes calculate the detrimental effects on the purchasing power of food for any population that is taken into consideration (Nelson *et al.*, 2014). Numerous indicators point to the immediate effects of both macro- and micro-level climate change to food affordability, and several climate change scenarios have shown that food costs will rise.

Household purchasing power affects food costs and, in turn, food access, both of which are severely harmed by climate change (Arnell, 2016; Tol, 2018). Climate change will probably affect the global food production geography, such as transference in suitable regions for crop and livestock production, which could have significant effects on cost of the food, businesses, and subsequently food access (Havlík *et al.*, 2014; Jones *et al.*, 2017). Through its effects on transportation networks, road infrastructure, and individual fortune, climate change also affects household's physical access to food (Tol, 2018). It is determined that because climate change affects food production systems and economic situations, it hurts food access.

2.4.3 Impacts on food utilization

Any moderate change in climate has a significant impact on how much food is used to meet dietary and nutritional requirements. Other secondary elements that also have an impact include the availability of potable water and sanitary services. There is a direct link between food consumption and climate change since climate variability impacts the availability of food items like drinking water (Duran-Encalada *et al.*, 2017). Seneviratne *et al.* (2012) state that inadequate sanitation infrastructure is a common concern in undeveloped areas, resulting in unsanitary conditions during extreme weather events such as droughts or floods. Consequently, various stomach disorders are brought on by a lack of adequate sanitation systems, lowering the intake of vital nutrients and consequently linked to temperature fluctuations. Dietary quality is threatened by climate change, which leads to an increase in demand for hygienic food and the need for sound techniques to prevent food contamination from any unfavorable environmental factors (Medina *et al.*, 2014). Higher costs for food are typically observed under climate change-induced events.

Two factors that are the primary determinants of food utilization are (i) food safety along the food supply chain and (ii) climate change. They may have direct or indirect health effects that impede dietary results. In particular, fisheries, fruits, and vegetables are anticipated to have lower food safety due to high- and low-temperature shocks because of increasing rates of microbial activities under altered climatic conditions (Hammond *et al.*, 2015; Galstyan *et al.*, 2019; Raza, 2020; Alegbeleye *et al.*, 2022). The food chain in the food system is impacted by climate change in a variety of dynamic ways, both directly and indirectly. These include diseases spread by vectors, temperature extremes, and natural disasters like floods and droughts that affect people's capacity to care for their families, their income, their nutrition, and their security. Human and cattle health is at risk due to the consequences of climate change on water resources, which include increasing frequency and severity of droughts and floods, contaminated water supplies, and reduced food quality (Uyttendaele *et al.*, 2015; Djekic *et al.*, 2016).

2.4.4 Impacts on food stability

Climate changes have a significant impact on food stability since they can have a long- or short-term variable effect on food price patterns (Myers *et al.*, 2017). Small disruptions in the food chain, whether on the supply or demand side, have affected prices during the past 10 years and frequently increased food costs (Haile and Wossen, 2016). Due to an increase in food volatility, climate change increases the unpredictability of the relationship between food supply and demand (Haile and Wossen, 2016).

Food stability is threatened by demand-side fluctuations brought on by climate change. A recent case study on food instability revealed that, in general, food crises were caused by inadequate policies, lower agricultural, livestock, fisheries, agroforestry productivity, and an absence of mitigation and adaptation strategies (both short- and long-term) against climate change and catastrophic occurrences. Food security and the environment, which is dependent on the availability of food and water, regulation of the climate, extreme weather, pests and diseases, and support of water and nutrient recycling, are linked both directly and indirectly. Extreme weather events and climate change are increasing the strain on ecosystems. Temperature fluctuations and changes in the frequency of extreme events are brought on by climate change, which reduces biodiversity and alters the interactions between and within groups (Mach *et al.*, 2016; Oppenheimer *et al.*, 2016). In the end, this jeopardizes food security and productivity in forestry and agriculture (Khoury *et al.*, 2014).

Many soil-, water-, livestock-, and crop-related problems are brought on by climate variability, and these problems ultimately affect human health and the ecosystem. They also cause soil degradation, changes in the properties of the soil, poor-quality surface and groundwater, problems with water availability, and a decrease in the amount and quality of food produced. Climate change thus puts the social and economic foundations of the food chain in jeopardy (Fig. 2.5).

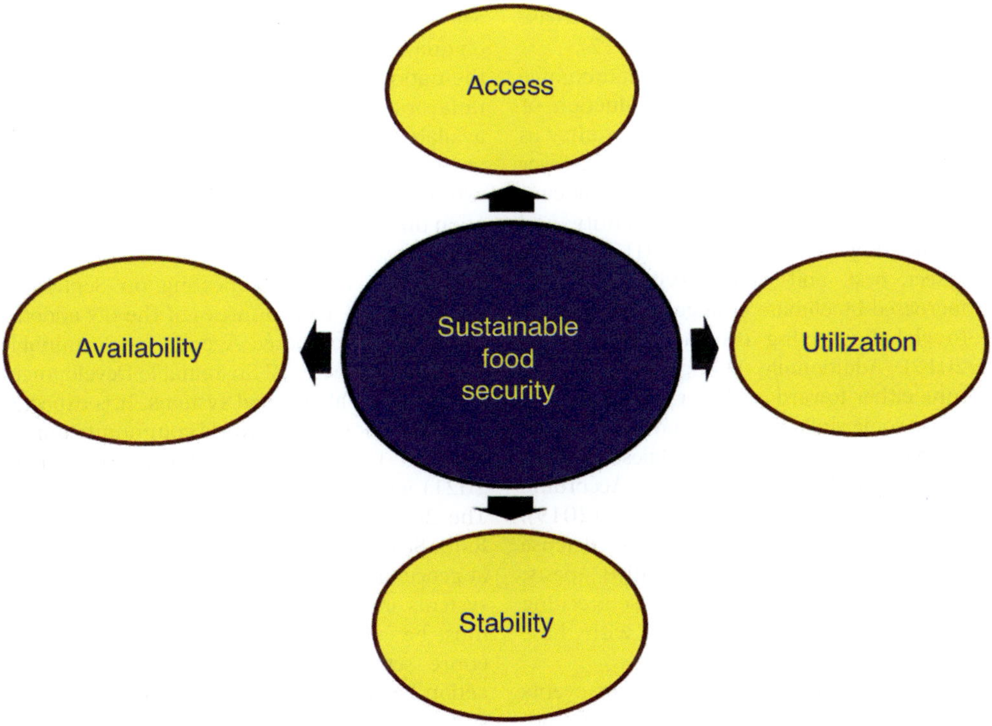

Fig. 2.5. Dimensions of food security affected by climate change.

2.5 Impacts on Overall Food Productivity

Climate predictions have shown with reasonable certainty that unpredictable and unequal climate fluctuations would lead to a fall in global food production, notwithstanding the inherent flaws in climate-crop modeling (Porter *et al.*, 2014; Martinich *et al.*, 2017).

According to recent IPCC climate estimates, this included some uncertainty.

1. The average productivity of the three main food crops—rice, maize, and wheat—would decrease by 3–10% for every 1°C increase in temperature (Challinor *et al.*, 2014, Challinor *et al.*, 2018).
2. Because of competition for natural resources, feed quantity and quality, livestock diseases, heat and cold stress, and biodiversity loss, climate change will influence livestock and fisheries productivity.
3. Furthermore, by the middle of the 21st century, it is predicted that the demand for livestock and fishery products will increase by 100% (Rojas-Downing *et al.*, 2017).
4. Through physiological stress, climate change can also directly affect the production of food for animals. Using milk and poultry as an example, when the temperature rises over the ideal range (≥30°C), it has a significant negative influence on both quantity and quality (Rojas-Downing *et al.*, 2017).
5. Insect, pest, and weed survival rates are increased by climate change, primarily due to global warming (Chidawanyika *et al.*, 2019). Additionally, changes in temperature, either toward colder or higher ranges, cause some shifts in the latitudinal range of the prevalence of diseases and pests (Bebber *et al.*, 2013; Deutsch *et al.*, 2018). According to Bebber (2015) and Bebber *et al.* (2019), indigenous crop varieties lack defense mechanisms against non-native pests; hence, it may be necessary to use breeding management strategies to deal with these new threats.
6. On the other hand, severe weather events can sometimes make crops and cattle more resilient to pests (Seidel, 2014). Due to their ability to create ideal conditions for the growth of weeds, pests, and diseases, climate change and extreme weather events reduce agricultural productivity (Rosenzweig *et al.*, 2014; Powell and Reinhard, 2015). Climate change is predicted to increase CO_2 concentration, which would alter the makeup of weeds and crop plants' defense systems against pests (Myers *et al.*, 2017). Increased CO_2 concentrations impair the effectiveness of herbicides and pesticides in controlling weeds and pests (Table 2.2; Varanasi *et al.*, 2016; Ramesh *et al.*, 2017).

Even though there is already enough food produced per unit area to satisfy the world's rapidly expanding population, 811 million people suffer from acute undernourishment, and there are indications that the drive to achieve zero hunger is waning (FAO, 2021). In the meantime, malnutrition is having a severe negative impact on both developed and developing nations. Over 2 billion adults, teenagers, and children worldwide are currently obese or overweight, despite a modest decline in stunted growth, low height, and decreasing weight with age (Ferreira *et al.*, 2020). As a result, the effects on quality of life, livelihoods, public health, and the nation's economy are getting worse. These unfavorable trends coexist with declining land availability, degraded soil and biodiversity, and an increase in the frequency and intensity of extreme weather events, which worsen the situation due to their combined effects with climate change.

During a special meeting on September 25, 2015, the 193 countries of the UN adopted the 2030 action-based Agenda for Sustainable Development with 17 Sustainable Development Goals (SDGs) about food systems. It is expected that the actions of the world community during the next 15 years (from 2016 to 2030; FAO, 2021) will be supported by these global goals. The 2030 Agenda of the Sustainable Growth Rate (SGR) presents a fairer picture of a world at peace in which no one is left behind. All food systems, including agroforestry and agriculture, must be completely fulfilled to accomplish the entire set of SDGs. Rural development and action-based investments in different farming systems, including crops, livestock, forestry, fisheries, and aquaculture, will receive special attention (Gregersen *et al.*, 2020). This will lead to more sustainable development and provide the entire society with more effective instruments to reduce hunger, poverty, and undernourishment.

Some of the 17 main components of SDGs, as determined by a United Nations (UN) agency's joint judgment, are as follows (FAO, 2021):

1. Putting an end to poverty everywhere and in all forms.
2. Putting an end to hunger, attaining food security objectives, enhancing wholesome eating, supporting sustainable agricultural systems ensuring wholesome lives, and fostering well-being in an inclusive manner.
3. Promoting long-term learning while guaranteeing inclusive, high-quality awareness and education for all communities.
4. Attaining gender parity and empowering everyone, particularly women and girls.
5. Utilizing sustainable management practices to guarantee the availability of sanitized water.
6. Ensuring that all communities have access to resources for clean, sustainable, inexpensive, and dependable energy.
7. Ensuring and fostering job possibilities, dependable working systems, and equitable, comprehensive, and sustainable economic growth.
8. Constructing more durable transit networks, housing, and market infrastructures, and encouraging innovative and sustainable industry development lowering disparities, both within and between countries.
9. Increasing the inclusiveness, safety, catastrophe resilience, and sustainability of cities.
10. Making sure that patterns of food production and consumption are sustainable.
11. Combating climate change vulnerabilities by taking essential and expeditious measures.
12. Preservation and guaranteeing the long-term utilization of maritime resources maintaining trees sustainably, preventing desertification, slowing down or stopping the process of land degradation and stopping the loss of biodiversity, and fostering social justice, harmony, and inclusivity.
13. Improving international collaborations for sustainable development initiatives.

2.6 Conclusions

There are still questions about how climatic variability affects humans, crops, animals, and forests, as well as the best strategies for adjusting to erratic and unpredictable weather. Even if these uncertainties predict exact changes in the food supply in the future, there is ample reason to be ready for a wider range of likely unfavorable results. This research also shows that several places are already facing climate change and have fewer resources available for mitigation and adaptation so are frequently disadvantaged by climatic unpredictability. It is essential to outline the local food priorities at the community level. For example, calculating food availability can assist in assessing how temperature variations, such as cold stress or elevated CO_2 and global warming, affect nutrition and health. This estimate, however, only provides information on the availability of food, not its actual availability and consumption among socioeconomically different communities, and it falls short of adequately characterizing the distribution of food at the local and national levels. There is a lack of knowledge and comprehension about future food consumption and access worldwide, the ways and degrees to which climate variability will impact agriculture, livestock, fisheries, and forest farming at local levels, and the potential health consequences of these factors. There is little question that climate change will have an impact on the food supply chain, which encompasses farm production, processing, storage, transportation, retail, and human consumption, but the precise nature of these effects is still up in the air. These uncertainties relate to the long-term climate as well as the exposure and response of the interdependent natural and human systems to long-term climate change. For instance, the consequences of climate change on agricultural labor, the livestock supply chain from farmgate to merchants, and the prices and consumption trends of animal products have not received much attention in the past. The consequences of climate change on the production and quality of animal feed, especially forages, have also received less attention than those of major staple grains. Moreover, the past is probably no longer a good indicator of the future due to the rapid changes in global socioeconomic and environmental variables. Further investigation will reduce this uncertainty.

Food production, security, and climate change are all interconnected and reliant on one another. A person's changes may have a direct or indirect detrimental effect on other people. For instance, concerns about food insecurity are brought on by a rapidly expanding population. Resolving these problems eventually exacerbates climate change because they increase food production through agricultural management practices (such as deforestation and heavy fertilizer and pesticide use). Furthermore, the energy needed for food processing, livestock rearing, fishing, forest farming, and agriculture increases the variability of the climate.

The heightened frequency and intensity of natural disasters brought on by climate change, their effects on the food system (crops, livestock, fisheries, and forest farming), as well as the variables influencing food security, will necessitate extensive interactive research-based analysis in the future. This is because modifications made to one of these domains may have several detrimental repercussions on the others. In light of climate change, it is critical to have a thorough understanding of all the factors affecting food security as well as the multi-sectoral effects that affect total food security. This is because, to ensure food security, it will also be necessary to increase mitigation and adaptation initiatives. A wide range of factors (political, economic, and social) should be taken into account in addition to climate change and extreme events that worsen food insecurity issues to further close the research gaps on the effects of climate change on food production (livestock, crops, fisheries, and forest farming), availability, access, utilization, and stability.

References

Ainsworth, E.A., Lemonnier, P. and Wedow, J.M. (2020) The influence of rising tropospheric carbon dioxide and ozone on plant productivity. *Plant Biology (Stuttgart, Germany)* 22 Suppl 1(Suppl 1), 5–11. DOI: 10.1111/plb.12973.

Alegbeleye, O., Odeyemi, O.A., Strateva, M. and Stratev, D. (2022) Microbial spoilage of vegetables, fruits and cereals. *Applied Food Research* 2(1), 100122. DOI: 10.1016/j.afres.2022.100122.

Arnell, N.W. (2016) The global-scale impacts of climate change: the QUEST-GSI project. *Climatic Change* 134(3), 343–352. DOI: 10.1007/s10584-016-1600-x.

Augustine, D.J., Blumenthal, D.M., Springer, T.L., LeCain, D.R., Gunter, S.A. *et al.* (2018) Elevated CO_2 induces substantial and persistent declines in forage quality irrespective of warming in mixed grass prairie. *Ecology Applied* 28, 721–735. DOI: 10.1002/eap.1680.

Bebber, D.P. (2015) Range-expanding pests and pathogens in a warming world. *Annual Review of Phytopathology* 53, 335–356. DOI: 10.1146/annurev-phyto-080614-120207.

Bebber, D.P., Ramotowski, M.A.T. and Gurr, S.J. (2013) Crop pests and pathogens move polewards in a warming world. *Nature Climate Change* 3(11), 985–988. DOI: 10.1038/nclimate1990.

Bebber, D.P., Field, E., Heng, G., Mortimer, P., Holmes, T. *et al.* (2019) Many unreported crop pests and pathogens are probably already present. *Global Change Biology* 25, 2703–2713. DOI: 10.1101/519223.

Bell, J.E., Herring, S.C., Jantarasami, L., Adrianopoli, C., Benedict, K. *et al.* (2016) Impacts of extreme events on health. In: *The Impacts of Climate Change on Human Health in the United States: A Scientific Assessment*. U.S. Global Change Research Program, Washington, DC, pp. 99–128. DOI: 10.7930/J0BZ63ZV.

Biglari, T., Maleksaeidi, H., Eskandari, F. and Jalali, M. (2019) Livestock insurance as a mechanism for household resilience of livestock herders to climate change: evidence from Iran. *Land Use Policy* 87, 104043. DOI: 10.1016/j.landusepol.2019.104043.

Bourguignon, M., Nelson, J.A., Carlisle, E., Ji, H., Dinkins, R.D. *et al.* (2015) Eco physiological responses of tall fescue genotypes to fungal endophyte infection, elevated temperature, and precipitation. *Crop Science* 55(6), 2895–2909. DOI: 10.2135/cropsci2015.01.0020.

Brown, A.L., Cavagnaro, T.R., Gleadow, R. and Miller, R.E. (2016) Interactive effects of temperature and drought on cassava growth and toxicity: implications for food security? *Global Change Biology* 22(10), 3461–3473. DOI: 10.1111/gcb.13380.

Campbell, B.M., Vermeulen, S.J., Aggarwal, P.K., Corner-Dolloff, C., Girvetz, E. *et al.* (2016) Reducing risks to food security from climate change. *Global Food Security* 11, 34–43. DOI: 10.1016/j.gfs.2016.06.002.

Challinor, A.J., Watson, J., Lobell, D.B., Howden, S.M., Smith, D.R. *et al.* (2014) A meta-analysis of crop yield under climate change and adaptation. *Nature Climate Change* 4(4), 287–291. DOI: 10.1038/nclimate2153.

Challinor, A.J., Müller, C., Asseng, S., Deva, C., Nicklin, K.J. *et al.* (2018) Improving the use of crop models for risk assessment and climate change adaptation. *Agricultural Systems* 159, 296–306. DOI: 10.1016/j.agsy.2017.07.010.

Chidawanyika, F., Mudavanhu, P. and Nyamukondiwa, C. (2019) Global climate change as a driver of bottom-up and top-down factors in agricultural landscapes and the fate of host-parasitoid interactions. *Frontiers in Ecology and Evolution* 7, 80. DOI: 10.3389/fevo.2019.00080.

CIRAD (2016) *Livestock Farming & Local Development*. CIRAD, Montpellier, France.

Connolly-Boutin, L. and Smit, B. (2016) Climate change, food security, and livelihoods in Sub-Saharan Africa. *Regional Environmental Change* 16(2), 385–399. DOI: 10.1007/s10113-015-0761-x.

Deutsch, C.A., Tewksbury, J.J., Tigchelaar, M., Battisti, D.S., Merrill, S.C. *et al.* (2018) Increase in crop losses to insect pests in a warming climate. *Science* 361(6405), 916–919. DOI: 10.1126/science.aat3466.

Djekic, I., Kuzmanović, J., Anđelković, A., Saraèević, M., Stojanović, M.M. *et al.* (2016) Relationships among hygiene indicators in take-away foodservice establishments and the impact of climatic conditions. *Journal of Applied Microbiology* 121(3), 863–872. DOI: 10.1111/jam.13211.

Duran-Encalada, J.A., Paucar-Caceres, A., Bandala, E.R. and Wright, G.H. (2017) The impact of global climate change on water quantity and quality: a system dynamics approach to the US–Mexican transborder region. *European Journal of Operational Research* 256(2), 567–581. DOI: 10.1016/j.ejor.2016.06.016.

Ebi, K.L., Balbus, J.M., Luber, G., Bole, A., Crimmins, A. *et al.* (2018) Human health. In: Reidmiller, D.R., Avery, C.W., Easterling, D.R., Kunkel, K.E. and Lewis, K.L.M. *et al.* (eds) *Impacts, Risks, and Adaptation in the United States: Fourth National Climate Assessment*, Vol. II. U.S. Global Change Research Program, Washington, DC, p. 544. DOI: 10.7930/NCA4.2018.CH14.

Fann, N., Brennan, T., Dolwick, P., Gamble, J.L., Ilacqua, V, *et al.* (2016) Air quality impacts. In: *The Impacts of Climate Change on Human Health in the United States: A Scientific Assessment*. U.S. Global Change Research Program, Washington, DC, pp. 69–98. DOI: 10.7930/J0GQ6VP6.

FAO (2009) Global agriculture towards 2050. Available at: https://www.fao.org/fileadmin/templates/wsfs/docs/Issues_papers/HLEF2050_Global_Agriculture.pdf (accessed 5 August 2024).

FAO (2021) *The State of Food Security and Nutrition in the World 2021*. FAO, Rome, Italy. DOI: 10.4060/CB4474EN.

FAOSTAT (2020) Food and agriculture data. Available at: https://www.fao.org/faostat/en/#home (accessed 5 August 2024).

Ferreira, H.S., Albuquerque, G.T., Santos, T.R., Barbosa, R.D.L., Cavalcante, A.L. *et al.* (2020) Stunting and overweight among children in Northeast Brazil: prevalence, trends (1992-2005-2015) and associated risk factors from repeated cross-sectional surveys. *BMC Public Health* 20(1), 736. DOI: 10.1186/s12889-020-08869-1.

Fleming, E., Payne, J., Sweet, W., Craghan, M., Haines, J. *et al.* (2018) Coastal effects. In: Reidmiller, D.R., Avery, C.W., Easterling, D.R., Kunkel, K.E. and Lewis, K.L.M. *et al.* (eds) *Impacts, Risks, and Adaptation in the United States: Fourth National Climate Assessment*, Vol. Volume II. U.S. Global Change Research Program, Washington, DC, pp. 322–352. DOI: 10.7930/NCA4.2018.CH8.

Flood, J. (2010) The importance of plant health to food security. *Food Security* 2(3), 215–231. DOI: 10.1007/s12571-010-0072-5.

Galstyan, A.G., Petrov, A.N., Illarionova, E.E., Semipyatniy, V.K., Turovskaya, S.N. *et al.* (2019) Effects of critical fluctuations of storage temperature on the quality of dry dairy product. *Journal of Dairy Science* 102(12), 10779–10789. DOI: 10.3168/jds.2019-17229.

Gartaula, H., Patel, K., Johnson, D., Devkota, R., Khadka, K. *et al.* (2017) From food security to food wellbeing: examining food security through the lens of food wellbeing in Nepal's rapidly changing agrarian landscape. *Agriculture and Human Values* 34(3), 573–589. DOI: 10.1007/s10460-016-9740-1.

Gleadow, R.M., Ottman, M.J., Kimball, B.A., Wall, G.W., Pinter, P.J. *et al.* (2016) Drought-induced changes in nitrogen partitioning between cyanide and nitrate in leaves and stems of sorghum grown at elevated CO_2 are age dependent. *Field Crops Research* 185, 97–102. DOI: 10.1016/j.fcr.2015.10.010.

Godde, C.M., Mason-D'Croz, D., Mayberry, D.E., Thornton, P.K. and Herrero, M. (2021) Impacts of climate change on the livestock food supply chain; a review of the evidence. *Global Food Security* 28, 100488. DOI: 10.1016/j.gfs.2020.100488.

Gregersen, H., El-lakany, H. and Frechette, A. (2020) Forests, forest people and UN 2030 agenda's ethical mandate: "LEAVE NO ONE BEHIND". Available at: https://rightsandresources.org/wp-content/uploads/2020/09/Forests-ForestPeople-UN2030AgendasEthicalMandate-LEAVENOONEBEHIND.pdf (accessed 5 August 2024).

Gupta, S., Pingali, P. and Pinstrup-Andersen, P. (2019) Women's empowerment and nutrition status: the case of iron deficiency in India. *Food Policy* 88, 101763. DOI: 10.1016/j.foodpol.2019.101763.

Haile, M.G. and Wossen, T. (2016) Impacts of climate and price changes on global food production. In: *Proceedings of the 2016 Fifth International Conference,* 246374. African Association of Agricultural Economists (AAAE), Addis Ababa, Ethiopia.

Hammond, S.T., Brown, J.H., Burger, J.R., Flanagan, T.P., Fristoe, T.S. *et al.* (2015) Food spoilage, storage, and transport: implications for a sustainable future. *BioScience* 65(8), 758–768. DOI: 10.1093/biosci/biv081.

Havlík, P., Valin, H., Herrero, M., Obersteiner, M., Schmid, E. *et al.* (2014) Climate change mitigation through livestock system transitions. *Proceedings of the National Academy of Sciences of the United States of America* 111(10), 3709–3714. DOI: 10.1073/pnas.1308044111.

Herrero, M., Havlík, P., Valin, H., Notenbaert, A.M., Rufino, M.C. *et al.* (2013) Biomass use, production, feed efficiencies, and greenhouse gas emissions from global livestock systems. *Proceedings National Academy Science of the United States of America* 110, 208–288. DOI: 10.1073/pnas.1308149110.-93.

Herrero, M., Wirsenius, S., Henderson, B., Rigolot, C., Thornton, P. *et al.* (2015) Livestock and the environment: what have we learned in the past decade? *Annual Review of Environment and Resources* 40(1), 177–202. DOI: 10.1146/annurev-environ-031113-093503.

Hughes, S. (2020) Principles, drivers, and policy tools for just climate change adaptation in legacy cities. *Environmental Science & Policy* 111, 35–41. DOI: 10.1016/j.envsci.2020.05.007.

IPCC (2012) *Managing the Risks of Extreme Events and Disasters to Advance Climate Change Adaptation*. IPCC, Geneva, Switzerland. DOI: 10.1017/cbo9781139177245.

IPCC (2014) *Climate Change 2014: Synthesis Report. Contribution of Working Groups I, II and III to the Fifth Assessment Report of the Intergovernmental Panel on Climate Change*. IPCC, Geneva, Switzerland.

IPCC (2021) *Climate Change 2021: The Physical Science Basis, the working Group I contribution to the Sixth Assessment Report*. Cambridge University Press, Cambridge, UK.

Jones, J.W., Antle, J.M., Basso, B., Boote, K.J., Conant, R.T. *et al.* (2017) Brief history of agricultural systems modeling. *Agricultural Systems* 155, 240–254. DOI: 10.1016/j.agsy.2016.05.014.

Khoury, C.K., Bjorkman, A.D., Dempewolf, H., Ramirez-Villegas, J., Guarino, L. *et al.* (2014) Increasing homogeneity in global food supplies and the implications for food security. *Proceedings of the National Academy of Sciences* 111(11), 4001–4006. DOI: 10.1073/pnas.1313490111.

Lee, Y.H., Sang, W.G., Baek, J.K., Kim, J.H., Shin, P. *et al.* (2020) The effect of concurrent elevation in CO_2 and temperature on the growth, photosynthesis, and yield of potato crops. *PloS One* 15(10), e0241081. DOI: 10.1371/journal.pone.0241081.

Lipton, D., Rubenstein, M.A., Weiskopf, S.R., Carter, S., Peterson, J. *et al.* (2018) Ecosystems, ecosystem services, and biodiversity. In: Reidmiller, D.R., Avery, C.W., Easterling, D.R., Kunkel, K.E. and Lewis, K.L.M. et al. (eds) *Impacts, Risks, and Adaptation in the United States: Fourth National Climate Assessment*, Vol. Volume II. U.S. Global Change Research Program, Washington, DC, pp. 268–321. DOI: 10.7930/NCA4.2018.CH7.

Livensperger, C., Steltzer, H., Darrouzet-Nardi, A., Sullivan, P.F., Wallenstein, M. *et al.* (2019) Experimentally warmer and drier conditions in an arctic plant community reveal microclimatic controls on senescence. *Ecosphere* 10(4), e02677. DOI: 10.1002/ecs2.2677.

Lobell, D.B., Schlenker, W. and Costa-Roberts, J. (2011) Climate trends and global crop production since 1980. *Science* 333(6042), 616–620. DOI: 10.1126/science.1204531.

Lobell, D.B., Hammer, G.L., McLean, G., Messina, C., Roberts, M.J. *et al.* (2013) The critical role of extreme heat for maize production in the United States. *Nature Climate Change* 3(5), 497–501. DOI: 10.1038/nclimate1832.

Loladze, I. (2014) Hidden shift of the ionome of plants exposed to elevated CO_2 depletes minerals at the base of human nutrition. *eLife* 3, e02245. DOI: 10.7554/eLife.02245.

Mach, K.J., Mastrandrea, M.D., Bilir, T.E. and Field, C.B. (2016) Understanding and responding to danger from climate change: the role of key risks in the IPCC AR5. *Climatic Change* 136(3–4), 427–444. DOI: 10.1007/s10584-016-1645-x.

Martinich, J., Crimmins, A., Beach, R.H., Thomson, A. and McFarland, J. (2017) Focus on agriculture and forestry benefits of reducing climate change impacts. *Environmental Research Letters* 12(6), 060301. DOI: 10.1088/1748-9326/aa6f23.

Masipa, T.S. (2017) The impact of climate change on food security in South Africa: current realities and challenges ahead. *Jamba* 9(1), 411. DOI: 10.4102/jamba.v9i1.411.

Medina, A., Rodriguez, A. and Magan, N. (2014) Effect of climate change on aspergillus flavus and aflatoxin B1 production. *Frontiers in Microbiology* 5, 348. DOI: 10.3389/fmicb.2014.00348.

Mekonnen, Z.A., Riley, W.J., Berner, L.T., Bouskill, N.J., Torn, M.S. *et al.* (2021) Arctic tundra shrubification: a review of mechanisms and impacts on ecosystem carbon balance. *Environmental Research Letters* 16(5), 053001. DOI: 10.1088/1748-9326/abf28b.

Mills, G., Pleijel, H., Malley, C.S., Sinha, B., Cooper, O.R. *et al.* (2018) Tropospheric ozone assessment report: present-day tropospheric ozone distribution and trends relevant to vegetation. *Elementa* 6, 47. DOI: 10.1525/elementa.302.

Molotoks, A., Smith, P. and Dawson, T.P. (2021) Impacts of land use, population, and climate change on global food security. *Food and Energy Security* 10(1), e261. DOI: 10.1002/fes3.261.

Myers, S.S., Zanobetti, A., Kloog, I., Huybers, P., Leakey, A.D.B. *et al.* (2014) Increasing CO_2 threatens human nutrition. *Nature* 510(7503), 139–142. DOI: 10.1038/nature13179.

Myers, S.S., Smith, M.R., Guth, S., Golden, C.D., Vaitla, B. *et al.* (2017) Climate change and global food systems: potential impacts on food security and undernutrition. *Annual Review of Public Health* 38, 259–277. DOI: 10.1146/annurev-publhealth-031816-044356.

Nelson, G.C., Valin, H., Sands, R.D., Havlík, P., Ahammad, H. *et al.* (2014) Climate change effects on agriculture: economic responses to biophysical shocks. *Proceedings of the National Academy of Sciences* 111(9), 3274–3279. DOI: 10.1073/pnas.1222465110.

Oppenheimer, M., Little, C.M. and Cooke, R.M. (2016) Expert judgement and uncertainty quantification for climate change. *Nature Climate Change* 6(5), 445–451. DOI: 10.1038/nclimate2959.

Peña-Lévano, L.M., Taheripour, F. and Tyner, W.E. (2019) Climate change interactions with agriculture, forestry sequestration, and food security. *Environmental and Resource Economics* 74(2), 653–675. DOI: 10.1007/s10640-019-00339-6.

Phiiri, G.K., Egeru, A. and Ekwamu, A. (2016) Climate change and agriculture nexus in sub-Saharan Africa: the agonizing reality for smallholder farmers. *International Journal of Current Research and Review* 8(2), 57–64.

Porter, J.J., Dessai, S. and Tompkins, E.L. (2014) What do we know about UK household adaptation to climate change? A systematic review. *Climatic Change* 127(2), 371–379. DOI: 10.1007/s10584-014-1252-7.

Powell, J.P. and Reinhard, S. (2016) Measuring the effects of extreme weather events on yields. *Weather and Climate Extremes* 12, 69–79. DOI: 10.1016/j.wace.2016.02.003.

Ramesh, K., Matloob, A., Aslam, F., Florentine, S.K. and Chauhan, B.S. (2017) Weeds in a changing climate: vulnerabilities, consequences, and implications for future weed management. *Frontiers in Plant Science* 8, 95. DOI: 10.3389/fpls.2017.00095.

Ray, D.K., West, P.C., Clark, M., Gerber, J.S., Prishchepov, A.V. *et al.* (2019) Climate change has likely already affected global food production. *PLoS One* 14(5), e0217148. DOI: 10.1371/journal.pone.0217148.

Raza, A. (2022) Metabolomics: a systems biology approach for enhancing heat stress tolerance in plants. *Plant Cell Reports* 41(3), 741–763. DOI: 10.1007/s00299-020-02635-8.

Raza, A., Razzaq, A., Mehmood, S.S., Zou, X., Zhang, X. *et al.* (2019) Impact of climate change on crops adaptation and strategies to tackle its outcome: a review. *Plants* 8(2), 34. DOI: 10.3390/plants8020034.

Raza, A., Tabassum, J., Kudapa, H. and Varshney, R.K. (2021) Can omics deliver temperature resilient ready-to-grow crops? *Critical Reviews in Biotechnology* 41(8), 1209–1232. DOI: 10.1080/07388551.2021.1898332.

Raza, A., Tabassum, J., Mubarik, M.S., Anwar, S., Zahra, N. *et al.* (2022) Hydrogen sulfide: an emerging component against abiotic stress in plants. *Plant Biology* 24(4), 540–558. DOI: 10.1111/plb.13368.

Rojas-Downing, M.M., Nejadhashemi, A.P., Harrigan, T. and Woznicki, S.A. (2017) Climate change and livestock: impacts, adaptation, and mitigation. *Climate Risk Management* 16, 145–163. DOI: 10.1016/j.crm.2017.02.001.

Rosenzweig, C., Elliott, J., Deryng, D., Ruane, A.C., Müller, C. *et al.* (2014) Assessing agricultural risks of climate change in the 21st century in a global gridded crop model intercomparison. *Proceedings of the National Academy of Sciences* 111(9), 3268–3273. DOI: 10.1073/pnas.1222463110.

Seidel, P. (2014) Extremwetterlagen und Auswirkungen auf Schaderreger–extreme Wissenslücken. *Gesunde Pflanzen* 66(3), 83–92. DOI: 10.1007/s10343-014-0319-8.

Selvaraj, S., Ganeshamoorthi, P. and Pandiaraj, T. (2013) Potential impacts of recent climate change on biological control agents in agro-ecosystem: a review. *International Journal Biodiversity Conservation* 5, 845–852.

Seneviratne, S.I., Nicholls, N., Easterling, D., Goodess, C.M., Kanae, S. *et al.* (2012) Changes in climate extremes and their impacts on the natural physical environment. In: Field, C.B., Barros, V., Stocker, T.F., Qin, D. and Dokken, D.J. (eds) *Managing the Risks of Extreme Events and Disasters to Advance Climate Change Adaptation. A Special Report of Working Groups I and II of the IPCC*. Cambridge University Press, Cambridge, UK, pp. 109–230.

Shah, A.A., Khan, W.U., Yasin, N.A., Akram, W., Ahmad, A. *et al.* (2020) Butanolide alleviated cadmium stress by improving plant growth, photosynthetic parameters and antioxidant defense system of brassica oleracea. *Chemosphere* 261, 127728. DOI: 10.1016/j.chemosphere.2020.127728.

Smith, M.R. and Myers, S.S. (2018) Impact of anthropogenic CO_2 emissions on global human nutrition. *Nature Climate Change* 8(9), 834–839. DOI: 10.1038/s41558-018-0253-3.

Syed, A., Raza, T., Bhatti, T.T. and Eash, N.S. (2022) Climate impacts on the agricultural sector of Pakistan: risks and solutions. *Environmental Challenges* 6, 100433. DOI: 10.1016/j.envc.2021.100433.

Tol, R.S.J. (2018) The economic impacts of climate change. *Review of Environmental Economics and Policy* 12(1), 4–25. DOI: 10.1093/reep/rex027.

Ullah, A., Bano, A. and Khan, N. (2021) Climate change and salinity effects on crops and chemical communication between plants and plant growth-promoting microorganisms under stress. *Frontiers in Sustainable Food Systems* 5, 618092. DOI: 10.3389/fsufs.2021.618092.

Uyttendaele, M., Liu, C. and Hofstra, N. (2015) Special issue on the impacts of climate change on food safety. *Food Research International* 68, 1–6. DOI: 10.1016/j.foodres.2014.09.001.

Varanasi, A., Prasad, P.V.V. and Jugulam, M. (2016) Impact of climate change factors on weeds and herbicide efficacy. *Advance Agronomical* 15, 107–146. DOI: 10.1016/bs.agron.2015.09.002.

Vose, R.S., Easterling, D.R., Kunkel, K.E., LeGrande, A.N. and Wehne, M.F. (2017) Temperature changes in the United States. In: Wuebbles, D.J., Fahey, D.W., Hibbard, K.A., Dokken, D.J. and Stewart, B.C. *et al.* (eds) *Climate Science Special Report: Fourth National Climate Assessment*, Vol. Volume I. U.S. Global Change Research Program, Washington, DC, pp. 197–199. DOI: 10.7930/J0N29V45.

Wu, W., Verburg, P.H. and Tang, H. (2014) Climate change and the food production system: impacts and adaptation in china. *Regional Environmental Change* 14(1), 1–5. DOI: 10.1007/s10113-013-0528-1.

Wuebbles, D.J., Fahey, D.W., Hibbard, K.A., Dokken, D.J., Stewart, B.C. *et al.* (2017) *Climate Science Special Report: Fourth National Climate Assessment*, Vol. Volume I. U.S. Global Change Research Program, Washington, DC. DOI: 10.7930/J0J964J6.

3 Sustainable Agriculture, Food Safety, and Climate Change

Akarsh Parihar[1], Jalpa Dobaria[2] and Radadiya Chirag[1*]
[1]Department Genetics and Plant Breeding, B. A. College of Agriculture, Anand Agricultural University, Anand, India; [2]B. A. College of Agriculture, Anand Agricultural University, Anand, India

Abstract
With the population growing so fast worldwide, food production is facing unprecedented barriers by present agriculture, mainly with malnutrition, adverse variations in climate, overexploitation of natural resources, loss of biodiversity, food contamination, and food loss. Our current food and agriculture systems are failing to address the key challenges, while millions still go hungry or malnourished. This chapter explores different aspects related to farming to food for a sustainable life for one and all. For a solution, the farming system should have a balanced ecosystem as well as economic benefits. It should also positively affect major biological and ecological processes for healthy soil and a pollution-free environment. All these efforts seem fulfilled when food on the plate is safe. That food channel must avoid foodborne disease, as various pathogens depend upon it for their survival and life cycle.

Keywords: Food Production, Malnutrition, Biodiversity, Ecosystem, Farming System

3.1 Introduction

There is a need for urgent effort in order to boost agricultural production, improve the global food supply chain, and ensure availability of nutritious food. People are suffering from hunger and malnutrition. If agriculture and food systems are unsustainable, fail to benefit those whose livelihoods depend on it, rely on outdated approaches and technologies, or if access to resources, inputs, and markets is limited, then it may cause a huge crisis worldwide. Various international committees believe that it is possible to eliminate hunger within the next few years and are working together to achieve this goal (Muhie, 2022). For example, a few years ago, the UN launched the zero-hunger challenge, which sought to ensure zero stunted children under the age of 2, 100% access to adequate food all year-round, increase in smallholder productivity and income, zero loss or waste of food, among others. Ending hunger and achieving food security and improved nutrition has interlinkages with supporting sustainable agriculture, empowering small landholder farmers, ending rural poverty, ensuring healthy food, tackling climate change and other issues. Beyond adequate calorie intake, proper nutrition has other dimensions, including micronutrient availability, vitamin-rich food, and balanced healthy diets (Singh *et al.*, 2014). Extreme poverty and hunger are predominantly rural,

*Corresponding author: radadiyachirag111@gmail.com

DOI: 10.1079/9781800625037.0003

with smallholder farmers and their families making up a very significant proportion of the poor and hungry. Thus, eliminating poverty and hunger is integrally linked to boosting food production and agricultural productivity.

3.2 Sustainable Agriculture

Agricultural systems must globally become more productive as well as less wasteful. Sustainable agricultural practices, including both production and consumption, must be pursued from an integrated perspective. Healthy soils, pollutant-free water, and superior plant genetic resources are key inputs into food production, and their growing scarcity in many parts of the world makes it imperative to use and manage them sustainably. The current acceleration of soil degradation is due to unsustainable land management practices, including deforestation, excessive use of agrochemicals, frequent tillage, overgrazing, and poor irrigation causing soil salinization. Furthermore, soil degradation depletes food of its nutritional value, which could additionally be contaminated by the uptake of heavy metals or by the deficit of essential micronutrients (Zdruli and Zucca, 2023). Finally, poor soil health has an impact on human health, accelerated by water and air contamination. For a solution, high yields on existing agricultural lands, including restoration of degraded lands, through sustainable agricultural practices would also relieve the pressure to clear forests for agricultural production (Singh *et al.*, 2022). Management of scarce water through improved irrigation and novel storage technologies, combined with development of new drought-resistant crop varieties, can also contribute to sustaining dry lands' productivity. Halting and reversing land degradation will also be critical to meeting future food needs. Given the current extent of land degradation globally, the potential benefits of land restoration for food security and for mitigating climate change are enormous. Sustainable agriculture and food safety depends directly on climate change. Considering expected changes in temperatures, precipitation, and pests associated with climate change, the global community is called upon to increase investment in research, development, and demonstration of technologies to improve the sustainability of food systems everywhere. Building resilience of local food systems will be critical to averting large-scale shortages in the future, ensuring food security and good nutrition for all.

3.3 Integrated Farming System

Integrated farming system (IFS) is a sustainable agricultural system that integrates livestock, crop production, fish, poultry, tree crops, plantation crops, and other systems that benefit each other. IFS approach is considered to be the most powerful tool for enhancing profitability of farming systems especially for small and marginal farmers to make them bountiful. It is hypothesized that these complex farms would be more productive at a system level and produce fewer negative externalities than simplified farms. Farming system research is an agricultural research and development approach that views the whole farm as a system and examines how these components interact with each other with respect to physical, biological, and socio-economic factors not under the household's control (Dey *et al.*, 2012). Efficiently managed IFSs are expected to be less risky, as they benefit from enterprise synergies, product diversity, and ecological reliability. This approach encourages ecological intensification and aims to reduce use of anthropogenic inputs with enhanced ecosystem functioning like nutrient recycling, soil formation, soil fertility improvement, and enhancing environmental performance (Fig. 3.1) (Salton *et al.*, 2014). High input of pesticides and fertilizer cause a huge impact on environmental degradation, leading to loss of soil fertility and a rise in pollution, thus causing irreversible damage to the environment. It can be rectified to some extent by adopting IFS. The higher net income required for IFS would be due to reduced production costs by recycling by-products and residue of different components within the system.

The input cost especially critical inputs like fertilizers, pesticides, and herbicides consumption can be reduced by the adoption of IFS through encouraging resource flow and integrated pest and nutrient management. The

Fig. 3.1. Positive benefits from an integrated farming system.

higher product diversification in IFS, especially from livestock (i.e. dairy and poultry), has the potential to generate daily income for small and marginal farmers. The livestock component such as dairy, goatery, poultry, and piggery will act as farm insurance at the time of crop failure. Rautaray *et al.* (2005) reported that the rice–fish system under lowland ecologies with vegetables, fruits, ornamental plants, and agroforestry components has the potential to produce 2.8 times higher income over rice crop alone. The coconut-based IFS study at ICAR-CPCRI, Kasaragod, produced 19,125 nuts, 9275 l milk, 526 kg poultry, 50 kg Japanese quail, and 400 kg fish from 1.04 ha (Reddy and Biddappa, 2000). This system highlighted that the increased residue recycling and tree components offset the negative effect of climate change by sequestrating more carbon into the soil and aboveground biomass of the trees. The IFS approach has the capacity to reduce methane (CH_4) absorption, as observed previously by Dong *et al.* (2000) and Chen *et al.* (2010), and reported 30% lesser CH_4 absorption under temperate plains due to increased nutrient recycling through organic farming practices thus improved the abundance and activity of methanotrophs in soil. Methanotrophs possibly decreased air diffusion, thus impairing CH_4 diffusion. Therefore, management practices such as nutrient management through composting and crop residues, use of legumes for nitrogen fixation, and change in cultivation practices can increase the resilience of crops to changing climate and also can also reduce greenhouse gas emission.

Farm enterprises can be integrated based on a variety of criteria. Mainly, it includes soil texture, climatic conditions, resource availability, land, labor, capital, present level of utilization of resources, and management skill of the farmer.

Optimal conditions for introduction of IFS:

1. The farmer is seeking to maximize profits on existing holding.
2. The farmer is looking to reduce chemical control methods.
3. The farm household is struggling to buy food or is below the poverty line.
4. The farmer wants to improve the soil quality and fertility.
5. The farmer wants to gain profits from allied activities.
6. The farm is being eroded by wind and/or water.
7. Chemical inputs are expensive or unavailable.
8. Soil fertility has decreased as a result of inorganic fertilizer use.
9. The farm has been suffering from water scarcity.
10. The land is fragmented.
11. The community wants to reduce pollution.
12. The community wants to maintain biodiversity.

3.3.1 Methodology to organize farming systems

Improved agronomic management and conservation of biodiversity lead to resilient, productive, and sustainable systems and can reduce environmental pollution. The impact of climate change can be reversed or reduced by carbon sequestration and nutrient availability through agroforestry, alley cropping, and intercropping. In general, monoculture crop systems affect soil biology through genetic erosion, reduction in groundwater availability, and several other environmental problems. Farmers choose crop diversity on small farms considering several factors, including increased nutrition, market diversification, and risk mitigation. Multi-enterprise schemes like IFS have the potential to enhance ecological function through biodiversity restoration, in addition to expanding whole-system economic and agronomic productivity. Agricultural diversification occurs when a farm or agricultural community adds more plants, plant varieties, or animal breeds. The IFS promotes growing multiple crops together as intercrops, mixed crops, or sequential crops, thereby providing ecosystem services from agriculture (nutrient recycling, improved soil quality, reduced economic losses due to crop failure, nitrogen fixation, water penetration, and pollination). It also stimulates soil microbial biodiversity through the addition of compost or manure or by duck droppings in rice–fish–duck culture. Structural variation in soil microbial diversity is observed through organic manure recycling with the production of plankton and macrobenthos in rice–fish–duck, rice–duck, and rice–fish systems over conventional rice production systems (Nayak *et al.*, 2018). Integration of livestock or fish with crops (i.e., cover cropping, fodder production, and rotational grazing) at the field scale results in heterogeneous landscapes which supports overall agrobiodiversity (Perfecto *et al.*, 2005).

Thus, having multiple enterprises encourages biodiversity conservation; provides feed, fodder, and fuel; and also reduces risks associated with crop failure. The well-integrated complementary IFS provides the dietary needs of farm families partially or fully from a small piece of land. It offers scope to utilize land and time to grow short-duration vegetable crops, pulses, and fodder for livestock. These systems are very critical for achieving future food and nutrition for the burgeoning population (Paramesh *et al.*, 2021).

Despite several advantages, there are several constraints in different regions so that farmers are unable to adopt the IFS systems: financial, biophysical, sociocultural, institutional, or policy problems. Lack of required finance due to the high cost of inputs is a major limitation in adopting crop-livestock integrated system, as a high initial investment is required for the establishment of animal sheds, purchase of livestock, and so on. Further, non-availability of quality planting material, lack of skills and knowledge of new crops such as fodder, and lack of access to veterinary services are also significant factors. The idiosyncratic character and attitude of farmers is also found as a major criterion in adopting an IFS system (Purnomo *et al.*, 2021), and farmers are resilient to change and were found to be laggards in the adoption of new technology, improved crops, and livestock breeds. Hence, based on suitable motivation and encouragement through training and demonstration, along with credit facilities and ensured supply of required quality planting materials, the farmer's attitude, and adoption of IFS could be modified appropriately. As farming is very closely associated with the environment, it will have a greater impact on soil, water, landscape, and biodiversity. Hence, there is a need for a specific policy to provide crop-specific price, insurance, and income support to insulate the farmers from market fluctuations and conserve the agroecological assets, maintaining ecological balance through sustainable use of natural resources. Future research should further investigate the well-being of laborers, farmers, and consumers, and their interaction with farm size and with other social and environmental outcomes.

IFS is adoptable, with reducing risks and raising profits being the primary goals. By recycling resources, about 90–95% of the nutritional needs are met on their own, lowering cultivation costs and boosting employment and profit margins. Maintaining the integration of crops, livestock, fisheries, apiculture, and even biogas on their farms can enhance the farmers' net income. Farmers seeking to improve their income to lift themselves out of poverty will have to focus on integrating more businesses of several kinds onto their farms (Sheikh *et al.*, 2021).

An advantage of IFS is that the production cost can be reduced by the use of the least expensive waste material from one component. Thus, mostly, there is a linkage established between the use of waste materials and the removal of intermediary interference. Secondly, natural components can be combined to create a variety of nutrition sources to provide a balanced diet. Waste materials are efficiently recycled in IFS, thus lowering pollution. Interaction of business crops, eggs, milk, mushrooms, honey, silkworm cocoons, and so on, provides the farmer with a consistent annual income stream. Through agricultural intensification and related businesses, IFS maximizes space usage and offers a chance to raise economic yield per unit area per unit time. By appropriate crop rotation and organic compost, farmers may be able to improve soil fertility, enhance soil physical structure, and reduce weeds, insect, pests, and diseases. In addition, IFS assures the restoration of system efficiency. IFS is also a solution to deforestation, thereby ensuring the ecosystem's sustainability. An efficient means of recycling biogas can be established from the system's organic wastes. Every piece of land area is effectively utilized by IFS, such as by the plantation of perennial legume fodder trees on the borders of the field. These procedures will significantly alleviate the issue of the animal component associated with poor-quality feed availability.

3.3.2 Recommended approaches for IFS

- Guarantee prompt delivery of high-quality inputs, including planting materials and seeds, among other agroinputs.
- Conduct strategic and fundamental research on production technologies to increase farming system mode efficiency in the use of agricultural resources.
- Conduct on-farm testing, verification, and improvement of technologies for system-based farm productivity.
- Evaluation and improvement of low-cost methods and instruments for value addition and storage of fruits, vegetables, milk, fish, and other farm goods on-site.
- Using decision support systems, resource inflow, and input-output linkages are optimized in various IFS models.
- Developing training and demonstration programs to increase the capability of stakeholders in IFS.
- Optimal nutrient supply and site-specific nutrient management strategies for intensive cropping systems.
- Options and interventions tailored to a particular location for crop diversification, involving high-value, low-volume crops and corresponding production technologies, through intensification, substitution, or interruption.
- The creation of integrated plant nutrient supply systems is necessary for increased productivity and resource sustainability,

particularly in areas that are extensively farmed.

- The creation of economically viable methods for managing agricultural residues *in situ*.
- Establishing an appropriate electronic communication channel between scientists and farmers.
- The identification and standardization of novel cropping systems suitable for protected agriculture and resource conservation methods.
- Enhancing the knowledge and abilities of researchers, extension agents, and rural business owners in significant fields of endeavor.
- Information on organic farming is shared via field trips, agriculture extension offices, and print, electronic, and online media.
- Assessment of credit delivery system and insurance in agriculture.
- The installation of block-level, farmer-friendly IFS information kiosks in local tongues.

3.4 Biological and Ecological Processes

Sustainable economic growth depends on the effective management of natural resources; the resource conversation agenda is threatened by the growing ecological footprints that are creating a biocapacity deficit. The management of natural resources is crucial to sustaining economic activity for both the current and upcoming generations. It is challenging to manage natural resources effectively due to widespread deforestation, depletion of mineral resources, reduction of energy and oil, and coal burning. Global resource deterioration is caused by a multitude of factors, including population growth, a lack of awareness about resource conservation, rising fuel use, incorrect financial markets, natural and technological disasters, deteriorating resource extraction processes, and a lack of green financing. Three main elements are thought to be responsible for the destruction of natural resources: (i) industrialization, (ii) ecological footprints, and (iii) rapid population increase. An increase in the number of foreign migrants brings with it greater difficulties, worsening the state of the economy and the environment. Because of the excessive use of natural and financial resources, massive population increase makes it more difficult to achieve goals related to resource sustainability. Royalties from natural resources can be raised, epidemic shocks can be prevented, and population expansion can be reined in to increase the environmental sustainability index (Lee *et al.*, 2021).

Climate change and globalization have decreased food security, as natural resources, including soil quality, biodiversity, and water quality, have been contaminated. Food security exists when all people, at all times, have access to sufficient, safe, and nutritious food to meet their dietary needs and food preferences for an active and healthy life (Flora, 2010). Considerable progress has been made in the previous 3 years in comprehending some of the major processes and how they are reflected in global land models. These include the impact of plant-microbe interactions on nutrient cycles, the transformation of inorganic soil phosphorus, and the release of nitrogen from rocks. The quantity of anthropogenic CO_2 emissions that would otherwise remain in the atmosphere has been reduced, since the land biosphere has absorbed over 25% of the total anthropogenic CO_2 emissions since the beginning of the industrial revolution. This carbon sink over land is presently slowing down anthropogenic global warming, together with the ocean's carbon sink. It's very unclear how long this terrestrial carbon sink will last into the future. The degree to which soil nutrients influence the strength of the land carbon sink is a significant source of uncertainty.

A healthy, functioning soil ensures nutrient cycling for optimum plant growth for agricultural production. Agricultural productivity is often limited by available soil nutrients, especially nitrogen. Nitrogen is not present in soil parent material, despite the fact that nitrogen content in the atmosphere is highest among all the atmospheric gases (Hedin *et al.*, 2009).

The reaction of biological nitrogen fixation (BNF) to rising atmospheric CO_2 and climate change is expected to have a major effect on the carbon balance of natural ecosystems. BNF is considered the biggest natural nitrogen input to the land biosphere, contributing 60–110 TgN/

year. Soil nitrogen input for plant nutrition and crop productivity largely depends on organic matter degradation, synthetic fertilizer applications, and BNF via nitrogenase, an enzyme that catalyzes the conversion of N_2 gas to ammonia (NH_3). This limited bioavailability of nitrogen and the escalating reliance of crop growth on nitrogen have created a colossal nitrogen-based fertilizer industry worldwide. Nitrogen makes up 50% of dry plant tissue, being engaged in one way or another and limits primary productivity in terrestrial ecosystems. As a vital component of nucleic acids, amino acids, and chlorophyll, it is essential to the production of agroecosystems. In order to sustain high yields, the majority of intensive agricultural production systems apply large amounts of synthetic nitrogen fertilizer, which significantly changes the global nitrogen cycle. Nitrogen is required to compensate for crop harvest losses, which cause agroecosystems to lose a significant amount of their nitrogen-rich organic matter. Through the conversion of atmospheric N_2 into biologically accessible forms of reactive nitrogen, the intense cultivation of nitrogen-fixing crops has also increased the amount of reactive nitrogen in cropping systems. Adding legume biomass to soils can supply 23–176 kg ha^{-1} of nitrogen, which is comparable to inorganic additions. Looking closer, one of the main objectives of sustainable agroecosystems is nitrogen cycling. It entails field-level management techniques to encourage equilibrium between the inorganic and organic nutrient pools in the soil.

The economic and environmental costs of the heavy use of chemical nitrogen fertilizers in agriculture are a global concern. BNF, a microbiological process which converts atmospheric nitrogen into a plant-usable form, offers a good alternative to synthetic fertilizers. More than 50% of the applied nitrogen-based fertilizer is used by the plants and the remaining can be subjected to losses like surface runoff and leaching, leading to nitrate contamination of soils and groundwater. In addition, manufacturing nitrogen-based fertilizers requires six times more energy than that needed to produce either phosphorous- or potassium-based fertilizers (Da Silva *et al.*, 1978). Nitrogen-fixing systems offer an economically attractive and ecologically sound means of reducing external inputs and improving internal resources. Symbiotic systems such as that of legumes and *Rhizobium* can be a major source of nitrogen in most cropping systems and that of *Azolla* and *Anabaena* can be of particular value to a flooded rice crop. Nitrogen fixation by associative, and free-living microorganisms can also be important. The diazotrophic microorganisms from bacteria or archaea domains are responsible for BNF, and only some prokaryotes are able to use atmospheric nitrogen through BNF by encoding nitrogenase. In fact, nearly 50% of the total nitrogen in crop fields is the contribution of BNF by diazotrophic bacteria (Ramirez-Puebla *et al.*, 2019). Moreover, fixed nitrogen can also be transferred to intercropped non-legumes in the case of mixed cropping systems, such as the soybean–wheat system, or the next season crops in crop rotation. Free-living diazotrophs correspond to a small fraction of the plant rhizospheres ecosystem, and they belong to alphaproteobacteria (i.e. Rhizobia, *Bradyrhizobia*, *Rhodobacteria*), betaproteobacteria (i.e. *Burkholderia*, *Nitrosospira*), gammaproteobacteria (i.e. *Pseudomonas*, *Xanthomonus*), firmicutes, and cyanobacteria. The symbiotic relationship between soil bacteria, collectively known as rhizobia, includes the genera *Rhizobium*, *Bradyrhizobium*, *Mesorhizobium*, and *Sinorhizobium*. Rhizobia and legume roots generate nodules that fix atmospheric nitrogen through the action of the nitrogenase enzyme. Thus, BNF is an extremely sensitive process influenced by nutrient and environmental conditions and enables a plant to supply all or part of its requirements through interactions with endosymbiotic, associative, and endophytic symbionts, thus offering a competitive advantage over any non–nitrogen-fixing plants (Mahmud *et al.*, 2020).

Plant growth and nutrient cycling depend on the recycling of nutrients from dead organic matter through microbe-mediated mineralization, as plants typically absorb nutrients from the soil in inorganic forms. Because soil microbes make up only a small portion of the total soil organic carbon and have a short turnover time, their role in nutrient mineralization has received little attention. Instead, the amount and quality of the substrate as well as environmental factors are typically taken into account when modelling this process. Plants absorb nutrients either directly from their roots or through symbiotic relationships with fungi called mycorrhizae, which help them extract nutrients from low-nutrient soils. In exchange for the photosynthetic carbon that plants produce, mycorrhizae mine nutrients that are inaccessible to plant roots. When soil nutrient concentration is high, the carbon cost of nutrient acquisition through mycorrhizae is much higher than direct root uptake; however, when soil nutrient availability is limited, the process becomes highly cost-effective. In the case of phosphorus, productivity can be restricted in agroecosystems and is a necessary component of DNA, RNA, ATP, and the phospholipids that make up cell membranes.

Phosphorus is essential for seed formation, flowering, fruiting, and root growth. Many plant species create symbiotic relationships with mycorrhizal fungi, which scavenge phosphorus from the soil and transfer it straight to root cells in order to deal with phosphorus limitation. Plant growth can be restricted by phosphorus, despite the fact that plants require very little phosphorus annually in comparison to the amount found in the soil. This restriction stems from two main factors: (i) most soil inorganic phosphorus is chemically or physically stable, making it inaccessible for direct biological uptake and (ii) the uneven distribution of soil phosphorus in space. In global land models, biochemical mineralization is frequently represented as a function of maximal biochemical mineralization rate, which is given a fixed value for a particular soil type or plant functional type (Wang and Goll, 2021). Biochemical mineralization is an important process for recycling organic soil phosphorus. Fires have two main effects on ecosystem nutrient cycling: directly, by changing the physical and chemical characteristics of soils and releasing nutrients from burning biomass, and indirectly, by changing the composition and structure of ecosystems. For many ecosystems, the impacts of fires on nitrogen cycling are well documented. The main indicator of how well plants absorb available nutrients in a cropping system is called nutrient utilization efficiency. NUE calculations typically consist of two main components: (i) the amount of the applied nutrient that was neither incorporated into plant biomass nor made available to subsequent crops and (ii) the recovery of the applied nutrient in the plant's harvestable biomass or in the total aboveground biomass. Since measuring residual nitrogen mineralization and absorption is challenging, most research focuses on fertilizer recovery in crops during a single growing season (Ladha *et al.*, 2005). Plant nutrient uptake and use efficiency may be influenced at the plant level by temperature, stand structure, insect impacts, and requirements for water and light. Nutrients that are lost to the environment at the field level and are not absorbed by plants or retained in stable soil reserves can cause both local and global pollution.

Our ability to feed people worldwide has increased significantly as a result of human modification of the nutrient cycles, though at a significant cost to the stability of earth system functions. Investing in field and landscape management techniques that increase yields per unit of input and reduce environmental losses is imperative in order to effectively tighten nutrient cycles in agroecosystems.

3.5 Management Practices to Maximize Nutrient Cycling and Nutrient Use Efficiency

The term "efficient use of all nutrient sources" refers to nutrient management, and the main obstacles to maintaining soil fertility are as follows:

1. Minimize the loss of nutrients
2. Preserve or enhance the ability of plants to store nutrients
3. Encourage the recycling of plant nutrients
4. Add extra nutrients in the right amounts

Furthermore, cultural practices that encourage the growth of robust, healthy root systems also lead to effective nutrient uptake and use. Establishing varied crop rotations, lowering tillage, managing and maintaining crop residue, cultivating cover crops, handling manure as a valuable nutrient source, composting and utilizing all available wastes or by-products, liming to maintain soil pH, adding supplemental fertilizers and conducting routine soil tests are just a few cultural practices that aid in achieving these goals. It is crucial to integrate the application of these beneficial management strategies and consider their effects on the entire soil-crop system, rather than just one part of it, because they have numerous implications on soil fertility and nutrient cycling. There are numerous effective farming methods, so various approaches or combinations of techniques are suitable for various systems to achieve comparable objectives.

The fact that yields for a crop cultivated in rotation with other crops are typically 5–15% higher than for that same crop grown continuously is known as the "rotation effect." Planting a range of crops in succession has several beneficial benefits on soil fertility, while the exact source of higher yields is sometimes unclear and most likely not singular. Deep-rooted crops alternate with shallower, fibrous-rooted species in a varied rotation to draw nutrients up from lower soil levels. By doing this, nutrients that could otherwise be lost from the system are captured. When a variety of different crops are cultivated, variations in plant rooting patterns, such as root density and root branching at different soil depths, also lead to more effective nutrient extraction from all soil layers. Sod-forming crops improve soil organic matter and reduce runoff and erosion-related losses of soil and nutrients by being planted in rotation with row crops. Planting legumes to fix atmospheric nitrogen improves the amount of nitrogen stored in organic matter for future crops and decreases the requirement for purchased fertilizer. The most effective utilization of biologically fixed nitrogen occurs in crop rotations, where crops with high nitrogen requirements come after legumes. By providing a variety of residue types and food sources, rotating crops also improves soil biodiversity and nutrient cycling capacity. It also breaks down disease and pest cycles by reducing the accumulation and carryover of soilborne disease organisms and insect pests, and it can aid in establishing ideal growing conditions for robust, healthy crop root systems. Soil and water conservation is also important. Topsoil, which has the highest concentration of organic matter and nutrients, is often lost due to soil erosion. Restricting runoff and erosion through the implementation of soil and water conservation measures reduces nutrient losses and maintains soil productivity. Soil erosion is primarily caused by tillage techniques, crop residue cover, soil topography, structure, and drainage. By limiting water transport over the soil and lowering soil particle dissociation from wind or raindrop contact, surface residue reduces erosion. The amount of crop residue remaining on the soil's surface is controlled by tillage techniques. Remainder coverage is maximized by little or no tillage. Because water travels quickly and is more erosive on steep slopes, it is best to minimize tillage, leave surface residue, cultivate sod crops, and plant in contour strips or on the contour itself. One way to think of cultivating cover crops and green manure crops is as a kind of crop rotation, where the growing season is extended by interspersing non–revenue-generating crops between yearly cash crops. As a result, many of the advantages are identical to those obtained through crop rotation.

Adding legume cover crops to the soil fixes nitrogen organically. By adding more plant diversity, cover crops encourage a wider range of soil microbes, improve the cycling of nutrients and carbon, and strengthen root health. Because the soil surface is covered for a greater portion of the year, runoff and erosion-related nutrient losses are minimized. Because the plants are growing for a longer time, they produce a larger amount of biomass, which increases the amount of organic matter that is added to the soil. Cover crops provide an extended growth period that prolongs root activity and increases the capacity of chemicals secreted from the roots to release insoluble soil nutrients (Tully and Ryals, 2017). A significant amount of the plant nutrients extracted from harvested crops are recycled when manure is returned to crop fields. A significant source of nutrient inputs for the entire farming system is manure applied to crop fields on farms, where livestock are given a lot of feed that is purchased off-farm. However, just as nutrients can be lost from the soil, they can also be lost from manure during handling, application, and storage. This can be inefficient financially as well as environmentally. Ammonia volatilization causes nitrogen to be easily lost during storage and when manure is not quickly integrated after field application. Applying manure at rates higher than what the crop needs results in nutritional losses as well. Farm fields can also benefit from the use of organic amendments, including biosolids, food processing wastes, animal by-products, yard waste, seaweed, and various types of composted materials, in addition to manure. Although biosolids are much cleaner now than they were

20 years ago and still contain most of the critical plant nutrients, rules governing their use on farms must be observed to avoid the risk of excessive trace metal accumulation. Furthermore, biosolids are not a suitable source of nutrients for production that is certified organic.

A decomposition process called composting is comparable to the organic matter's natural breakdown in soil. By integrating mobile nutrients into organic forms where they are more stable and less likely to be lost, proper composting preserves volatile and soluble nitrogen as well as other mobile nutrients found in waste products. Composting reduces the majority of organic waste and increases the viability of transporting and using various waste products in the field. Manure and other farm wastes can be handled more easily if they are composted on the property. The majority of organic materials are compostable, with all of it containing ingredients that are beneficial to plants. Recycling of any waste or by-products that are appropriate for use in soil-crop systems by direct field application or composting it should be encouraged. These methods increase the amount of organic matter in the soil and offer a steady supply of slowly released nutrients. In addition, certain composts have qualities that prevent disease, promoting healthy root development. In addition, inorganic wastes can be recycled through the soil to provide nutrients for plants. Regional variations exist in the materials available; nonetheless, among the waste items that can be beneficially reused include wood ash, rock dust from quarries, gypsum from scrubbers in power plants using high-sulfur coal, and waste lime from water treatment plants.

3.6 Integrated Pest Management

Pesticide use has significantly affected soil and irrigation water, as their residue remains in the cultivated area. There is an increased chance of sperm abnormality, infertility, spontaneous abortion, birth defects, and retarded child development at certain pesticide exposure levels. Negative reproductive outcomes, such as stillbirths, miscarriages, neonatal deaths, and congenital malformations, which are signs of chromosomal abnormalities in germ cells, were observed in cotton field workers exposed to pesticides. Numerous studies have discovered a positive correlation between fetal death and occupational pesticide exposure. Exposure to phthalates can damage the male reproductive system. Most reproductive problems are caused by monoester metabolites, which are produced in the liver. The human disorder known as testosterone deficiency syndrome, which is characterized by low sperm counts, cryptorchidism, hypospadias, and testicular cancer, is brought on by phthalate exposure. The complex chemical combinations that are employed and formed during the production process are exposed to workers in the plastics sector. Infertility, poor reproductive outcomes, and miscarriages were found in a study of women employed in the plastics industry. There are developmental, reproductive, genetic, acute, and sub-chronic toxicity signs associated with exposure in its inorganic form (Skakkebek *et al.*, 2001; Thompson *et al.*, 2009; DeMatteo *et al.*, 2013). Therefore, it must be replaced by other substances. The United Nations recommends agroecology both as a new approach to development and as a comprehensive alternative to the massive and hazardous use of pesticides, in order to meet the challenges of food rights and human rights. The many harmful consequences of the massive use of pesticides are faced by farmers, consumers, and society in general; biodiversity is being eroded, particularly that of insects and birds (Deguine *et al.*, 2021).

Integrated pest management (IPM) is a leading complement and alternative to synthetic pesticides and a form of sustainable intensification, with particular importance for tropical smallholders (Fig. 3.2). IPM systems may also deliver an array of ecosystem goods and services beyond pest control, increasing general resilience at farm and landscape scales (Pretty and Bharucha, 2015). Alam *et al.* (2016) found that there were significant differences between the treatment and the observed percentage of dead heart, grain weight, and yield in rice by IPM. Plant biostimulants like humic acid, fulvic acids, seaweed extracts, botanicals, protein hydrolysates, chitosan, and biopolymer are major substances that enhance plant nutrition (du Jardin, 2015). They promote plant growth, quality of plant products, and tolerance to abiotic stress. This includes cell

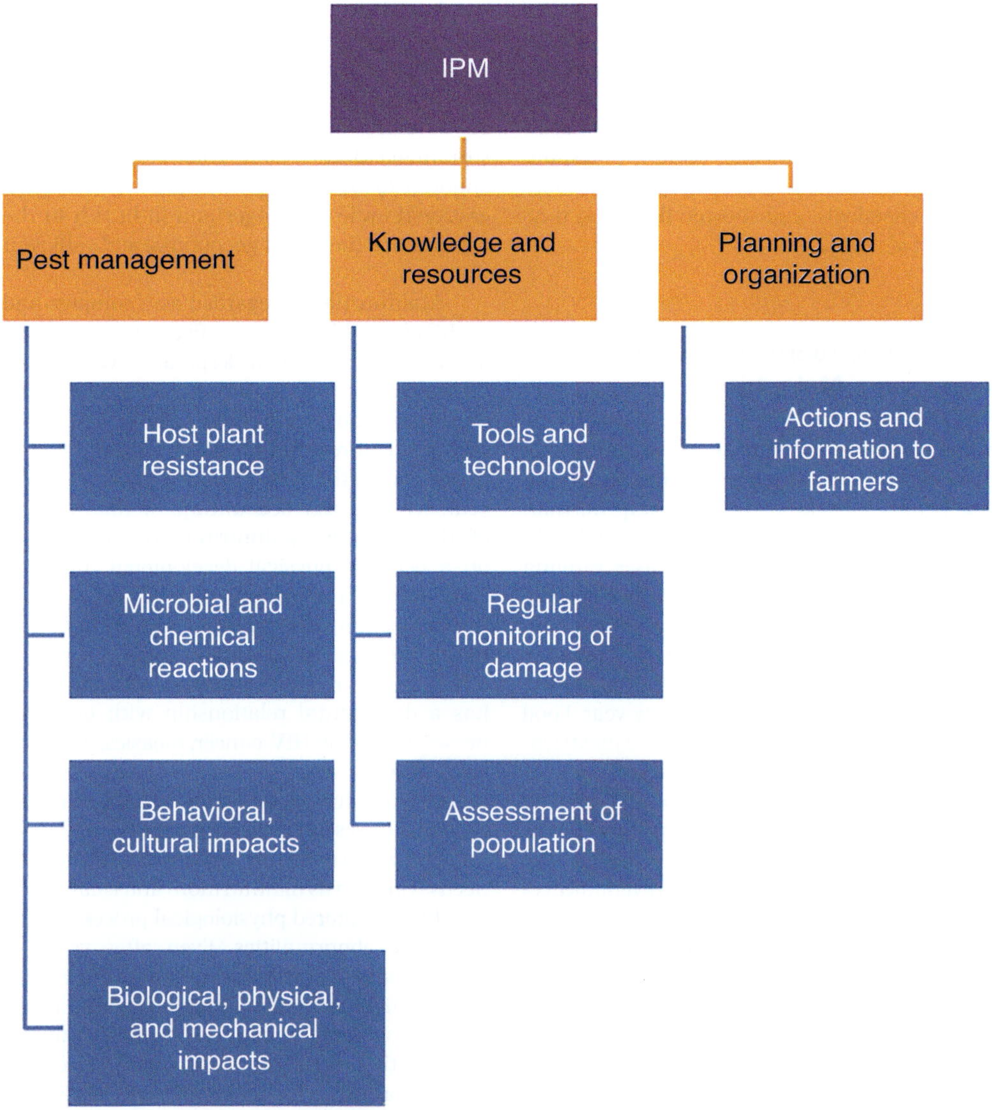

Fig. 3.2. Components of an integrated pest management system (IPM).

wall modification, osmoregulation, thermal regulation, antioxidant protection; interactions with symbionts, pathogen and herbivore response, and protection against heavy metals' toxicity, and thus can be used in farming for benefit to plants via ecological interactions in the cultivated ecosystem (Pilon-Smits *et al.*, 2009). Biofertilizers increase nutrient use efficiency and open new routes of nutrient acquisition by plants. Microbial biostimulants include mycorrhizal and non-mycorrhizal fungi, bacterial endosymbionts (like *Rhizobium*), and Plant Growth Promoting Rhizobacteria (PGPR) (Tiwari *et al.*, 2021; Singh *et al.*, 2023). Thus, microorganisms applied to plants can have a dual function of biocontrol agents and biostimulants. Organic farming and agroecology promote the use of such biological solutions and materials of biological origin. Biofertilizers have been found to boost plants' growth and yield by 10–40%.

When applied to the rhizosphere and the interior of the plant, they colonize the environment to promote plant growth and also add nutrients to the soil to improve soil fertility and crop yield and protect the plant against pests and diseases. They have been shown to enhance seedling survival, extend the root system's life, eliminate harmful chemicals, and shorten flowering time (Daniel *et al.*, 2022).

3.7 Foodborne Diseases and Malnutrition

Access to sufficient amounts of safe and nutritious food is key to sustaining life and promoting good health. Unsafe food, containing harmful bacteria, viruses, parasites, or chemical substances, causes more than 200 diseases, ranging from diarrhea to cancers. Almost 1 in 10 people globally fall ill after eating contaminated food, and 4.20 lakh die every year. Children under 5 years of age carry 40% of the foodborne disease burden, with 1.25 lakh deaths every year. Food safety, nutrition, and food security are inextricably linked. Unsafe food creates a vicious cycle of disease and malnutrition. DALYs attributed to these agents are most devastating to children younger than 5 years of age, the elderly, and the sick. To ensure food safety and to prevent unnecessary foodborne illnesses, rapid and accurate detection of pathogenic agents is essential.

Micronutrient deficiencies are associated with infectious disease, cognitive impairment, and growth retardation in infants and children (Black *et al.*, 2013). In addition, children in low- and middle-income families have high susceptibility to intestinal infections and parasitic infestations, which decrease nutrient absorption and appetite. Because of limited gastric capacity and increased nutrient requirements of infants, complementary foods should be diverse, with frequent consumption of nutrient-dense foods such as meat, milk, eggs, and fruits and vegetables (WHO, 2001). During pregnancy, the elevated demand for micronutrients puts pregnant women at an even higher risk of micronutrient deficiency. Some micronutrients are relatively well known; women at a reproductive age are recommended to take folic acid supplementation for the prevention of neural tube defect. If food is safe and uncontaminated, it will ensure good health of mothers and children.

It is well accepted that one of the best starting places for both intellectual and physical development is nutrition. This is a crucial sustainable development goal. For women and children to live, thrive, and end the intergenerational cycle of hunger and ill health in the community, they must get optimal and sufficient nutrition.

Malnutrition is regarded as a complex and multifaceted illness that affects people of all ages. It has received widespread coverage and grown significantly in importance as one of the most important health issues of our day. Based on statistical research, around 12 million children between the ages of 0 and 5 pass away each year, with hunger accounting for the majority of these deaths. Malnutrition has an impact on a person's physical development, mobility, cognitive function, length of hospital stay, institutionalization, low quality of life, and physical frailty among hospitalized patients, among other health-related outcomes. Malnutrition has a detrimental relationship with other illnesses, including HIV, cancer, measles, and diarrhea. This increases the number of deaths from malnutrition and causes enormous human and financial losses in developing nations. A state of imbalance between an organism's intake and needs is known as malnutrition, and it can cause weight loss, altered physiological processes, and metabolic abnormalities. Even after years of research and innumerable attempts, the burden of malnutrition remains frightening and is still regarded as a major cause of death worldwide. Malnutrition accounts for about 45% of the 12 million children under the age of 5 who die each year, the majority of whom are from poor nations (Fan *et al.*, 2022). Malnutrition causes significant clinical effects, including mortality, increased hospital visits, poor quality of life, physical weakness, and socioeconomic problems. It also forms connections with other infections.

Food by nature is biological. It is capable of supporting the growth of microbes that are potential sources of foodborne diseases. Microbes are the creators of epidemics and pandemics. Viruses are more responsible for the majority of foodborne illnesses, but hospitalizations and deaths associated with foodborne infections are due to bacterial agents. Illnesses range from mild gastroenteritis to neurologic, hepatic, and renal syndromes caused by either toxin from the disease-causing microbe. Foodborne bacterial agents are the leading cause of severe and fatal foodborne illnesses. Over 90% of food poisoning illnesses are caused by species of Staphylococcus, Salmonella, Clostridium, Campylobacter, Listeria, Vibrio, Bacillus, and E. coli. In the US and France in the 1990s, Salmonella was the most frequent cause of bacterial foodborne illness, accounting for 5700–10,200 cases, followed by Campylobacter for 2600–3500 cases and Listeria for 304 cases (Fung *et al.*, 2018).

Foodborne bacteria include *Bacillus cereus*, *Campylobacter*, *Clostridium perfringens*; Enterohemorrhagic *Escherichia coli*, Salmonella, *Shigella*, *Staphylococcus aureus*, as well as various types of Vibrio, especially *Vibrio cholerae* and *Vibrio parahaemolyticus*. Illnesses caused by enteric viruses are less well investigated, such as hepatitis A and E viruses, *caliciviruses* (e.g. norovirus), rotavirus, and possibly adenoviruses. Sources of salmonellosis include eggs, poultry, and other products of animal origin. Symptoms include fever, headache, nausea, vomiting, abdominal pain, and diarrhea. Foodborne *Campylobacter* is caused by raw milk, raw or undercooked poultry, and drinking water. *E. coli* are associated with unpasteurized milk, undercooked meat, and fresh fruits and vegetables. Listeria infection increases the risk of spontaneous abortions and stillbirths. Listeria is also found in unpasteurized dairy products and various ready-to-eat foods and can grow at refrigeration temperatures (Vaillant *et al.*, 2005). *Vibrio cholerae* infects through contaminated water or food. Symptoms include abdominal pain, vomiting, and profuse watery diarrhea, which may lead to severe dehydration and possibly death. Norovirus infections are characterized by nausea, explosive vomiting, watery diarrhea, and abdominal pain. Hepatitis A typically spreads through raw or undercooked seafood or contaminated raw produce (Todd and Todd, 2014). Food handlers infected with Hepatitis A virus are a common source of contamination.

Contaminants are introduced due to soil, air, and water pollution caused by human activity. The human body is primarily exposed to pollutants through three main routes: ingestion, oral absorption, and inhalation. Air pollution is generally caused by dangerous materials that come from both natural and artificial sources. Car emissions, power stations, burning trash, chemical businesses, and volcanic eruptions are the primary sources of pollutants. Many pollutants are emitted into the air that people breathe in, including sulfur dioxide, carbon monoxide, nitrogen oxides, heavy metals, biological contaminants, ozone, tobacco smoke, and so on. Ingestion of these contaminants disrupts the body's internal processes, leading to illnesses such as respiratory, cardiovascular, reproductive, and prenatal central nervous system problems, as well as cancer. Illnesses are caused by tobacco smoke, which contains toxic substances such as formaldehyde, benzene, cadmium, arsenic, and nicotine. Environmental pollutants cause oxidative stress and suppress the synthesis of antioxidant enzymes such as glutathione peroxidase, catalase, and superoxide dismutase. As a result, ROS are produced, which damage cells' lipids, proteins, carbohydrates, and DNA. Male infertility is caused by apoptosis and anomalies in germ cells, both of which are brought on by oxidative stress. The impacted sperm may transfer genetic damage to the growing baby, leading to early miscarriage or other developmental issues. Reproductive organs' capacity to function as antioxidants is reduced when mothers are exposed to PM2.5 (particulate matter) either before or during pregnancy. By generating reactive oxygen species, it impairs reproductive health and oxidative stress response. When granulocytes and oocytes are exposed to PM2.5, they may undergo apoptosis induction, activation of the apoptotic pathway, and mitochondrial dysfunction. When PM2.5 enters the bloodstream, chemokines and proinflammatory substances are generated. It causes an increase in free radical production and an increase in adhesion factors and cytokines produced by inflammatory cells. These factors amplify damage and set off a series of events that culminate in systemic immunological and inflammatory responses. Exposure to PM2.5 alters the amount of DNA methylation in the non-coding repetitive sections of the genome. This modification affects gene expression or genomic instability and adds to oxidative stress, inflammation, and health issues (Wang *et al.*, 2021). Numerous environmental pollutants possess carcinogenic and mutagenic qualities. Among all cancer types, 70–80% have been associated with environmental pollutants. The molecular processes of pollution-related carcinogenesis have received little attention in studies. Nonetheless, a number of inverse models show how pollution impacts cancer and cell division. A meta-analysis has confirmed the nonlinear dose-response relationship between DNA adducts and air pollution. There are multiple phases of carcinogenesis, including initiating, promoting, and progressing. Although the effects of pollutants on the transformation of cancer cells have been extensively researched and understood, they are affected by both time-dependent and individual doses. Certain carcinogens and their combinations cause chromosome instability, tumor suppressor gene inactivation, oncogene activation, somatic cell proliferation, inhibition of apoptosis, altered cell cycle dependent on TP53 activation, activation of energetic dysregulation, and other molecular processes. Users of herbicides like acetochlor had a higher chance of developing lung cancer than nonusers. According to certain research, the plastic addition of bisphenol A raises the risk of metabolic disorders, endometrial hyperplasia, recurrent miscarriages, polycystic ovarian syndrome, breast and prostate cancer, and infertility. Most agricultural insecticides are designed to affect the nervous system of the pest. Because of the similarities in neurochemical processes, exposure to pesticides causes neurotoxicity in humans. It is common to see neurodevelopmental anomalies such as learning disabilities, attention deficit hyperactivity disorder, autism spectrum disorders, developmental delays, and behavioral issues. Pesticides such as ethylene bismuth, carbamates and organophosphates, and herbicides containing thiocarbamates and chlorophenoxy may cause neurodevelopmental damage (Shetty *et al.*, 2023).

Further, there are several negative impacts of heavy metal toxicity on the human neurological system and brain, especially among children. However, these poisons only make up a small portion of environmental risks that could impair people's cognitive function. More recently, a growing body of evidence has shown that children are at high risk for impaired cognitive development and functioning, including IQ deficiencies and subpar academic performance, even having blood lead concentrations less than 10 μg/dL. It is particularly concerning because prenatal exposure to lead through vertical transmission from mother to child, even at concentrations below 10 μg/dL, might negatively impact a child's and infant's IQ and cognitive function (Jedrychowski *et al.*, 2009).

Another heavy element that seriously endangers children's health is mercury. The fact that fish and shellfish contain mercury, in the form of methyl mercury, which is easily ingested by children, is quite concerning. However, higher hair mercury levels were linked to better visual-spatial abilities in an adolescent study. This could be because eating fish generally has benefits (such as increased protein and omega-3 fatty acids) that outweigh the risks of mercury accumulation. Given that children's developing neurocognitive systems are more susceptible to damage during this time, it is probable that the effects of mercury are more noticeable at this time. Certain studies conducted during pregnancy have made a distinction between the impact of elevated mercury levels and the impact of increased fish diet by the mother. A higher mercury level in early infants is linked to a lower IQ when the exposure came from the atmosphere rather than their diet. Adults are adversely affected by lead as well. Higher levels of lead have been linked in several studies to deteriorations in cognitive function. In addition, there is concern that lead exposure could eventually cause dementia disorders, including Alzheimer's disease. Low levels of mercury deposition seem to impair children's cognitive function; there are less data to suggest that adults are similarly affected by damage (Shetty *et al.*, 2023). In a study of people who were exposed to mercury at work, the effects were negligible and hard to identify on an individual basis. It is believed that the brain plays a major role in the harmful health effects of lead and mercury exposure. Lead is thought to increase oxidative stress, trigger neuronal death, and have an impact on neurotransmitter release. Lead can either flow through the bloodstream or build up in the bones, where it can leak into the bloodstream when there is a bone turnover. Lead can replace calcium in the blood, allowing it to pass through the blood-brain barrier and interfere with other calcium-based functions. It is possible that these systems have a direct impact on cognitive function. Studies on brain imaging in rats and monkeys have revealed alterations in the hippocampus's structural flexibility and a rise in amyloid plaques in the frontal lobe. Dietary methylmercury can build up in the brains of adults and fetuses alike. It gradually changes into inorganic mercury.

The possible consequences of air pollution on children are a primary area of study. Given the ongoing neurological and physical changes that children experience, it is possible that they are more vulnerable to the deleterious effects of pollutants. Adults, on the other hand, have more developed and mature neural systems, making them more resilient to damage. A significant portion of pediatric research focuses on nitrogen dioxide's (NO_2) effects and how it affects the respiratory system. NO_2 can impair lung function in those with pulmonary disease and primarily works as an irritant on the mucosa of the nose, throat, and eyes as well as the rest of the respiratory tract. According to Morales *et al.* (2009), children who were exposed to higher levels of NO_2 were also more likely to experience the symptoms of attention deficit hyperactivity disorder. For adults, air pollution may be just as harmful to their matured and mature neural systems. The possible health effects of air pollution may be more likely to affect the elderly. It is believed that air pollution raises proinflammatory signals throughout the body, which eventually causes inflammation in the brain and impacts the immune system as well. Inflammation and alterations in immune system cells, including microglia, can cause direct brain damage or disruption to the blood-brain barrier, which may facilitate the entry and accumulation of pollutants in the brain (Liu and Lewis, 2014).

Food insecurity is made even more acute by the increasing scarcity and degradation of natural resources. Low soil fertility is a consistent problem faced by agriculturalists and herders. The dominant international thrust to increase food production has been to stimulate trade, enhanced by technology and its transfer. While international bodies acknowledge the importance of small farmers, they operate as if improving the technologies, trade regimes, and value chains that are characteristic of industrial agriculture will have the same results in local ecosystems in developing countries. Price volatility makes access to purchased inputs riskier for smallholders and for the governments that subsidize those inputs. In contrast, a systems approach that focuses on sustainability of the local ecosystem, social and cultural relationships, and economic security can be as, or more, productive than industrial agriculture and can have a much better opportunity to increase food security in developing countries.

Resource-conserving agriculture has been shown to increase yields in developing countries. Priority should be given to developing technologies that follow the systems principles of sustainable agriculture, integrating biological and ecological processes (such as nutrient cycling, nitrogen fixation, soil regeneration, and biodiversity) into the production processes, minimizing the use of non-renewable inputs that utilize the knowledge and skills of farmers and their collective capacities to work together to solve common problems. New farming technology, improved irrigation system, and cropping pattern are important, as well as newly improved varieties, genetically modified genotypes, and biofertilizers are waiting to break the policy barriers. Implementation of novel agricultural approaches and practices like climate-smart agriculture, organic farming, biodynamic agriculture and sustainable intensification, and remote sensing in farming and food systems should be implemented for ecological sustainability and food security as a safeguard for the alarmingly increasing human population (Muhie, 2022). A boost to integrated decision making processes at national and regional levels is needed to achieve synergies and adequately address trade-offs among agriculture, water, energy, land, and climate change.

3.8 Conclusions

Achieving a level of production that meets our needs from an already seriously depleted natural resource base will be impossible without profound changes in our food and agriculture systems. We need to expand and accelerate the transition to sustainable food and agriculture that ensures world food security, provides economic and social opportunities, and protects the ecosystem systems on which agriculture depends. Future research should be focused on the efficient and strategic use of IFS, IPM, planned agroforestry, artificial symbioses, associative nitrogen fixation in non-legume plants, and prevalence of micronutrient deficiencies. The food producers, distributors, handlers, and vendors bear primary responsibility, while consumers must remain vigilant and literate. Government agencies must enforce food safety laws to safeguard public and individual health, especially among infants, children, and elderly people. Medical providers must remain passionate about preventing foodborne illnesses and may consider treating diseases with safe diet therapy under proper medical supervision. There are many elements of traditional farmer knowledge that, enriched by the latest scientific knowledge, can support productive food systems through sound and sustainable soil, land, water, nutrient, and pest management, and the more extensive use of organic fertilizers.

References

Alam, M.Z., Crump, A.R., Haque, Md.M., Islam, Md.S., Hossain, E. *et al.* (2016) Effects of integrated pest management on pest damage and yield components in a rice agro-ecosystem in the Barisal region of Bangladesh. *Frontiers in Environmental Science* 4, 22. DOI: 10.3389/fenvs.2016.00022.

Black, R.E., Victora, C.G., Walker, S.P., Bhutta, Z.A., Christian, P. *et al.* (2013) Maternal and child undernutrition and overweight in low-income and middle-income countries. *Lancet* 382(9890), 427–451. DOI: 10.1016/S0140-6736(13)60937-X.

Chen, W., Wolf, B., Yao, Z., Brüggemann, N., Butterbach-Bahl, K. *et al.* (2010) Annual methane uptake by typical semiarid steppe in Inner Mongolia. *Journal of Geophysical Research* 115(D15), D15108. DOI: 10.1029/2009JD013783.

Da Silva, J.G., Serra, G.E., Moreira, J.R., Conçalves, J.C. and Goldemberg, J. (1978) Energy balance for ethyl alcohol production from crops. *Science* 201(4359), 903–906. DOI: 10.1126/science.201.4359.903.

Daniel, A.I., Fadaka, A.O., Gokul, A., Bakare, O.O., Aina, O. *et al.* (2022) Biofertilizer: the future of food security and food safety. *Microorganisms* 10(6), 1220. DOI: 10.3390/microorganisms10061220.

Deguine, J.-P., Aubertot, J.-N., Flor, R.J., Lescourret, F., Wyckhuys, K.A.G. *et al.* (2021) Integrated pest management: good intentions, hard realities. A review. *Agronomy for Sustainable Development* 41(3), 38. DOI: 10.1007/s13593-021-00689-w.

DeMatteo, R., Keith, M.M., Brophy, J.T., Wordsworth, A., Watterson, A.E. *et al.* (2013) Chemical exposures of women workers in the plastics industry with particular reference to breast cancer and reproductive hazards. *New Solutions* 22(4), 427–448. DOI: 10.2190/NS.22.4.d.

Dey, S.A., Chandra, U.N. and Bhatt, B.P. (2012) Integrated farming system for improving agricultural productivity. In: Bhatt, B.P., Sikka, A.K., Mukherjee, J., Islam, A. and Dey, A. (eds) *Status of Agricultural Development in Eastern India*. ICAR Research Complex for Eastern Region, Bihar, India, pp. 205–230.

Dong, Y.S., Zhang, S., Qi, Y.C. and Geng, Y.B. (2000) Diurnal variation of CO_2, CH_4 and N_2O in a typical steppe in Inner Mongolia. *Chinese Science Bulletin (in Chinese)* 45(3), 318. DOI: 10.3321/j.issn:0023-074X.2000.03.018.

du Jardin, P. (2015) Plant biostimulants: definition, concept, main categories and regulation. *Scientia Horticulturae* 196, 3–14. DOI: 10.1016/j.scienta.2015.09.021.

Fan, Y., Yao, Q., Liu, Y., Jia, T., Zhang, J. *et al.* (2022) Underlying causes and co-existence of malnutrition and infections: an exceedingly common death risk in cancer. *Frontiers in Nutrition* 9, 814095. DOI: 10.3389/fnut.2022.814095.

Flora, C.B. (2010) Food security in the context of energy and resource depletion: sustainable agriculture in developing countries. *Renewable Agriculture and Food Systems* 25(2), 118–128. DOI: 10.1017/S1742170510000177.

Fung, F., Wang, H.-S. and Menon, S. (2018) Food safety in the 21st century. *Biomedical Journal* 41(2), 88–95. DOI: 10.1016/j.bj.2018.03.003.

Hedin, L.O., Brookshire, E.N.J., Menge, D.N.L. and Barron, A.R. (2009) The Nitrogen Paradox in tropical forest ecosystems. *Annual Review of Ecology, Evolution, and Systematics* 40(1), 613–635. DOI: 10.1146/annurev.ecolsys.37.091305.110246.

Jedrychowski, W., Perera, F.P., Jankowski, J., Mrozek-Budzyn, D., Mroz, E. *et al.* (2009) Very low prenatal exposure to lead and mental development of children in infancy and early childhood: Krakow prospective cohort study. *Neuroepidemiology* 32(4), 270–278. DOI: 10.1159/000203075.

Ladha, J.K., Pathak, H., Krupnik, T.J. and Six, J. (2005) Efficiency of fertilizer nitrogen in cereal production: retrospects and prospects. *Advances in Agronomy* 87, 85–156.

Lee, T.C., Anser, M.K., Nassani, A.A., Haffar, M., Zaman, K. *et al.* (2021) Managing natural resources through sustainable environmental actions: a cross-sectional study of 138 countries. *Sustainability* 13(22), 12475. DOI: 10.3390/su132212475.

Liu, J. and Lewis, G. (2014) Environmental toxicity and poor cognitive outcomes in children and adults. *Journal of Environmental Health* 76(6), 130–138.

Mahmud, K., Makaju, S., Ibrahim, R. and Missaoui, A. (2020) Current progress in nitrogen fixing plants and microbiome research. *Plants* 9(1), 97. DOI: 10.3390/plants9010097.

Morales, E., Julvez, J., Torrent, M., de Cid, R., Guxens, M. *et al.* (2009) Association of early-life exposure to household gas appliances and indoor nitrogen dioxide with cognition and attention behavior in preschoolers. *American Journal of Epidemiology* 169(11), 1327–1336. DOI: 10.1093/aje/kwp067.

Muhie, S.H. (2022) Novel approaches and practices to sustainable agriculture. *Journal of Agriculture and Food Research* 10, 100446. DOI: 10.1016/j.jafr.2022.100446.

Nayak, P.K., Nayak, A.K., Panda, B.B., Lal, B., Gautam, P. *et al.* (2018) Ecological mechanism and diversity in rice based integrated farming system. *Ecological Indicators* 91, 359–375. DOI: 10.1016/j.ecolind.2018.04.025.

Paramesh, V., Chakurkar, E.B., Bhagat, T., Sreekanth, G.B., Kumar, H.B.C. *et al.* (2021) Impact of integrated farming system on residue recycling, nutrient budgeting and soil health. *The Indian Journal of Agricultural Sciences* 91(1), 44–48. DOI: 10.56093/ijas.v91i1.110923.

Perfecto, I., Vandermeer, J., Mas, A. and Pinto, L.S. (2005) Biodiversity, yield, and shade coffee certification. *Ecological Economics* 54(4), 435–446. DOI: 10.1016/j.ecolecon.2004.10.009.

Pilon-Smits, E.A.H., Quinn, C.F., Tapken, W., Malagoli, M. and Schiavon, M. (2009) Physiological functions of beneficial elements. *Current Opinion in Plant Biology* 12(3), 267–274. DOI: 10.1016/j.pbi.2009.04.009.

Pretty, J. and Bharucha, Z.P. (2015) Integrated pest management for sustainable intensification of agriculture in Asia and Africa. *Insects* 6(1), 152–182. DOI: 10.3390/insects6010152.

Purnomo, S.H., Sari, A.I., Emawati, S. and Rahayu, E.T. (2021) Factors influencing the adoption of integrated crop-livestock to support land conservation of organic agriculture in Mojosongo area, Karanganyar, Indonesia. *IOP Conference Series: Earth and Environmental Science* 724(1), 12049. DOI: 10.1088/1755-315/724/1/012049.

Ramirez-Puebla, S.T., Ormeno-Orrillo, E., Rogel, M.A., Lopez-Guerrero, M.G., Lopez-Lopez, A. *et al.* (2019) La diversidad de rizobios nativos de Mexico a la luz de la genómica. *Revista Mexicana de Biodiversidad* 90(0), 902681. DOI: 10.22201/ib.20078706e.2019.90.2681.

Rautaray, S.K., Das, P.C., Sinhababu, D.P. and Singh, B.N. (2005) Increasing farm income through rice-fish based integrated farming system in rainfed lowlands of Assam. *Indian Journal of Agricultural Science* 75(2), 79–82.

Reddy, D.V.S. and Biddappa, C.C. (2000) Coconut based cropping/farming systems in India. *Journal of Plantation Crops* 28(1), 1–18.

Salton, J.C., Mercante, F.M., Tomazi, M., Zanatta, J.A., Concenço, G. *et al.* (2014) Integrated crop-livestock system in tropical Brazil: toward a sustainable production system. *Agriculture, Ecosystems & Environment* 190, 70–79. DOI: 10.1016/j.agee.2013.09.023.

Sheikh, M.M., Riar, T.S. and Pervez, A.K.M. (2021) Integrated farming systems: a review of farmers friendly approaches. *Asian Journal of Agricultural Extension, Economics & Sociology* 39(4), 88–99. DOI: 10.9734/ajaees/2021/v39i430564.

Shetty, S.S., Deepthi, D., Harshitha, S., Sonkusare, S., Prashanth, P. *et al.* (2023) Environmental pollutants and their effects on human health. *Heliyon* 9(9), e19496. DOI: 10.1016/j.heliyon.2023.e19496.

Singh, P., Rayeen, F., Singh, R., Pathak, N., Singh, R.P. *et al.* (2023) Enhancing agricultural sustainability through microbial-mediated abiotic stress tolerance. *Journal of Ecophysiology and Occupational Health* 23(4), 233–247. DOI: 10.18311/jeoh/2023/34777.

Singh, P.K., Singh, P., Singh, R.P. and Singh, R.L. (2022) Transgenesis in plants: principle and methods. In: Singh, R.L., Mondal, S., Parihar, A. and Singh, P.K. (eds) *Plant Genomics for Sustainable Agriculture*. Springer Nature, Singapore, pp. 41–70.

Singh, R.L., Vishwakarma, S.P. and Singh, P. (2014) Vitamins and minerals, roles and plant sources. In: Prakash, D. (ed.) *Phytochemicals of Nutraceutical Importance and Their Role in Human Health*. CAB International, Wallingford, UK, pp. 248–265. DOI: 10.1079/9781780643632.0310.

Skakkebek, N.E., Rajpert-De Meyts, E. and Main, K.M. (2001) Testicular dysgenesis syndrome, supplement, in: an increasingly common developmental disorder with environmental aspects. *Human Reproduction* 16(5), 972–978. DOI: 10.1093/humrep/16.5.972.

Thompson, R.C., Moore, C.J., vom Saal, F.S. and Swan, S.H. (2009) Plastics, the environment and human health: current consensus and future trends. *Philosophical Transactions of the Royal Society of London. Series B, Biological Sciences* 364(1526), 2153–2166. DOI: 10.1098/rstb.2009.0053.

Tiwari, A., Singh, D., Goswami, S.G., Singh, P.K., Singh, P. *et al.* (2021) Role of plant growth promoting microorganism for improved plant growth: a review. *Chhattisgarh Journal of Science and Technology* 18(3), 8–11.

Todd, E.C.D. and Todd, E. (2014) Overview of biological hazards and food borned diseases. *Encyclopedia of Food Safety* 1, 221–242. DOI: 10.1016/B978-0-12-378612-8.00071-8.

Tully, K. and Ryals, R. (2017) Nutrient cycling in agroecosystems: balancing food and environmental objectives. *Agroecology and Sustainable Food Systems* 41(7), 761–798. DOI: 10.1080/21683565.2017.1336149.

Vaillant, V., de Valk, H., Baron, E., Ancelle, T., Colin, P. *et al.* (2005) Foodborne infections in France. *Foodborne Pathogens and Disease* 2(3), 221–232. DOI: 10.1089/fpd.2005.2.221.

Wang, L., Luo, D., Liu, X., Zhu, J., Wang, F. *et al.* (2021) Effects of PM2.5 exposure on reproductive system and its mechanisms. *Chemosphere* 264(Pt. 1), 128436. DOI: 10.1016/j.chemosphere.2020.128436.

Wang, Y.P. and Goll, D.S. (2021) Modelling of land nutrient cycles: recent progress and future development. *Faculty Reviews* 10, 53. DOI: 10.12703/r/10-53.

WHO (2001) Complementary feeding, report of the global consultation, summary of guiding principles. Available at: https://iris.who.int/bitstream/handle/10665/42739/924154614X.pdf?sequence=1 (accessed 24 July 2023).

Zdruli, P. and Zucca, C. (2023) *Restoring Land and Soil Health to Ensure Sustainable and Resilient Agriculture in the Near East and North Africa region—State of Land and Water Resources for Food and Agriculture Thematic Paper*. FAO, Cairo, Egypt. DOI: 10.4060/cc1137en.

4 Improving Sustainable Agriculture and Food Security

Pranabesh Ghosh[1]*, Sirshendu Chatterjee[2], Abhishek Konar[1] and Amrita Chakraborty[2]

[1]School of Life Sciences, Seacom Skills University, Santiniketan, Bolpur, Birbhum, India; [2]Department of Biotechnology, Techno India University, West Bengal, Kolkata, India

Abstract

Agriculture plays a key role in global food security, stability, medicine, and prosperity. A nation's economy depends heavily on its agricultural sector, which is also the most susceptible to the effects of climate change. An addition to some bio-resources such as biodiesel, bio-hydrogen, bio-methane, and so on promotes the efficiency of agriculture and increases the water storage capacity of plants for cultivation. Long-term sustainability might necessitate a progressive shift away from synthetic inputs and toward legume-based crop rotation, including continuous cover cropping systems, where they may be feasible in some regions of the world depending on climatic circumstances. This chapter discusses sustainable agriculture toward food security, discusses perspectives and insights into the sustainability of food production and security through sustainable agriculture toward the future using traditional and recent advancements of technologies and strategies by sustainable production through plant and animal origin; productivity growth through pest management; and sustainable food security.

Keywords: Agriculture, Food Security, Sustainable Growth, Bio-Resources, Ecosystem

4.1 Introduction

Climate, as described geologically, is the characteristic weather experienced for the long term at a particular place that creates a major impact on the agriculture, ecosystem, livelihood, and habitat of that region. The yearly weather fluctuations provide a statistical idea about the extremity or seasonal variation of the climate of that zone (Kumari *et al.*, 2020). It has been observed over the last few decades that there has been an impactful change in the climate of the earth as shown through geological evidence like the melting of ice caps and rising sea level. A remarkable and rapid rise in the temperature has made scientists think about global warming, which has been further supported by the facts of changes in ocean currents, sudden volcanic eruptions, floods, solar activities, and many more natural phenomena (Aggarwal, 2008). The productivity and output of India's agriculture are suffering due to climate change. According to estimates from the Intergovernmental Panel on Climate Change (IPCC) India's temperatures are predicted to rise by 3–4°C by the end of the 21st century, resulting in a loss of 3–26% in net agricultural income. The lack of mitigation and

*Corresponding author: doctorpranabeshghosh@gmail.com

DOI: 10.1079/9781800625037.0004

adaptation strategies could cause farm revenue to drop by 12–40% in the upcoming years.

According to the 2007 IPCC report, crop productivity in the southern hemisphere is predicted to drop by as much as 20%, with less developed nations experiencing the worst consequences. The main and threatening climatic changes are being caused by human activities. Environmental studies and research have stated that the devastation that is being caused by human beings can and will last not only for a few decades but for several centuries and has added fuel to the global climate change (Kumari *et al.*, 2020). There has been an increase in the concentration of several gases in the atmosphere like CO_2, CFC, H_2, SO_2, N_2O, and many other gases resulting in ozone layer depletion.

Moreover, cutting down trees for increasing locality has resulted in deforestation, which has been a major cause of soil erosion and less rainfall. This has caused and is causing a major threat to agricultural production and food security (Pathak *et al.*, 2010). About 18% of the greenhouse gases are emitted from agricultural production in India. The increase in temperature causes reduced yield of crops, leaf abscission, photo-oxidative stress, and invasion of several pests and obnoxious weeds that ultimately can cause major damage to agriculture and food production (Fig. 4.1). Scientists and researchers further state that by the end of the 21st century, there will be an increase in rainfall by 15–40% resulting in floods and damaging agricultural practices causing reduced or no food availability (Aggarwal, 2008).

4.2 Climatic Changes' Impact on Agriculture in Several Direct and Indirect Ways

4.2.1 Impact on soil

Climatic changes affect soil majorly. The soil of a particular region where the magnitude of climate change is intense may show a change in its pH that can retard crop production or may favor weed growth. Reduction in the carbon content of soil due to an increase in temperature and CO_2 emission can lead to lower nutrient and water-holding capacity of the soil (Narayanan, 2018; Kumari *et al.*, 2020).

4.2.2 Impact on forest

Higher risks of drought, flood, or precipitation can be an alarming outcome of climate change.

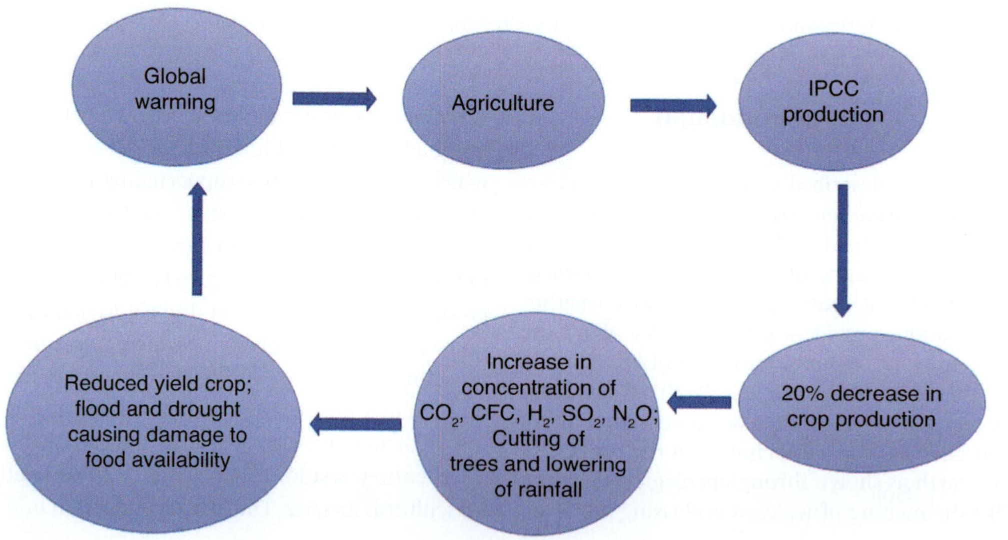

Fig. 4.1. Schematic representation of how global warming is related to loss of food availability.

This can cause a scarcity of water required for the growth of plants in forests. Excessive drought-like situations may result in forest fires due to the dryness of leaves and small shrubs. Drought retards sap growth in plants, resulting in invasion of some destructive insects like pine beetle (Sawalhah *et al.*, 2019).

4.2.3 Impact on land

The untimely and intense melting of ice caps and glaciers resulting in a rise in the sea level can cause river deltas to shift to higher ground levels. This might cause a conflict situation among the migrants and settled communities for access to sufficient land and water (Kumari *et al.*, 2020; Konar *et al.*, 2022).

4.2.4 Impact on water resources

The Monsoon in India has been a major reason for protecting its rich agricultural practices and biodiversity. Researchers have warned people that by 2050s an increase in average 2–4°C temperature, decreasing 15–20 days of rainfall in western and central India, while an increase in 5–10 days in the foothills of the Himalayas and Northeast India will overall affect the climatic conditions and trends. Moreover, a decreased summer rainfall will have a huge impact on agriculture and food production (Pathak *et al.*, 2010; Mahmoud and Gan, 2018).

4.2.5 Impact on crop yield

An economic impact is observed due to a variable change in climate (Fig. 4.2). An increase in average temperature by 3–6°C can adversely affect food availability, food utilization, food accessibility, and food system stability (Kumari *et al.*, 2020; Konar *et al.*, 2022). Several studies have shown various alarming declines in food crop yields. With the increase in temperatures up to 40°C there has been a decrease in rice production of up to 41% in Tamil Nadu, while in Punjab, the rise of minimum temperature by 1–3°C has decreased rice and wheat production by 3% and 10%, respectively. With the increase in atmospheric CO_2 concentration, along with temperature rise, a 20% increase in rice production and a 31% decline in wheat production has been observed. A negative impact has been seen in the growth of gram and ragi production due to the rise in maximum temperature. Extreme monsoon variation has an extensive effect on the jowar production in Karnataka that affects food security and incomes. Non-food grain crops like sugarcane, cotton, mustard, and sesamum crops are negatively affected by increasing temperature

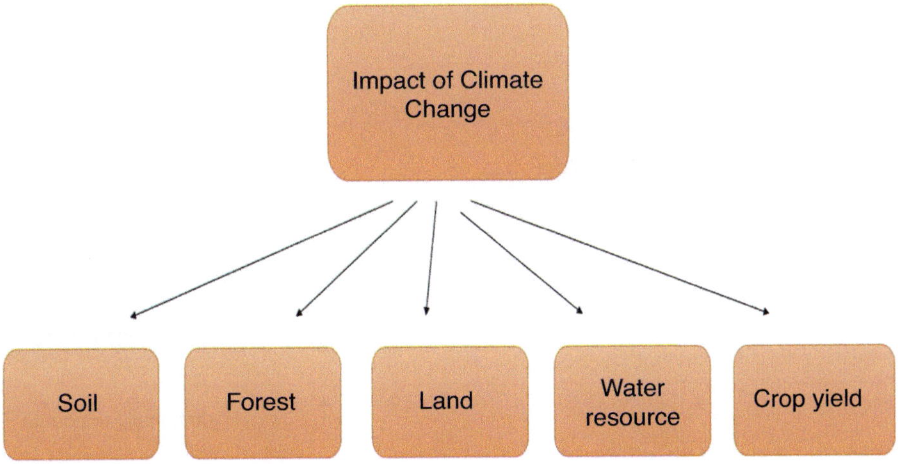

Fig. 4.2. Impact of climate change on different factors directly and indirectly.

along with a fluctuation in rainfall (Aggarwal, 2008; Mukherjee *et al.*, 2022).

4.3 Sustainable Agriculture and Its Key Principles

The word "sustain" has been taken from the Latin word "*sustinere*" where "sus" means from below and "tinere" means to hold. So, altogether, the meaning is to keep in existence or maintain long-term support. Thus, sustainable agriculture can be explained as a method of producing large quantities of food crops abundantly without destroying the ecological and environmental balance or polluting the environment by any means (Tiwari *et al.*, 2021; Singh *et al.*, 2022, 2023). The main idea of it is to create a system where the livestock or crops are self-sustainable. It can also be greatly helpful in food production as well as the economic well-being of several people involved with agricultural practices. Instead of pertaining to an infant stage, it should be taken into account to make this method of agriculture implemented in today's life (Pathak *et al.*, 2010; Mahmoud and Gan, 2018; Narayanan, 2018; Kumari *et al.*, 2020; Miśniakiewicz *et al.*, 2021; Konar *et al.*, 2022).

The principles of sustainable agriculture are as follows:

1. Increase in farm productivity
2. Any type of impact caused on natural or environmental resources is minimized
3. Use of chemicals in agriculture is minimized
4. Social benefits from agriculture are maximized
5. Well-equipped and flexible farming systems are adapted to manage probable risks

The major characteristics of sustainable agricultural practices are (Fig. 4.3):

1. Economically viable
2. Adaptable
3. Socially justifiable
4. Ecologically sound
5. Resource optimal
6. Nutritionally adequate

4.4 Climate-Smart Agriculture (CSA)

Recent studies are indicating a major decline in the yield of crops in India due to changes in climatic conditions between 2010 and 2039, which can be as high as 9%. This has given rise to the necessity of embracing the notion of climate-smart agriculture to minimize the

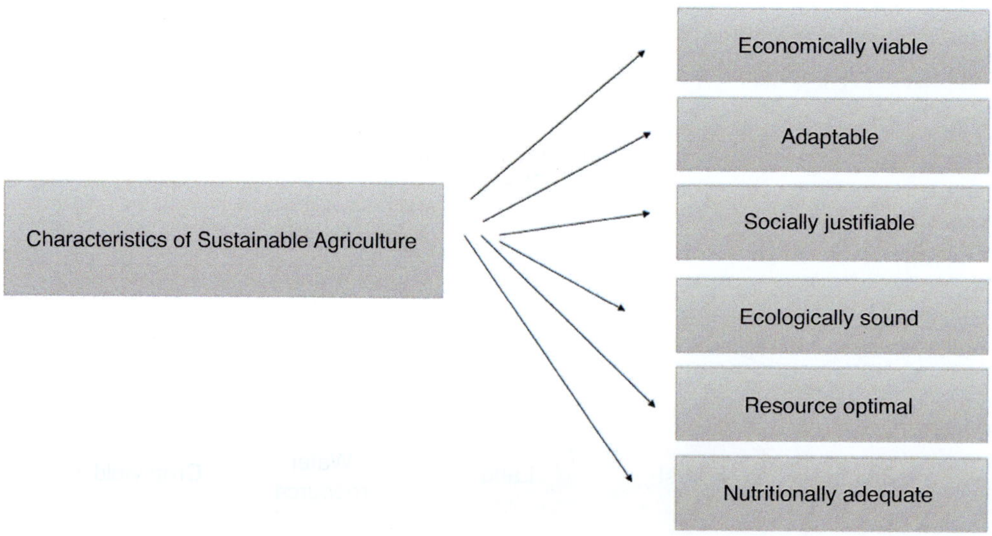

Fig. 4.3. Schematic representation of the major characteristics of sustainable agricultural practices.

negative impact of climate change on agriculture. Tied with sustainable agricultural practices, climate-smart agriculture can help contribute to the Millennium Development Goals of reducing hunger and improving environmental management (Caminade *et al.*, 2019).

Agricultural practices transforming agriculture systems to support food and security during climate change are collectively called "climate-smart agriculture." This approach integrates climatic change into the implementation of sustainable agriculture practices and aims mainly to increase the proficiency of agriculture, despite variable climatic variation and enable adaptation to changes in climate by reducing the contribution of agriculture to global warming (Cheung *et al.*, 2019; Aryal *et al.*, 2020; Negassa *et al.*, 2020; Coulibaly *et al.*, 2021).

4.4.1 Main characteristics of CSA

1. Systematic integration of climatic changes for planning and developing sustainable agricultural systems.
2. Increased production, enhanced resilience, and reduced emissions create an interrelated concern.
3. Maintaining ecosystem services.
4. CSA is a concept amenable to adaptation rather than providing a rigid set of particular practices, technologies, and methodologies.

4.4.2 Governmental initiatives on CSA

1. National Innovations on Climate Resilient Agriculture (NICRA)
2. National Mission on Sustainable Agriculture (NMSA)
3. The National Adaptation Fund for Climate (NAFCC)
4. Climate Smart Village (CSV)
5. Pradhan Mantri Krishi Sinchayee Yojana (PMSKY)
6. Pradhan Mantri Fasal Bima Yojana (PMFBY)
7. Soil Health Card Scheme (SHC)
8. National Water Mission (NWM)
9. Biotech-KISAN

4.4.3 Policy interventions

Agriculture remains a state subject under the Constitution of India. Though the central government controls the overall approach to agricultural policies, it is completely upon the choice of the state to accept or decline its policy. Immense cooperation between the state level and central government has proven successful in taking up existing policy challenges and their needful solutions (Kerdsriserm *et al.*, 2016; Dara, 2019; Ataei *et al.*, 2021; Lynch *et al.*, 2021).

The Central Government has shifted its focus from a production-centered approach during the Green Revolution to an income-centered approach that aims to "achieve high productivity, reduced cost of cultivation and remuneration price on the production, with a view to earn higher profits from farming."

Certain government policies and their targets are stated below (Fig. 4.4):

4.4.3.1 National Mission for Sustainable Agriculture (NMSA)

It aims to make agriculture sustainable in a comprehensive manner. It was the first among the eight missions outlined under the National Action Plan on Climate Change (Kumari *et al.*, 2020).

4.4.3.1.1 RAINFED AREA DEVELOPMENT. This scheme promotes mixed cropping, fishery, horticulture, livestock, and practices. Out of 140.1 mha of cultivated food grain production, 68.3 mha is irrigated, while the remaining 71.7 mha is rainfed.

4.4.3.1.2 SUB-MISSION ON AGROFORESTRY (SMAF). Following the National Agroforestry Policy (NAP), it aims to increase tree plantation. This is beneficial to the income generation capacity of marginal and small farmers.

4.4.3.1.3 SOIL HEALTH CARD SCHEME (SHC). This scheme aims at "promoting location as well as crop specific sustainable soil health management creating and linking soil fertility maps with macro-micro nutrient management, judicious application of fertilizers and organic farming practices."

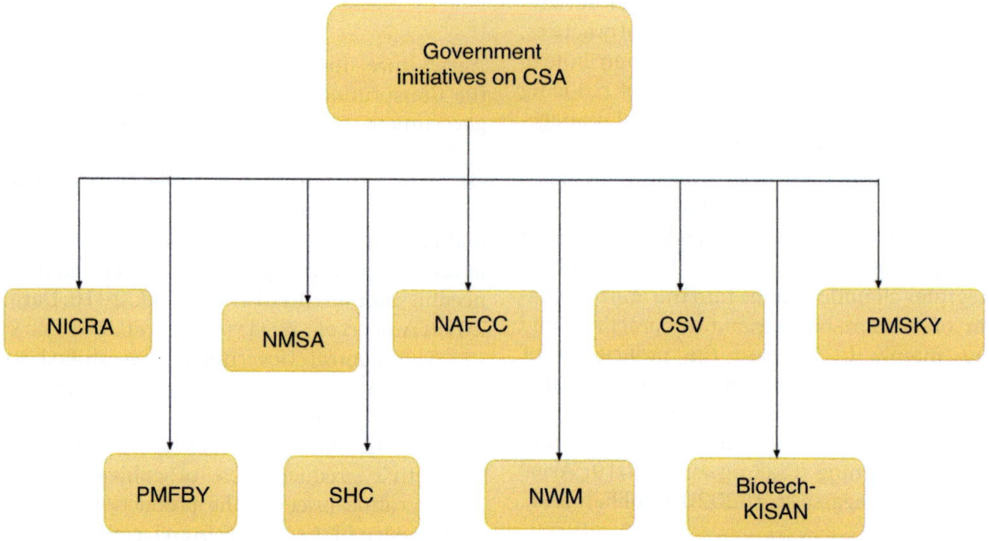

Fig. 4.4. Schematic presentation of various government initiatives on climate smart agriculture (CSA).

4.4.3.1.4 PARAMPARAGAT KRISHI VIKAS YOJANA (PKVY). This policy had a target of converting 2 lakh ha of cropping area into organic farming. Against this plan, it has brought 4.3 lakh ha area under it (Pathak *et al.*, 2010).

4.4.3.2 Pradhanmantri Krishi Sinchayee Yojana (PMKSY)

It is a highly publicized central government scheme focusing on expanding the infrastructure of irrigation in the country, especially in rainfed regions, and also improving water use in agriculture.

4.4.3.3 Pradhanmantri Fasal Bima Yojana (PMFBY)

This policy ensures coverage to farmers pre-sowing or post-harvesting to offset losses against non-preventable natural risks.

Several other announcements under agriculture sectors are:

1. INR 1 lakh crore agricultural infrastructure fund for filling up critical gaps in the agricultural value chain.
2. INR 20,000 crores under Pradhan Mantri Matsya Sampad Yojana.
3. Creating resilient agricultural supply chains using 50% subsidy on transportation to 50% on storage facilities.

4.5 Future Perspectives

Globally, agriculture and climate change are closely intertwined, impacting and shaping one another. Climate change affects soil fertility, crop quality, and productivity, and it may seriously jeopardize the security of food and nutrition (Duncan and Claeys, 2018). The difficulty facing Indian agriculture is the need to adapt to possible variations in temperature, precipitation, and extreme weather without sacrificing output and food security. With regard to the objectives of environmental policy, these modifications might make it necessary to accept current regulations (Aggarwal, 2008). Even though research is being conducted in new directions and strategies to lessen the negative effects of climate change are being developed, there is still a need to prioritize allocating enough funds to support the development of technology, dissemination of information, and creation of policies for climate change adaptation and mitigation (Liu *et al.*, 2019; Li *et al.*, 2021).

4.5.1 Case study of Global North

4.5.1.1 *Urban areas of Global North food security is an important relevant topic*

Contrary to rural communities in the Global South, Americans rely less on agriculture and food production for livelihoods (Opitz *et al.*, 2016). The United States has 3143 counties and 50 states (Parker, 2015). Since the 1960s, food security has become an increasingly relevant issue for the Global North (Opitz *et al.*, 2016). A century of food system modernization has resulted in the food system that exists in the United States today. In the early 1900s, 41% of the labor force was employed in small, diversified farms that produced an average of five commodities. These farms were engaged in labor-intensive agriculture, and at this time, 2% of the people lived on farms (Dimitri *et al.*, 2005; Raj *et al.*, 2022). The 1960s "Green Revolution," or the rise of farm mechanization, had a particularly significant impact on how family farming operated (Lobao and Meyer, 2001). Farm numbers have decreased, from 6.8 million in 1935 to 2.1 million in 2002. However, farms are more productive now than they were in the past because agricultural inputs such as chemicals, fertilizer, pesticides, and herbicides are more readily available and are used more often to achieve the current levels of productivity. A new approach to agricultural policy emerged in the 1970s when Secretary of Agriculture Earl Butz started to support large, capital-intensive farming operations rather than small, family farms (Dimitri *et al.*, 2005; Spittler *et al.*, 2011; Raj *et al.*, 2022). In addition to initiatives such as "Know Your Farmer, Know Your Food," and the 2008 Farm Bill (H.R. 6124), the US government has made a concerted effort to reintegrate consumers and producers of food (Parker, 2015).

In general, the United States exports more food than it consumes. On average, the country produces more than enough food to meet the dietary needs of its entire population (Maxwell, 2017). Effect of climate changes on food production alone will not immediately affect consumption patterns or levels of food security in American communities because of how far apart food production and consumption are in the American food system (Opitz *et al.*, 2016).

4.5.1.1.1 Factors that contribute to food insecurity in the US. High levels of food insecurity are caused by a number of factors in rural areas of the United States, including structural injustices, socioeconomic dynamics of rural areas, and policy oversight of rural food systems. Three indicators of farm production diversity, including the Simpson's Index, a metric accounting for both species richness and evenness, were calculated along with two indicators of dietary diversity: a modified Household Dietary Diversity Score (HDDS) and the Food Consumption Score (FCS). The majority of people who live in rural areas, even those who work in agriculture, buy a lot of their food from food retailers rather than farming for subsistence. As a result, supermarkets were also distributed closer to each other in urban counties as compared to rural counties; there was one supermarket every 75 km^2 in an urban county as opposed to 686 km^2 in a rural county. Small- and medium-sized grocery stores, more prevalent in rural areas, offer limited choices of healthy produce: two-thirds of retailers in the study did not sell vegetables, and one-third did not sell fruit (Morris *et al.*, 1992; Raj *et al.*, 2022).

Approximately 98% of US counties are food deserts, mostly in nonmetropolitan areas with fewer than 10,000 people per county (Morton and Blanchard, 2007). According to the USDA, this pattern of food distribution leads to large swaths of neighborhoods without supermarkets or grocery stores within 16 km, called "food deserts" (Morton and Blanchard, 2007; Sharkey, 2009; Raj *et al.*, 2022). Having access to food requires financial capital. There is a higher prevalence of poverty in rural than in urban areas, and the rate of poverty declined more significantly in urban and metro areas than in rural and isolated areas. Between 2015 and 2017, while real incomes in metro counties increased, they declined in completely rural and non-metro adjacent counties (Kusmin, 2012; Raj *et al.*, 2022). During the

2018 Farm Bill, the US government attempted to provide food safety nets to households and individuals in poverty, including Supplemental Nutritional Assistance Programs (SNAP) and a variety of smaller nutrition programs, including Women, Infants, and Children (WIC) (Mozaffarian *et al.*, 2019).

In rural areas, proper amounts of food at market price are not available. But, compared to urban areas where Meals on Wheels, soup kitchens, food pantries, and banks are available, rural areas have fewer such social safety nets. SNAP and WIC are safety nets that rural residents are able to access, but these services are limited in how they can be redeemed (Smith and Morton, 2009). In rural areas of the US, food costs more, and it is significantly more expensive to buy food in rural counties with persistent poverty than it is to receive food stamps under the thrifty food plan. Generally, in the US, fruit and vegetables cost more than dense fats, sweets, and grains (cheap calories). Rural areas create inflation due to a lack of food supply and high supply costs. Therefore, nutritional food supply is difficult in rural areas, and rural people face a nutritional challenge (Hendrickson *et al.*, 2006; Ramadurai *et al.*, 2012; Raj *et al.*, 2022). Another study found that Hispanics and Latinx people living in rural areas are more likely to have food insecurity than those who live in urban areas. The other article highlights that food level security is associated with time in the United States in rural areas. However, subsistence pathways, smallholder diets, and nutrition can be improved by expanding production diversity on their farms by introducing new crop and livestock species (Jones *et al.*, 2014; Sibhatu and Qaim, 2017; Raj *et al.*, 2022).

4.5.2 Case study of Global South

4.5.2.2 Indian food system

According to the Food and Agricultural Organization and the World Food Summit (WFS), "food security exists when all people have physical and economic access to sufficient, safe, and nutritious food that meets their dietary and food preferences for an active life at all times" (Pillay and Kumar, 2018). Currently, 1.37 billion people live in India, and it is one of the most populated countries in the world. India is spread across 3.3 million square kilometers and has 28 states and 8 union territories. Both states and union territories consist of 718 districts, which are further divided into rural villages and urban municipalities (Raj *et al.*, 2022). In India, 65% of the people belong in rural areas. India continues to be the world's largest producer of milk, pulses, and spices despite slowing agricultural growth. India is the country with the most land that is farmed worldwide for wheat, rice, and cotton. In addition, India supplies the world with fruits, vegetables, wheat, cotton, sugarcane, rice, sheep, goats, and farmed fish (Mukherjee *et al.*, 2022; Raj *et al.*, 2022). Grain production grew during 1966–1967 and 1970–1971, contributing to the crop revolution known as the "Green Revolution." Under the direction of MS Swaminathan, the Green Revolution of India was declared during the "plan holiday" (1966–1969) (Mukherjee *et al.*, 2022). Today, India produces 104 million tons of rice and 93 million tons of wheat annually (Pillay and Kumar, 2018). The livelihoods of rural Indians are closely linked to agricultural production and poverty (Raj *et al.*, 2022).

India's agriculture concentrated on developing high-yielding rice and wheat varieties and increasing chemical inputs, such as pesticides and fertilizers, after a green revolution in the middle of the 1960s. This increased output per hectare without expanding the amount of land under cultivation. The primary goal of the Green Revolution, which was implemented from 1966 to 1977, was to produce high-yielding seed varieties (Chakravarti, 1973; Mukherjee *et al.*, 2022; Raj *et al.*, 2022). The famine conditions under British rule contributed to the Green Revolution. Before India gained independence in 1945, the British exported agricultural goods from India to fund their empire and military operations abroad, while millions of Indians suffered from starvation (Sen, 1982). With less than two hectares of land, the majority of farmers in India are smallholder or subsistence farmers. Since the Green Revolution, agricultural productivity has increased, and India is now able to produce enough food to meet its population's caloric needs and has become self-sufficient in grain production since the 1970s (Chakravarti,

1973; Miśniakiewicz *et al.*, 2021; Mukherjee *et al.*, 2022; Raj *et al.*, 2022).

4.5.2.2.1 FACTORS THAT CONTRIBUTE TO FOOD INSECURITY IN INDIA. Despite being the world's top producer of several important agricultural products, India's average per capita net food seed availability has fluctuated and been uneven across states. The Indian government implemented a public distribution system to provide more equitable access across states. The PDS, the largest social protection program in the world, gives 800 million people in India access to subsidized cereals, which are available for purchase at more than 500,000 fair-price stores (Pillay and Kumar, 2018; Raj *et al.*, 2022). Another illustration of initiatives to create food security is the Indian ration card system. The card serves as a means of establishing the holders' identity and entitles them to the purchase of food and fuel subsidies through the PDS (liquefied petroleum gas and kerosene are two common fuels used by the impoverished for cooking) (Pillay and Kumar, 2018). The PDS has yielded varying outcomes. There are five distinct layers of food insecurity, ranging from "very high" to "very low," for the 17 major states of India. West Bengal, along with five other states (Haryana, Rajasthan, Uttar Pradesh, Assam, and Tamil Nadu), has a "moderate" level of food insecurity (Basu, 2020). In spite of having the necessary paperwork, 20% of the sample households in the Ali *et al.* (2012) study of Uttar Pradesh were unable to receive food from the PDS. In a similar vein, it was discovered that just 1 kg of PDS rice, out of 5.43 kg distributed, reached individuals in need; similarly, the distribution of sugar was even less effective, with only 1 kg of sugar distributed reaching those in need in Tamil Nadu. Some also mention corrupt practices impeding households' ability to purchase through the PDS, misclassifying households as above the poverty line as a justification for their exclusion from the program, and subpar grain quality at fair-price stores (Ali *et al.*, 2012; Dhanaraj and Gade, 2012; Kasim, 2012; Pillay and Kumar, 2018).

In 2030, water levels in India are predicted to drop below 50% of demand at current rates of use, severely taxing the country's river basins. Additionally, groundwater is becoming less abundant, particularly in India's northwest, particularly Punjab and Haryana states, which produce the majority of the country's wheat and rice. The rural poor in India depend on agriculture for approximately 75% of their income, and any future loss of grain will contribute to an escalation in food insecurity (Ahmad *et al.*, 2011; Merriott, 2016; Raj *et al.*, 2022).

Discrimination based on caste is another factor contributing to low income in rural India. The social and financial networks and capital that upper-caste landowners have access to are not available to small-scale and marginal farmers, who are members of lower and landless castes (Ali *et al.*, 2012; Goli *et al.*, 2021; Raj *et al.*, 2022). Ali *et al.* (2012) reported that food insecurity was found to be more severe in lower-caste households than in upper-caste households (Ali *et al.*, 2012). Ten years later, in Uttar Pradesh, Goli *et al.* (2021) discovered comparable outcomes. Food insecurity is four times more prevalent in households with no or marginal landholdings and three to four times more prevalent in lower-caste households compared to higher caste and medium to large agricultural landholding households, according to a study done on over 5000 households in the Uttar Pradesh state (Goli *et al.*, 2021). Caste prejudice is still pervasive in areas of India, which limits access to resources, credit, and education (Raj *et al.*, 2022).

4.6 Global South-North Comparison of Determinants of Food Security and Adaptations

There are similarities in food security production between India and the United States (US), despite differences in geography, socioeconomics, and culture. In both countries, rural areas face challenges when it comes to attracting development that would improve their quality of life. Although the nature of the constraints differs between the two countries, physical access is a key challenge for both. Farmers in India face obstacles in reaching aggregators due to inadequate infrastructure and utilities, which hinders their physical access to markets. Both regions' rural areas have had difficulty receiving government policy responses that offer effective safety nets to reduce food insecurity. A different rural landscape exists in the United States, where agriculture is not the prime source of livelihood, and farming losses are not as detrimental to food security as they are in rural India.

COVID-19 affected food security in both the Global North and South, irrespective of climate. Around 12 million cases and 162,000 deaths were recorded in India from COVID-19, while about 30 million cases and 550,000 deaths were recorded in the US from the same disease. We watch two radically different stories play out in Madera County, California, and Kerala, India. The lockdown imposed by the pandemic in Kerala is having an immediate impact on food security and is affecting every aspect of the food system, including manufacturing, production, retail, supply chain, and processing, partly because the lockdown fell during the busiest periods for harvesting (Ghosh, 2020; Pothan *et al.*, 2020; Raj *et al.*, 2022).

According to recent research by Lusk and Chandra (2021), Madera County has one of the highest COVID-19 rates among migrant workers nationwide. Those in the aforementioned high-risk Madera groups may not have received the US government's response because of their immigration status and tax-filing status, even though unemployment benefits were increased and stimulus checks were mailed to tax-filing citizens as a safety net (Lusk and Chandra, 2021). The second wave of COVID-19 raging in India reached small and marginal farmers of lower castes, who were likely more vulnerable to the pandemic's effects because of their limited capacity to cope (Ghosh, 2020). The two case studies included in this study highlight the necessity of context-based adaptation plans in both the Global North and South in order to protect food security from major disasters like climate change. We observe that the majority of proposals aimed at enhancing food security typically concentrate on food production and the availability aspect of food security.

4.7 Improvement in Next-Generation Plants

Climate change and food security are explored through various qualitative strategies in this paper (Islam and Wong, 2017). We chose to use both content and discourse analysis to gather the information necessary for this study on food production systems, climate change, and food security. In this article, our examination, we found various academic journals and research papers about this.

By increasing food production, we can reduce the vulnerability of the food system:

4.7.1 By root structure

There is some evidence as to how plants may be able to adapt to climate change based on some of their traits. In some regions, climate change or increased climate variability may result in less rainfall, resulting in an improved selection of genotypes that utilize limited supplies of water stored in soils. The efficient use of water by crops is influenced by a variety of plant characteristics and crop management strategies (Gregory, 2004). However, due in large part to the difficulty of measuring root systems, relatively little attention has been paid to root characteristics that may allow for greater water exploitation or efficiency (Porter and Semenov, 2005).

However, various genotypes exist, like the diameter and length of root and leaf, that can increase food production in difficult situations (O'toole and Bland, 1987). We know that water plays a vital role in crop productivity. Nowadays, the hydrological cycle is affected by climate change, industrial growth, pollution, and the threat to food production (Aerts and Droogers, 2004). In this situation, worldwide scientists search for various genotype plants that will produce higher crop-yielding crops. Scientists and farmers use crop water productivity to evaluate water savings and investments. In dry areas, genotypes help the adaptation possible under conditions of changed climate. As an example, crops in the Mediterranean environment of northern Syria rely heavily on the growing season rainfall. Another study showed that this genotype consistently had longer roots in the 3–4 weeks leading up to anthesis compared with other genotypes like Beecher with deeper roots.

In this condition, faster rates of root growth were reflected in a higher crop-yielding crop food. As a result of the development of DNA-based molecular markers, it has been possible to identify the genes (quantitative trait loci) behind different root traits. Champoux *et al.* (1995) developed drought-tolerant rice (*Oryza sativa* L.). An extensive root system is one of the important adaptive strategies for drought tolerance. Researchers studied the root growth, water use, and yield of the local landrace (*Arabic abiad*) at sites typically receiving less than 350 mm of rainfall per year. They found that this genotype consistently had longer roots than other genotypes, such as the variety Beecher, at depths below 15 cm (Champoux *et al.*, 1995; Babu *et al.*, 2001).

4.7.2 A new transpiration module

Transpiration is a type of water loss from plant parts. Transpiration is most important for the plant, but it is a major issue for survival in drought areas. Transpiration occurs in stomata through Lenticels (Jia and Zhang, 2008; Konar, 2021). However, lack of water supply, or drought, is a major issue in the field of agriculture. That is why scientists are searching for a new transportation module including G-protein regulation, vapor pressure deficit, and stomatal response which is used to regulate atmospheric and edaphic water stress on crops. For crop production, minimum transpiration rate could play an important role in water savings.

4.7.2.1 Transpiration at elevated vapor pressure deficit

A limit on maximum transpiration rate could be particularly important in non-irrigated crop production. Worldwide, great focus has been given to limiting transpiration at elevated vapor pressure deficit (VPD), also called "limited transpiration rate" trait, which is involved in both heat and drought tolerance (Sinclair *et al.*, 2005; Lobell *et al.*, 2013; Watson *et al.*, 2017). Transpiration from produce is controlled by the difference between the vapor pressures of the air and the surface evaporating. Differences in vapor pressure are referred to as vapor pressure deficits. Genotype variation of low transpiration rate has been observed in various crops like sorghum, soybean, maize, and groundnut (Fletcher *et al.*, 2007; Devi *et al.*, 2010; Yang *et al.*, 2012). Collins *et al.* reported that limited rate of transpiration demand has the potential to improve drought adaptation, crop water productivity, and food security (Collins *et al.*, 2021). Variations in LTR are commonly expressed as a segmented response of TR for each unit of leaf area in response to VPD. Genotypic differences in wheat were found in the VPD breakpoint, the "second" slope (for higher VPD), and the "first" slope (for VPD less than the VPD breakpoint) (Schoppach and Sadok, 2012).

However, Medina *et al.* (2019) found durum wheat to have low and unique VPD breakpoint (~1.1 kPa) genotypes and segmented TR–VPD responses, which is close to the breakpoint of 1.3 kPa found in a recent study, in an examination with a relatively low number of evaluation performed at different levels of controlled VPD (Medina *et al.*, 2019). Certain contemporary genotypes that are frequently grown in the northeast seem to have a high transpiration efficiency and a low transpiration rate trait. Hence, the findings of various reports of the recent study state that genetic material carrying the low transpiration rate trait has the potential to improve wheat germplasm for yield beyond the recent levels demonstrated in Australian breeding programs, especially in northeast Australia (Collins *et al.*, 2021).

4.7.2.2 G-protein regulation

A plant's transpiration depends on stomatal movements. Stomata regulate the main function of water loss and gas exchange (Jia and Zhang, 2008; Nilson and Assmann, 2010; Konar, 2021). Worldwide, cucumber (*Cucumis sativus* L.) is one of the most popular vegetables, and drought stress causes a major decrease in the growth, quality, and yield of cucumber plants (Sun *et al.*, 2016; Liu *et al.*, 2021; Zait *et al.*, 2021). In plants (rice and cucumber) G-proteins have a central role in the development of abiotic stress. A recent study found that stomata were regulated by G-protein or G-protein's subunit.

In the active state, the G-protein works as a trimer and consists of an α-subunit (Gα) bound to GDP, a β-subunit (Gβ), and a γ-subunit (Gγ). Currently, new GPCR-type G-proteins (GTG1 and GTG2) have been identified in Arabidopsis, and they are one type of abscisic acid (ABA) receptors, which is hormonal and environmental signaling, with stomatal aperture regulations (Assmann, 2002; Joo *et al.*, 2005; Pandey *et al.*, 2006; Fan *et al.*, 2008; Nilson and Assmann, 2010). We have found in various papers that the α-subunit of the heterotrimeric G-protein, GPA1, is a regulator of Transpiration efficiency (TE). There has been evidence that GPA1, a molecular on/off switch for signaling, plays an important role in morphogenesis and abiotic stress response. In drought stress, ABA concentration plays a role in the leaves, including stomatal closure, and inhibits stomatal opening (Schroeder *et al.*, 2001; Nilson and Assmann, 2010; Jangam *et al.*, 2016). Few experiments have found that ABA of G-protein mutants suggests that heterotrimeric G-proteins may regulate the water status of the whole plant (Maren *et al.*, 2022).

A key mechanism of ABA's regulation of stomatal opening is its inhibition of inward-rectifying K^+ channels, which causes a reduction in K^+ influx and a reduction in water ingress into cells (Schroeder *et al.*, 2001). In soil water conditions equivalent to abundant water (90%) or drought (30%) stress of the soil water-carrying capacity, GPA1 regulation of stomatal movements, and ABA signaling TE was measured on GPA1–3, GPA1–4, and the Columbia ecotype (Wang *et al.*, 2001; Pandey *et al.*, 2006). Under abundant water conditions, *gpa1* mutants had an approximate 12% increase in TE, compared to Col (P=0.0107 for *gpa1–3*, P<0.0001 for *gpa1–4*). Under drought stress, *gpa1* mutants saw a 14% increase in TE compared to Col (P<0.0001 for both *gpa1* alleles) (Chalise *et al.*, 2019).

However, to directly examine ABA's role in regulating TE by *gpa1*, we exogenously treated rosettes of *gpa1* and Col with ABA and studied their carbon isotope ratios. To directly assess the role of ABA in the regulation of TE by *gpa1*, the rosettes of *gpa1* and Col were exogenously treated with ABA, and carbon isotope analysis was performed. *Gpa1* mutants are hyposensitive to ABA inhibition of stomatal opening, *gpa1* mutants had decreased avoidance against ^{13}C in rosette tissue compared with Col, even in this research where 25 μM ABA was employed directly to the leaf (P<0.001 for *gpa1–3* and P=0.004 for *gpa1–4*). Unexpectedly, given that *gpa1* mutants are hyposensitive to ABA inhibition of stomatal opening (albeit wild type for ABA promotion of stomatal closure), *gpa1* mutants had reduced discrimination against ^{13}C in rosette tissue compared with Col, even in this experiment where 25 μM ABA was applied directly to the leaves (P<0.001 for *gpa1–3* and P=0.004 for *gpa1–4*) (Gregory *et al.*, 2009; Nilson and Assmann, 2010).

4.7.2.3 Stomatal response against climate change

Globally, climate change in the context of including drought and heat is expected to increase in many agricultural areas. Drought and heat stress both have a great impact on the agricultural field and strongly threaten food security (Serna, 2022). The world's most cultivated crop is maize, with 1.1 billion tons of grain produced in 2021. It can avoid drought and heat stress damage through a change in stomatal number and/or function (Gregory *et al.*, 2009; Gupta *et al.*, 2020; Serna, 2022). Transpiration decreases leaf temperature. Therefore, stomatal opening prevents leaf heating. However, water deficit increases by 40–50% due to stomatal size, density, and opening (Zheng *et al.*, 2013; Maren *et al.*, 2022).

However, researchers are searching how maize can avoid a high amount of water loss and grow in drought and high temperatures. Stomata of maize consist of two dumbbell-shaped guard cells. By reducing transpiration, associated with high water-use efficiency, we can expect lower rates of CO_2 assimilation and diminished yields, possibly through a direct decrease in CO_2 uptake and increased leaf temperatures that negatively impact protein synthesis. Under elevated CO_2 levels, C_4 species experience less drought damage due to reduced stomatal transpiration and increased water-use efficiency and soil water content (Long *et al.*, 2004; Ghannoum, 2009). Therefore, maize may benefit from increased CO_2 levels in drought conditions (Chalise *et al.*, 2019).

4.7.3 MicroRNAs in plant drought tolerance

During drought stress, microRNA (miRNA) differential expression has been found in plants (Fig. 4.5). This miRNA can help the plant survive in low water supply conditions (Kantar *et al.*,

2010; Chen *et al.*, 2012; Ferdous *et al.*, 2015). In 1993, miRNA were first discovered in the nematode *Caenorhabditis elegans*. First, animal miRNA was identified, and after 10 years, plant miRNA was identified. miRNA are non-coding RNA with a single strand, sized usually between 20 and 24 nucleotides, that acts as a gene regulator in a wide range of organisms including roots, stems, leaves, and flower parts (Lee *et al.*, 1993; Chen, 2004; Ferdous *et al.*, 2015). A growing body, including roots, stems, leaves, and flower parts, suggests that miRNA plays a key role in a plant's response to biotic and abiotic stresses (Chen, 2004; Ferdous *et al.*, 2015).

Numerous genes and metabolites, such as dehydrins, vacuolar acid invertase, glutathione S-transferase, and abscisic acid (ABA)-inducible genes, have been shown to change in expression when under drought stress. Embryo abundance, cold regulation, ABA responsiveness, helicase, proline, carbohydrates, and Rubisco (5-bisphosphate carboxylase oxygenase) play important roles in the plant stress system (Nezhadahmadi *et al.*, 2013; Ferdous *et al.*, 2015). Various research articles reported drought-responsive miRNA in many plant species such as tobacco, Arabidopsis, rice, cowpea, soybean, *Phaseolus vulgaris*, as given in Table 4.1 (Chen, 2004; Frazier *et al.*, 2011; Kulcheski *et al.*, 2011; Ferdous *et al.*, 2015).

Transgenic miR393 in rice showed hyposensitivity to auxin treatments, suggesting that native miR393 may be controlled by auxin signaling and under drought stress decrease the plant growth. Transgenic overexpression of miR393 in rice resulted in hyposensitivity to synthetic auxin analog treatments, suggesting that native miR393 may regulate auxin signaling and would thus reduce plant growth under drought stress (Xia *et al.*, 2012; Ferdous *et al.*, 2015; Maleksaeidi and Keshavarz, 2019). Another study reported that overexpression of miR169 c in tomatoes reduced stomatal transportation and water loss than non-transgenic tomatoes with drought tolerance (Zhang *et al.*, 2011; Ferdous *et al.*, 2015). miR393 is a regulator of root adaptation under drought stress and has two auxin receptors (TIR1 and AFB2), For drought-inhibited lateral root growth to occur, miR393-guided cleavage is required (Chen, 2004; Ferdous *et al.*, 2015; López-Vicente and Wu, 2019). Two auxins of miR167 target response factors (ARFs) that play a key role in root

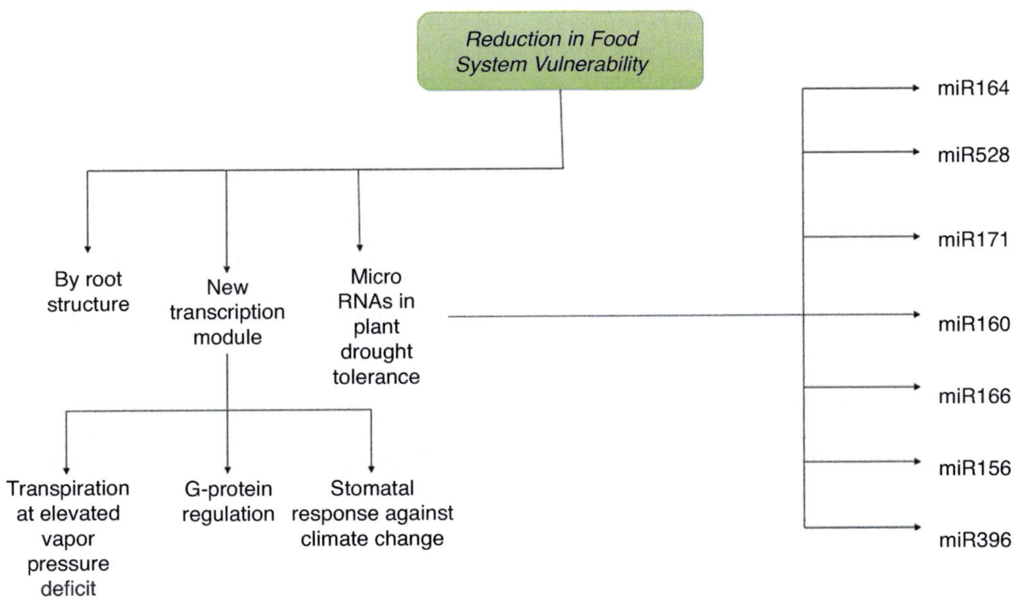

Fig. 4.5. Different methods showing reduction in food system vulnerability.

Table 4.1. Drought-responsive miRNAs in different plant species.

miRNA	Target name with functions	Species[a]	References
miR164	NAC domain TF—lateral root development	*Brachypodium distachyon* ↑	Shuai *et al.*, 2013
miR528	Peroxidase Elimination of ROS	*Zea mays* ↓	Wei *et al.*, 2009
miR171	Growth-regulating factor of floral development and transcription factors of response in abiotic pressure	*Triticum dicoccoides* ↓	Kantar *et al.*, 2011; Ferdous *et al.*, 2015
miR160	ARF 10, ARF 16 and ARF 17-seed germination and postgermination stages	*Populus tomentosa* ↓	Jones-Rhoades *et al.*, 2006; Ferdous *et al.*, 2015
miR166	Class III homeodomain leucine zipper transcription factor of axillary leaf and vascular development	*Glycine max* ↑	Kantar *et al.*, 2011
miR156	SBP family of transcription factors—promote phase transitions, flowering time.	*Hordeum vulgare* ↑	Zhao *et al.*, 2007
miR396	Growth-regulating factor, transcription factors of leaf and cotyledon development.	*Oryza sativa* ↑	Wang *et al.*, 2005

[a]Upregulation by drought ↑; downregulation by drought ↓

development. ABA plays a key role in seed dormancy. During seed germination or under drought conditions, miR159 levels of *Arabidopsis* increase exogenous ABA. Recent research found drought and ABA upregulated miR169, miR319, miR397, miR2118, miR393, miR398, and miR399 in tobacco, *Phaseolus vulgaris* and rice plants (Reyes and Chua, 2007; Ferdous *et al.*, 2015).

4.8 Conclusions

As can be seen above, climate change is causing an increase in the frequency of heat and cold waves, droughts, and floods worldwide. They have a devastating effect on farmers' means of subsistence, particularly in India, given our economy is more reliant on farming. It is interesting to note that opposite-nature weather extremes, such as heat and cold waves, floods, and droughts, can occur in the same year in different parts of the same region or elsewhere, and they are probably going to get worse in the coming decades. Significant crop and human losses are probably in store. The imbalance between greenhouse gases and aerosols produced by humans, which is linked to the entire climate change process, may cause uncertainty in the behavior of the monsoon from year to year across India.

As a result, both industrialized and emerging nations should make a concerted effort to reduce the amount of greenhouse gases that industrialization releases into the atmosphere. In a similar manner, programs to raise knowledge about climate change and its implications on a range of industries, including infrastructure, water, forestry, fisheries, agriculture, health, and sea level, as well as the role that human activities play in climate change, should be given top attention. People's habits should also be altered during this process to prevent adding CFCs and greenhouse gases to the atmosphere and endangering the earth's atmosphere.

From the perspective of agriculture, the regional effects of extreme weather events on crops should be recorded so that planners can use the information to mitigate the negative consequences of similar reoccurring events. Additionally, in order to reduce risk in the agricultural sector, farmers must be informed about the expected effects of climate change and made aware of potential mitigation and adaptation strategies.

Acknowledgments

The authors of the article are grateful to Seacom Skills University for all kinds of support to complete the review work.

Conflict of Interest

The author declares no conflict of interest.

References

Aerts, J.C.J.H. and Droogers, P. (2004) *Climate Change in Contrasting River Basins. Adaptation Strategies for Water, Food and Environment*. CAB International, Wallingford, UK. DOI: 10.1079/9780851998350.0000.

Aggarwal, P.K. (2008) Global climate change and indian agriculture: impacts, adaptation and mitigation. *Indian Journal of Agricultural Sciences* 78, 911.

Ahmad, J., Alam, D. and Haseen, M.S. (2011) Impact of climate change on agriculture and food security in India. *International Journal of Agriculture, Environment and Biotechnology* 4(2), 129–137.

Ali, M., Rehman, H. and Husain, S.M. (2012) Status of food insecurity at household level in rural India: a case study of Uttar Pradesh. *International Journal of Physical and Social Sciences* 2(8), 227–244.

Aryal, J.P., Khatri-Chhetri, A., Sapkota, T.B., Rahut, D.B. and Erenstein, O. (2020) Adoption and economic impacts of laser land leveling in the irrigated rice-wheat system in Haryana, India using endogenous switching regression. *Natural Resources Forum* 44(3), 255–273. DOI: 10.1111/1477-8947.12197.

Assmann, S.M. (2002) Heterotrimeric and unconventional GTP binding proteins in plant cell signaling. *The Plant Cell* 14(Suppl), S355–73. DOI: 10.1105/tpc.001792.

Ataei, P., Gholamrezai, S., Movahedi, R. and Aliabadi, V. (2021) An analysis of farmers' intention to use green pesticides: the application of the extended theory of planned behavior and health belief model. *Journal of Rural Studies* 81, 374–384. DOI: 10.1016/j.jrurstud.2020.11.003.

Babu, R.C., Shashidhar, H.E., Lilley, J.M., Thanh, N.D., Ray, J.D. *et al.* (2001) Variation in root penetration ability, osmotic adjustment and dehydration tolerance among accessions of rice adapted to rainfed lowland and upland ecosystems. Journal of Plant Breeding 120(3), 233–238. DOI: 10.1046/j.1439-0523.2001.00578.x.

Basu, R. (2020) The Saga of food security in West Bengal. *International Journal of Research-Granthalayah* 8(1), 189–202. DOI: 10.29121/granthaalayah.v8.i1.2020.267.

Caminade, C., McIntyre, K.M. and Jones, A.E. (2019) Impact of recent and future climate change on vector-borne diseases. *Annals of the New York Academy of Sciences* 1436(1), 157–173. DOI: 10.1111/nyas.13950.

Chakravarti, A.K. (1973) Green revolution in India. *Annals of the Association of American Geographers* 63(3), 319–330. DOI: 10.1111/j.1467-8306.1973.tb00929.x.

Chalise, D., Kumar, L. and Kristiansen, P. (2019) Land degradation by soil erosion in Nepal: a review. *Soil Systems* 3(1), 12. DOI: 10.3390/soilsystems3010012.

Champoux, M.C., Wang, G., Sarkarung, S., Mackill, D.J., O'Toole, J.C. *et al.* (1995) Locating genes associated with root morphology and drought avoidance in rice via linkage to molecular markers. *Theoretical and Applied Genetics* 90(7–8), 969–981. DOI: 10.1007/BF00222910.

Chen, H., Li, Z. and Xiong, L. (2012) A plant microRNA regulates the adaptation of roots to drought stress. *FEBS Letters* 586(12), 1742–1747. DOI: 10.1016/j.febslet.2012.05.013.

Chen, X. (2004) A microRNA as A translational repressor of APETALA2 in Arabidopsis flower development. *Science* 303(5666), 2022–2025. DOI: 10.1126/science.1088060.

Cheung, W.W.L., Lam, V.W.Y., Sarmiento, J.L., Kearney, K., Watson, R. *et al.* (2009) Projecting global marine biodiversity impacts under climate change scenarios. *Fish and Fisheries* 10(3), 235–251. DOI: 10.1111/j.1467-2979.2008.00315.x.

Collins, B., Chapman, S., Hammer, G. and Chenu, K. (2021) Limiting transpiration rate in high evaporative demand conditions to improve Australian wheat productivity. *In Silico Plants* 3(1), 006. DOI: 10.1093/insilicoplants/diab006.

Coulibaly, T.P., Du, J. and Diakité, D. (2021) Sustainable agricultural practices adoption. *Agriculture (Pol'nohospodárstvo)* 67(4), 166–176. DOI: 10.2478/agri-2021-0015.

Dara, S.K. (2019) The new integrated pest management paradigm for the modern age. *Journal of Integrated Pest Management* 10(1), 12. DOI: 10.1093/jipm/pmz010.

Devi, M.J., Sinclair, T.R. and Vadez, V. (2010) Genotypic variation in peanut for transpiration response to vapor pressure deficit. *Crop Science* 50(1), 191–196. DOI: 10.2135/cropsci2009.04.0220.

Dhanaraj, S. and Gade, S. (2012) Universal PDS: efficiency and equity dimensions. Available at: https://ssrn.com/abstract=2020130 (accessed 13 August 2024).

Dimitri, C., Effland, A. and Conklin, N.C. (2005) The 20th century transformation of US agriculture and farm policy. Economic Information Bulletin 59390, United States Department of Agriculture, Economic Research Service. DOI: 10.22004/ag.econ.59390.

Duncan, J. and Claeys, P. (2018) Politicizing food security governance through participation: opportunities and opposition. *Food Security* 10(6), 1411–1424. DOI: 10.1007/s12571-018-0852-x.

Fan, L.M., Zhang, W., Chen, J.G., Taylor, J.P., Jones, A.M. *et al.* (2008) Abscisic acid regulation of guard-cell K+ and anion channels in Gβ-and RGS-deficient arabidopsis lines. *Proceedings of the National Academy of Sciences* 105(24), 8476–8481.

Ferdous, J., Hussain, S.S. and Shi, B.J. (2015) Role of micro RNA in plant drought tolerance. *Plant Biotechnology Journal* 13(3), 293–305. DOI: 10.1111/pbi.12318.

Fletcher, A.L., Sinclair, T.R. and Allen, L.H. (2007) Transpiration responses to vapor pressure deficit in well watered 'slow-wilting' and commercial soybean. *Environmental and Experimental Botany* 61(2), 145–151. DOI: 10.1016/j.envexpbot.2007.05.004.

Frazier, T.P., Sun, G., Burklew, C.E. and Zhang, B. (2011) Salt and drought stresses induce the aberrant expression of microrna genes in tobacco. *Molecular Biotechnology* 49(2), 159–165. DOI: 10.1007/s12033-011-9387-5.

Ghannoum, O. (2009) C4 photosynthesis and water stress. *Annals of Botany* 103(4), 635–644. DOI: 10.1093/aob/mcn093.

Ghosh, J. (2020) A critique of the indian government's response to the COVID-19 pandemic. *Journal of Industrial and Business Economics* 47(3), 519–530. DOI: 10.1007/s40812-020-00170-x.

Goli, S., Rammohan, A. and Reddy, S.P. (2021) The interaction of household agricultural landholding and caste on food security in rural Uttar Pradesh, India. *Food Security* 13(1), 219–237. DOI: 10.1007/s12571-020-01109-9.

Gregory, P.J., Bengough, A.G., Grinev, D., Schmidt, S., Thomas, W.B. *et al.* (2009) Root phenomics of crops: opportunities and challenges. *Functional Plant Biology* 36(11), 922–929. DOI: 10.1071/FP09150.

Gupta, A., Rico-Medina, A. and Caño-Delgado, A.I. (2020) The physiology of plant responses to drought. *Science* 368(6488), 266–269. DOI: 10.1126/science.aaz7614.

Hendrickson, D., Smith, C. and Eikenberry, N. (2006) Fruit and vegetable access in four low-income food deserts communities in Minnesota. *Agriculture and Human Values* 23(3), 371–383. DOI: 10.1007/s10460-006-9002-8.

Islam, M.S. and Wong, A.T. (2017) Climate change and food in/security: a critical nexus. *Environments* 4(2), 38. DOI: 10.3390/environments4020038.

Jangam, A.P., Pathak, R.R. and Raghuram, N. (2016) Microarray analysis of rice d1 (RGA1) mutant reveals the potential role of G-protein alpha subunit in regulating multiple abiotic stresses such as drought, salinity, heat, and cold. *Frontiers in Plant Science* 7, 11. DOI: 10.3389/fpls.2016.00011.

Jia, W. and Zhang, J. (2008) Stomatal movements and long-distance signaling in plants. *Plant Signaling & Behavior* 3(10), 772–777. DOI: 10.4161/psb.3.10.6294.

Jones, A.D., Shrinivas, A. and Bezner-Kerr, R. (2014) Farm production diversity is associated with greater household dietary diversity in Malawi: findings from nationally representative data. *Food Policy* 46, 1–12. DOI: 10.1016/j.foodpol.2014.02.001.

Jones-Rhoades, M.W., Bartel, D.P. and Bartel, B. (2006) MicroRNAS and their regulatory roles in plants. *Annual Review of Plant Biology* 57, 19–53. DOI: 10.1146/annurev.arplant.57.032905.105218.

Joo, J.H., Wang, S., Chen, J.G., Jones, A.M. and Fedoroff, N.V. (2005) Different signaling and cell death roles of heterotrimeric G-protein α and β subunits in the Arabidopsis oxidative stress response to ozone. *The Plant Cell* 17(3), 957–970. DOI: 10.1105/tpc.104.029603.

Kantar, M., Unver, T. and Budak, H. (2010) Regulation of barley miRNAs upon dehydration stress correlated with target gene expression. *Functional & Integrative Genomics* 10(4), 493–507. DOI: 10.1007/s10142-010-0181-4.

Kantar, M., Lucas, S.J. and Budak, H. (2011) MiRNA expression patterns of triticum dicoccoides in response to shock drought stress. *Planta* 233, 471–484.

Kasim, C.M. (2012) Food security and nutrition in Kerala: an exploratory approach. *Journal of Rural Development* 31(4), 513–534.

Kerdsriserm, C., Suwanmaneepong, S. and Mankeb, P. (2016) Factors affecting adoption of organic rice farming in sustainable agriculture network, Chachoengsao Province, Thailand. *International Journal of Agricultural Technology* 12(7.1), 1227–1237.

Konar, A. (2021) Regulatory effect of G-protein in Transpiration. *International Journal of Scientific Research in Engineering and Management (IJSREM)* 5(12), 1–7.

Konar, A., Mukherjee, K. and Kaur, J. (2022) A critical approach of plant tissue culture for rural economy. *International Journal of Current Science* 12(4), 32–743.

Kulcheski, F.R., de Oliveira, L.F., Molina, L.G., Almerão, M.P., Rodrigues, F.A. *et al.* (2011) Identification of novel soybean microRNAs involved in abiotic and biotic stresses. *BMC Genomics* 12, 1–17. DOI: 10.1186/1471-2164-12-307.

Kumari, S., George, S.G., Meshram, M.R., Esther, D.B. and Kumar, P. (2020) A review on climate change and its impact on agriculture in India. *Current Journal of Applied Science and Technology* 39(44), 58–74. DOI: 10.9734/cjast/2020/v39i4431152.

Kusmin, L. (2012) Rural America at a glance. USDA-ERS Economic Brief, p. 21. DOI: 10.2139/ssrn.2202862.

Lee, R.C., Feinbaum, R.L. and Ambros, V. (1993) The C-elegans heterochronic gene lin-4 encodes small RNAs with antisense complementarity to lin-14. *Cell* 75(5), 843–854. DOI: 10.1016/0092-8674(93)90529-y.

Li, F., Zhang, K., Ren, J., Yin, C., Zhang, Y. *et al.* (2021) Driving mechanism for farmers to adopt improved agricultural systems in China: the case of rice-green manure crops rotation system. *Agricultural Systems* 192, 103202. DOI: 10.1016/j.agsy.2021.103202.

Liu, X., Pattanaik, N., Nelson, M. and Ibrahim, M. (2019) The choice to go organic: evidence from small US farms. *Agricultural Sciences* 10(12), 1566–1580. DOI: 10.4236/as.2019.1012115.

Liu, Y., Du, Q., Bai, L., Sun, M., Li, Y. *et al.* (2021) Interference of *CsGPA1*, the α-submit of G protein, reduces drought tolerance in cucumber seedlings. *Horticultural Plant Journal* 7(3), 209–220. DOI: 10.1016/j.hpj.2021.02.003.

Lobao, L. and Meyer, K. (2001) The great agricultural transition: crisis, change, and social consequences of twentieth century US farming. *Annual Review of Sociology* 27(1), 103–124. DOI: 10.1146/annurev.soc.27.1.103.

Lobell, D.B., Hammer, G.L., McLean, G., Messina, C., Roberts, M.J. *et al.* (2013) The critical role of extreme heat for maize production in the United States. *Nature Climate Change* 3(5), 497–501. DOI: 10.1038/nclimate1832.

Long, S.P., Ainsworth, E.A., Rogers, A. and Ort, D.R. (2004) Rising atmospheric carbon dioxide: plants FACE the future. *Annual Review of Plant Biology* 55, 591–628. DOI: 10.1146/annurev.arplant.55.031903.141610.

López-Vicente, M. and Wu, G.-L. (2019) Soil and water conservation in agricultural and forestry systems. *Water* 11(9), 1937. DOI: 10.3390/w11091937.

Lusk, J.L. and Chandra, R. (2021) Farmer and farm worker illnesses and deaths from COVID-19 and impacts on agricultural output. *PLoS One* 16(4), e0250621. DOI: 10.1371/journal.pone.0250621.

Lynch, J., Cain, M., Frame, D. and Pierrehumbert, R. (2021) Agriculture's contribution to climate change and role in mitigation is distinct from predominantly fossil CO_2-emitting sectors. *Frontiers in Sustainable Food Systems* 4, 518039. DOI: 10.3389/fsufs.2020.518039.

Mahmoud, S.H. and Gan, T.Y. (2018) Impact of anthropogenic climate change and human activities on environment and ecosystem services in arid regions. *Science of the Total Environment* 633, 1329–1344. DOI: 10.1016/j.scitotenv.2018.03.290.

Maleksaeidi, H. and Keshavarz, M. (2019) What influences farmers' intentions to conserve on-farm biodiversity? An application of the theory of planned behavior in fars province, Iran. *Global Ecology and Conservation* 20, e00698. DOI: 10.1016/j.gecco.2019.e00698.

Maren, N.A., Duan, H., Da, K., Yencho, G.C., Ranney, T.G. *et al.* (2022) Genotype-independent plant transformation. *Horticulture Research* 9, uhac047. DOI: 10.1093/hr/uhac047.

Maxwell, M.J. (2017) US farmers feed the world. Available at: https://archive-share.america.gov/u-s-farmers-feed-world/

Medina, S., Vicente, R., Nieto-Taladriz, M.T., Aparicio, N., Chairi, F. *et al.* (2018) The plant-transpiration response to vapor pressure deficit (VPD) in durum wheat is associated with differential yield performance and specific expression of genes involved in primary metabolism and water transport. *Frontiers in Plant Science* 9, 1994. DOI: 10.3389/fpls.2018.01994.

Merriott, D. (2016) Factors associated with the farmer suicide crisis in India. *Journal of Epidemiology and Global Health* 6(4), 217–227. DOI: 10.1016/j.jegh.2016.03.003.

Miśniakiewicz, M., Łuczak, J. and Maruszewska, N. (2021) Improvement of organic farm assessment procedures on the example of organic farming in Poland—Recommendations for organic farming in Poland. *Agronomy* 11(8), 1560. DOI: 10.3390/agronomy11081560.

Morris, P.M., Neuhauser, L. and Campbell, C. (1992) Food security in rural America: a study of the availability and costs of food. *Journal of Nutrition Education* 24(1), 52S–58S. DOI: 10.1016/S0022-3182(12)80140-3.

Morton, L.W. and Blanchard, T.C. (2007) Starved for access: life in rural America's food deserts. *Rural Realities* 1(4), 1–10.

Mozaffarian, D., Griffin, T. and Mande, J. (2019) The 2018 Farm Bill-implications and opportunities for public health. *JAMA* 321(9), 835–836. DOI: 10.1001/jama.2019.0317.

Mukherjee, K., Konar, A. and Ghosh, P. (2022) Organic farming in India: a brief review. *International Journal of Research in Agronomy* 5(2), 113–118. DOI: 10.33545/2618060X.2022.v5.i2b.120.

Narayanan, S. (2018) Effects of high temperature stress and traits associated with tolerance in wheat. *Open Access Journal of Science* 2(3), 177–186. DOI: 10.15406/oajs.2018.02.00067.

Negassa, M.D., Mallie, D.T. and Gemeda, D.O. (2020) Forest cover change detection using Geographic Information Systems and remote sensing techniques: a spatio-temporal study on Komto protected forest priority area, East Wollega Zone, Ethiopia. *Environmental Systems Research* 9(1), 1. DOI: 10.1186/s40068-020-0163-z.

Nezhadahmadi, A., Prodhan, Z.H. and Faruq, G. (2013) Drought tolerance in wheat. *The Scientific World Journal* 2013(1), 1–12. DOI: 10.1155/2013/610721.

Nilson, S.E. and Assmann, S.M. (2010) The α-subunit of the Arabidopsis heterotrimeric G- protein, GPA1, is a regulator of transpiration efficiency. *Plant Physiology* 152(4), 2067–2077. DOI: 10.1104/pp.109.148262.

Opitz, I., Berges, R., Piorr, A. and Krikser, T. (2016) Contributing to food security in urban areas: differences between urban agriculture and peri-urban agriculture in the Global North. *Agriculture and Human Values* 33(2), 341–358. DOI: 10.1007/s10460-015-9610-2.

O'toole, J.C. and Bland, W.L. (1987) Genotypic variation in crop plant root systems. *Advances in Agronomy* 41, 91–145.

Pandey, S., Chen, J.G., Jones, A.M. and Assmann, S.M. (2006) G-protein complex mutants are hypersensitive to abscisic acid regulation of germination and postgermination development. *Plant Physiology* 141(1), 243–256. DOI: 10.1104/pp.106.079038.

Parker, T. (2015) Updated ERS county economic types show a changing rural landscape. Amber Waves: The Economics of Food, Farming, Natural Resources, and Rural America (11). Available at: https://www.ers.usda.gov/amber-waves/2015/december/updated-ers-county-economic-types-show-a-changing-rural-landscape (accessed 22 March 2021).

Pathak, H., Jain, N., Bhatia, A., Patel, J. and Aggarwal, P.K. (2010) Carbon footprints of Indian food items. *Agriculture, Ecosystems & Environment* 139(1–2), 66–73. DOI: 10.1016/j.agee.2010.07.002.

Pillay, D.P.K. and Kumar, T.K.M. (2018) Food security in india: evolution, efforts and problems. *Strategic Analysis* 42(6), 595–611. DOI: 10.1080/09700161.2018.1560916.

Porter, J.R. and Semenov, M.A. (2005) Crop responses to climatic variation. *Philosophical Transactions of the Royal Society of London. Series B, Biological Sciences* 360(1463), 2021–2035. DOI: 10.1098/rstb.2005.1752.

Pothan, P.E., Taguchi, M. and Santini, G. (2020) *Local Food Systems and COVID-19: A Glimpse on India's Responses*. FAO, Rome, Italy.

Raj, S., Roodbar, S., Brinkley, C. and Wolfe, D.W. (2022) Food security and climate change: differences in impacts and adaptation strategies for rural communities in the global south and north. *Frontiers in Sustainable Food Systems* 5, 691191.

Ramadurai, V., Sharf, B.F. and Sharkey, J.R. (2012) Rural food insecurity in the United States as an overlooked site of struggle in health communication. *Health Communication* 27(8), 794–805. DOI: 10.1080/10410236.2011.647620.

Reyes, J.L. and Chua, N.H. (2007) ABA induction of miR159 controls transcript levels of two MYB factors during Arabidopsis seed germination. *The Plant Journal* 49(4), 592–606. DOI: 10.1111/j.1365-313X.2006.02980.x.

Sawalhah, M.N., Holechek, J.L., Cibils, A.F., Geli, H.M. and Zaied, A. (2019) Rangeland livestock production in relation to climate and vegetation trends in new mexico. *Rangeland Ecology and Management* 72, 832–845.

Schoppach, R. and Sadok, W. (2012) Differential sensitivities of transpiration to evaporative demand and soil water deficit among wheat elite cultivars indicate different strategies for drought tolerance. *Environmental and Experimental Botany* 84, 1–10.

Schroeder, J.I., Allen, G.J., Hugouvieux, V., Kwak, J.M. and Waner, D. (2001) Guard cell signal transduction. *Annual Review of Plant Biology* 52(1), 627–658.

Sen, A. (1982) *Poverty and Famines: An Essay on Entitlement and Deprivation*. Oxford University Press. DOI: 10.1093/0198284632.001.0001.

Serna, L. (2022) Maize stomatal responses against the climate change. *Frontiers in Plant Science* 13, 952146. DOI: 10.3389/fpls.2022.952146.

Sharkey, J.R. (2009) Measuring potential access to food stores and food-service places in rural areas in the US. *American Journal of Preventive Medicine* 36(4 Suppl), S151–5. DOI: 10.1016/j.amepre.2009.01.004.

Shuai, P., Liang, D., Zhang, Z., Yin, W. and Xia, X. (2013) Identification of drought-responsive and novel populus trichocarpa microRNAs by high-throughput sequencing and their targets using degradome analysis. *BMC Genomics* 14, 233. DOI: 10.1186/1471-2164-14-233.

Sibhatu, K.T. and Qaim, M. (2017) Rural food security, subsistence agriculture, and seasonality. *PLoS One* 12(10), e0186406. DOI: 10.1371/journal.pone.0186406.

Sinclair, T.R., Hammer, G.L. and Van Oosterom, E.J. (2005) Potential yield and water-use efficiency benefits in sorghum from limited maximum transpiration rate. *Functional Plant Biology* 32(10), 945. DOI: 10.1071/FP05047.

Singh, P., Rayeen, F., Singh, R., Pathak, N., Singh, R.P. *et al.* (2023) Enhancing agricultural sustainability through microbial-mediated abiotic stress tolerance. *Journal of Ecophysiology and Occupational Health* 23(4), 233–247. DOI: 10.18311/jeoh/2023/34777.

Singh, P.K., Singh, P., Singh, R.P. and Singh, R.L. (2022) Transgenesis in plants: principle and methods. In: Singh, R.L., Mondal, S., Parihar, A. and Singh, P.K. (eds) *Plant Genomics for Sustainable Agriculture*. Springer Nature, Singapore, pp. 41–70.

Smith, C. and Morton, L.W. (2009) Rural food deserts: low-income perspectives on food access in minnesota and iowa. *Journal of Nutrition Education and Behavior* 41(3), 176–187. DOI: 10.1016/j.jneb.2008.06.008.

Spittler, J., Ross, R. and Block, W. (2011) The economic impact of agricultural subsidies in the United States. *The Journal of Social, Political, and Economic Studies* 36(3), 301.

Sun, Y., Wang, H., Liu, S. and Peng, X. (2016) Exogenous application of hydrogen peroxide alleviates drought stress in cucumber seedlings. *South African Journal of Botany* 106, 23–28. DOI: 10.1016/j.sajb.2016.05.008.

Tiwari, A., Singh, D., Goswami, S.G., Singh, P.K., Singh, P. *et al.* (2021) Role of plant growth promoting microorganism for improved plant growth: a review. *Chhattisgarh Journal of Science and Technology* 18(3), 8–11.

Wang, J.W., Wang, L.J., Mao, Y.B., Cai, W.J., Xue, H.W. *et al.* (2005) Control of root cap formation by microRNA-targeted auxin response factors in Arabidopsis. *Plant Cell* 17(8), 2204–2216. DOI: 10.1105/tpc.105.033076.

Wang, X.Q., Ullah, H., Jones, A.M. and Assmann, S.M. (2001) G-protein regulation of ion channels and abscisic acid signaling in Arabidopsis guard cells. *Science* 292(5524), 2070–2072. DOI: 10.1126/science.1059046.

Watson, J., Zheng, B., Chapman, S. and Chenu, K. (2017) Projected impact of future climate on water-stress patterns across the Australian Wheatbelt. *Journal of Experimental Botany* 68(21–22), 5907–5921. DOI: 10.1093/jxb/erx368.

Wei, L., Zhang, D., Xiang, F. and Zhang, Z. (2009) Differentially expressed mirnas potentially involved in the regulation of defense mechanism to drought stress in maize seedlings. *International Journal of Plant Sciences* 170(8), 979–989. DOI: 10.1086/605122.

Xia, K., Wang, R., Ou, X., Fang, Z., Tian, C. *et al.* (2012) OsTIR1 and osAFB2 downregulation via osmiR393 overexpression leads to more tillers, early flowering and less tolerance to salt and drought in rice. *PLoS ONE* 7(1), e30039. DOI: 10.1371/journal.pone.0030039.

Yang, Z., Sinclair, T.R., Zhu, M., Messina, C.D., Cooper, M. *et al.* (2012) Temperature effect on transpiration response of maize plants to vapour pressure deficit. *Environmental and Experimental Botany* 78, 157–162. DOI: 10.1016/j.envexpbot.2011.12.034.

Zait, Y., Ferrero-Serrano, Á. and Assmann, S.M. (2021) The α subunit of the heterotrimeric G-protein regulates mesophyll CO_2 conductance and drought tolerance in rice. *New Phytologist* 232(6), 2324–2338. DOI: 10.1111/nph.17730.

Zhang, X., Zou, Z., Gong, P., Zhang, J., Ziaf, K. *et al.* (2011) Over-expression of microRNA169 confers enhanced drought tolerance to tomato. *Biotechnology Letters* 33(2), 403–409. DOI: 10.1007/s10529-010-0436-0.

Zhao, B., Liang, R., Ge, L., Li, W., Xiao, H. *et al.* (2007) Identification of drought-induced microRNAs in rice. *Biochemical and Biophysical Research Communications* 354(2), 585–590. DOI: 10.1016/j.bbrc.2007.01.022.

Zheng, Y., Xu, M., Hou, R., Shen, R., Qiu, S. *et al.* (2013) Effects of experimental warming on stomatal traits in leaves of maize (Zea may L.). *Ecology and Evolution* 3(9), 3095–3111. DOI: 10.1002/ece3.674.

5 Climate-Neutral Agriculture

Abhijith P., Aditya K.A., Saronik Bose and Geetha N.*
Department of Studies in Biotechnology, University of Mysore, Mysuru, India

Abstract
Sustainability has taken center stage both as an issue and practice in the 21st century; this is due to the agroclimatic and socioeconomic changes in the world that started affecting food production, which is the cornerstone for sustaining the world's population. As population increases and land-use changes, net available cropping land decreases, along with this, climate change and issues arising from biotic and abiotic stresses in plant health contribute toward a steady decrease in food production across the world. This furthers the cause for sustainability and sustainable carbon-neutral agriculture, as our society is highly addicted to chemical-based cultivation, which is causing deterioration in soil quality, which again causes reduction in food production, causing a feedback loop that is affecting human population across the land. Issues, causes, preventive measures, and the steps for stabilizing global, sustainable, and carbon-neutral agriculture are discussed in this chapter in depth.

Keywords: Agroclimatic, Food Production, Abiotic Stresses, Agriculture, Global Sustainable

5.1 Introduction

Agriculture plays a pivotal role in global sustainability, providing food, livelihoods, and raw materials. However, the conventional practices associated with agriculture contribute significantly to greenhouse gas emissions, primarily in the form of carbon dioxide (CO_2), methane (CH_4), and nitrous oxide (N_2O). The urgent need to address climate change has brought forth the concept of carbon-neutral agriculture, aiming to balance the emissions produced by agricultural activities with equivalent carbon removal or sequestration. This chapter explores the global challenges associated with current agricultural practices and outlines key principles for achieving carbon-neutral agriculture. The imperative to mitigate climate change has spurred the emergence of the concept of carbon-neutral agriculture, which seeks to offset the emissions generated by agricultural activities through the equivalent removal or sequestration of carbon. Achieving carbon-neutral agriculture is a complex and multifaceted challenge that requires a holistic approach to address the interconnected issues associated with agricultural production, land use, and environmental impact. One of the key challenges in current agricultural practices lies in the overreliance on synthetic fertilizers, particularly nitrogen-based fertilizers, which contribute significantly to nitrous oxide emissions. A study by Leip *et al.* (2015) underscores the importance of adopting precision agriculture techniques to optimize fertilizer use, thereby reducing nitrogen losses and associated greenhouse gas emissions. Furthermore, the expansion of agricultural land at the expense of natural ecosystems contributes to deforestation and loss of carbon sinks. Sustainable land management practices,

*Corresponding author: geetha@appbot.uni-mysore.ac.in

DOI: 10.1079/9781800625037.0005

such as agroforestry and reforestation, play a crucial role in sequestering carbon and enhancing the overall resilience of agricultural systems (Minang *et al.*, 2015).

Livestock farming, a prominent component of agriculture, is another major source of emissions, primarily due to enteric fermentation in ruminants and manure management. The adoption of improved feeding practices and the integration of manure into nutrient management strategies can mitigate emissions from livestock (Herrero *et al.*, 2016). Additionally, advancements in livestock genetics and breeding for improved feed efficiency contribute to reducing the environmental footprint of animal agriculture. In the pursuit of carbon-neutral agriculture, a comprehensive strategy must encompass the promotion of agroecological approaches, which prioritize biodiversity, soil health, and ecosystem resilience. Agroecological practices, as advocated by Altieri (2018), emphasize the integration of natural processes and biological diversity into agricultural systems, fostering sustainable and regenerative practices that enhance carbon sequestration. In conclusion, transitioning to carbon-neutral agriculture requires a paradigm shift in the way we approach and practice agriculture. The integration of precision agriculture, sustainable land management, and agroecological principles can play a pivotal role in mitigating greenhouse gas emissions from the agricultural sector. By adopting these strategies, we can work toward a more sustainable and resilient global food system.

5.2 Global Challenges in Sustainable Agriculture

5.2.1 Greenhouse gas emissions

Current agricultural practices contribute a substantial share of global greenhouse gas emissions. Emissions from the use of synthetic fertilizers, enteric fermentation in livestock, and land-use change are significant contributors to climate change (Smith *et al.*, 2014b). Synthetic fertilizers, particularly nitrogen-based fertilizers, release nitrous oxide (N_2O), a potent greenhouse gas, into the atmosphere during application and subsequent soil processes. This process significantly contributes to the agricultural sector's carbon footprint, as highlighted in the study by Leip *et al.* (2015), which emphasizes the need for precision agriculture techniques to optimize fertilizer use and minimize nitrogen losses.

Livestock farming is another major source of greenhouse gas emission, primarily due to enteric fermentation in ruminant animals. Methane (CH_4) released during digestion represents a significant portion of emissions from the livestock sector. Herrero *et al.* (2016) stress the importance of addressing emissions from livestock through improved feeding practices and the integration of manure management into sustainable nutrient strategies. These measures can contribute to mitigating emissions from the livestock sector, aligning with the goals of carbon-neutral agriculture. Furthermore, land-use change associated with agricultural expansion, deforestation, and conversion of natural ecosystems into cropland contributes to carbon dioxide (CO_2) emissions. The study by Smith *et al.* (2014a) underscores the interconnectedness of land-use change and greenhouse gas emissions, emphasizing the need for sustainable land management practices to preserve carbon sinks and biodiversity.

To address the challenges posed by greenhouse gas emissions in agriculture, it is crucial to adopt a multifaceted approach. This includes implementing precision agriculture to optimize fertilizer use, improving livestock management practices, and promoting sustainable land-use practices that prioritize carbon sequestration and biodiversity conservation.

5.2.2 Soil degradation

Intensive farming methods, such as monoculture and the excessive use of agrochemicals, contribute significantly to soil degradation, posing long-term threats to agricultural productivity (Lal, 2015). Monoculture, the practice of cultivating a single crop over large expanses of land, can lead to increased soil erosion and a decline in overall soil health. This is due to the lack of crop diversity, which disrupts the natural balance of the soil ecosystem. Lal's comprehensive study highlights the need for diversified cropping systems to mitigate the negative impacts of monoculture on soil structure and fertility (Lal, 2015). The excessive use of agrochemicals, including synthetic fertilizers and pesticides, further exacerbates soil degradation. These chemicals can disrupt the soil microbiota, decrease microbial diversity, and lead to the loss of beneficial soil organisms. Additionally, the application of synthetic fertilizers can contribute to nutrient imbalances, affecting the long-term fertility of the soil. Lal (2015) emphasizes the importance of adopting sustainable agricultural practices—such as agroecology—to maintain soil health and prevent degradation caused by chemical-intensive farming.

Soil erosion, a critical consequence of intensive farming, results from the removal of protective vegetation cover and the disruption of natural soil structures. The work of Montgomery (2007) underscores the historical impact of unsustainable farming practices on soil erosion and the subsequent decline of civilizations. Implementing conservation practices, such as contour plowing, cover cropping, and agroforestry, becomes essential to mitigate soil erosion and preserve the integrity of agricultural landscapes (Montgomery, 2007). Loss of organic matter, another consequence of intensive farming, leads to decreased soil fertility and resilience. Continuous cultivation without adequate organic matter input depletes the soil's ability to retain moisture and nutrients and support diverse microbial communities. Lal (2015) emphasizes the importance of incorporating organic farming practices, such as cover cropping and crop residue management, to enhance soil organic matter content and promote sustainable agricultural systems.

To conclude it should be said that, addressing soil degradation resulting from intensive farming requires a paradigm shift toward sustainable and regenerative agricultural practices. Diversifying cropping systems, reducing reliance on agrochemicals, implementing conservation measures, and promoting organic farming practices are crucial steps toward preserving soil health and ensuring long-term agricultural productivity.

5.2.3 Water scarcity

The agricultural sector stands as a significant consumer of water, and the inefficiencies in irrigation practices contribute substantially to global water scarcity issues. According to UN Water (2021), inefficient water management in agriculture exacerbates the strain on water resources. The demand for water in agriculture arises not only from crop irrigation but also from livestock production and other farming activities. Inefficient irrigation practices, such as flood irrigation and the use of outdated technologies, lead to substantial water wastage. This not only diminishes water availability for other sectors but also contributes to environmental problems such as soil salinization. The work of Molden *et al.* (2007) emphasizes the need for adopting water-efficient irrigation technologies, including drip and sprinkler systems, to enhance water productivity in agriculture. By implementing these technologies, water-use efficiency can be significantly improved, mitigating the impact of agricultural activities on water resources.

Climate change adds another layer of complexity to the challenge of water scarcity in agriculture. Altered precipitation patterns, a consequence of climate change, affect the timing and availability of water resources. The Intergovernmental Panel on Climate Change (IPCC) highlights the increasing frequency and intensity of extreme weather events, including droughts and floods, which disrupt agricultural water availability (IPCC, 2019). These changes necessitate adaptive water management strategies in agriculture to ensure resilience in the face of a changing climate. To address the water scarcity issues in agriculture, integrated water management practices are essential. This involves the optimization of irrigation systems, the adoption of water-efficient technologies, and the implementation of sustainable water-use policies. Furthermore, promoting soil moisture conservation techniques, such as mulching and conservation tillage, can enhance the water-holding capacity of soils, reducing the overall demand for irrigation.

In these aspects arriving to conclusions is not that easy, yet it could be said that the agricultural sector's substantial water consumption and inefficient irrigation practices contribute significantly to global water scarcity challenges. Addressing this issue requires a comprehensive approach, integrating water-efficient technologies, sustainable management practices, and climate-resilient strategies to ensure the long-term sustainability of agriculture and water resources.

5.2.4 Loss of biodiversity

Modern agricultural practices frequently entail the conversion of diverse ecosystems into monoculture landscapes, leading to significant consequences for biodiversity and ecosystem services (Tscharntke *et al.*, 2012). The shift toward monoculture reduces the variety of plant and animal species that can thrive in a given area. This phenomenon is particularly concerning as biodiversity plays a crucial role in maintaining ecosystem resilience and functionality (Cardinale *et al.*, 2012). Beyond the 2012 reference, recent studies further underscore the negative impact of monoculture on biodiversity and ecological sustainability. Recent research by Dainese *et al.* (2021) highlights the continued threat posed by monoculture to pollinators, such as bees and butterflies. Monoculture landscapes often lack the diverse floral resources needed to support a variety of pollinator species, leading to declines in their populations. This decline has cascading effects on plant reproduction and the overall health of ecosystems.

Moreover, the work of Liu *et al.* (2020) emphasizes the role of agricultural intensification, often associated with monoculture, in driving biodiversity loss. The study highlights that as agricultural practices intensify, the negative impacts on biodiversity become more pronounced. Monoculture not only diminishes plant diversity but also disrupts habitats, affecting the fauna and microbial communities that contribute to ecosystem health. In light of these findings, there is a growing recognition of the importance of transitioning toward more sustainable and biodiversity-friendly agricultural practices. Agroecological approaches, as advocated by Altieri and Nicholls (2020) , prioritize the integration of diverse crops, cover cropping, and other ecological principles to promote biodiversity and enhance ecosystem resilience. These practices not only mitigate the adverse effects of monoculture but also contribute to sustainable and resilient agricultural systems.

In summary, the conversion of diverse ecosystems into monoculture landscapes in modern agriculture continues to pose a threat to biodiversity and ecosystem services. Recent research underscores the urgent need for sustainable agricultural practices that prioritize biodiversity conservation, emphasizing the role of agroecology in fostering resilient and ecologically sound food production systems.

5.2.5 Chemical toxicity in soil

Chemical toxicity in soil is a pressing concern arising from non-sustainable agricultural practices, particularly the excessive use of agrochemicals such as pesticides and synthetic fertilizers. These substances, while aiding in pest control and enhancing crop productivity in the short term, can have detrimental effects on soil health and ecosystem sustainability. Recent studies underscore the negative impact of chemical toxicity in soil resulting from non-sustainable agricultural practices. For instance, research by Jiao *et al.* (2021) reveals that the persistent use of certain pesticides can lead to the accumulation of toxic residues in the soil, adversely affecting soil microbial communities and biodiversity. Prolonged exposure to these chemicals disrupts the balance of beneficial microorganisms crucial for nutrient cycling, ultimately compromising soil fertility and long-term agricultural productivity.

Furthermore, synthetic fertilizers, a cornerstone of conventional farming, contribute to chemical toxicity in soil. The study by Giller *et al.* (2020) highlights that the excessive application of nitrogen-based fertilizers can result in soil acidification and nutrient imbalances. This not only hampers plant growth but also leads to the leaching of excess nutrients into water bodies, contributing to water pollution and eutrophication. To address the issue of chemical toxicity in soil, sustainable agricultural practices are imperative. Recent research by Liu *et al.* (2022a) emphasizes the importance of agroecological approaches in reducing reliance on chemical inputs. Agroecology promotes the integration of organic farming practices, cover cropping, and diversified cropping systems, which not only enhance soil fertility but also minimize the environmental impact of agricultural activities.

To say the least, chemical toxicity in soil due to non-sustainable agricultural practices poses a significant threat to soil health, biodiversity, and long-term agricultural sustainability. Recent studies highlight the urgent need for a shift toward more sustainable farming methods to mitigate the adverse effects of agrochemicals on soil ecosystems. This makes it the most important aspect of study scientists must embark on to ensure our future generations do not repeat our mistakes. We have only one planet to live on and saving it for the next generation is our duty.

5.3 Principles for Pursuing Carbon-Neutral Agriculture

5.3.1 Sustainable soil management

Healthy soils play a pivotal role in the pursuit of carbon-neutral agriculture, and sustainable practices contribute to enhancing soil health. Conservation tillage, a key component of soil management, involves minimizing soil disturbance during planting and cultivation. Recent studies, including the work of Six *et al.* (2020), highlight that conservation tillage practices sequester carbon by reducing soil erosion and preserving soil structure. This approach enhances soil organic carbon content, promoting long-term carbon storage and overall soil health. Cover cropping is another valuable practice in carbon-neutral agriculture, wherein cover crops are planted during periods of the year when the main cash crops are not growing. The study by Poeplau *et al.* (2019) demonstrates that cover crops contribute to soil organic carbon sequestration through root biomass and residue incorporation. Additionally, cover crops help reduce soil erosion, enhance nutrient cycling, and provide ground cover, fostering a conducive environment for soil microorganisms that contribute to carbon sequestration. Agroforestry, the integration of trees and shrubs into agricultural systems, also holds significant promise for carbon-neutral agriculture. Research by Nair *et al.* (2021) emphasizes that agroforestry enhances soil organic carbon stocks by providing organic inputs through tree litter and root biomass. The shade and windbreak effects of trees in agroforestry systems contribute to improved microclimates, reducing soil moisture loss and promoting water retention. These combined effects enhance soil fertility and contribute to carbon sequestration. The practices such as conservation tillage, cover cropping, and agroforestry are integral components of carbon-neutral agriculture. These strategies not only enhance soil organic carbon content but also improve soil fertility, water retention, and overall resilience. Implementing these practices at scale can contribute to sustainable and climate-resilient agricultural systems.

5.3.2 Agroecological approaches

Adopting agroecological principles involves designing farming systems that mimic natural ecosystems, fostering sustainability and resilience. Diversifying crops is a fundamental aspect of agroecology, promoting biodiversity within agricultural landscapes. Recent research by Perfecto *et al.* (2021) demonstrates that diverse cropping systems enhance ecosystem services, such as pest control and pollination, contributing to overall farm productivity. By cultivating a variety of crops, farmers can reduce the risk of crop failure and enhance the adaptability of their systems to changing environmental conditions. Integration of livestock within agroecological farming systems is another key principle that has gained recognition for its multifaceted benefits. Livestock provides valuable ecosystem services, such as nutrient cycling and weed control, contributing to improved soil fertility. The work of van Zanten *et al.* (2021) highlights the potential of integrating livestock into agroecological practices to enhance resource use efficiency and minimize environmental impact. This integration not only supports sustainable agriculture but also contributes to diversified income sources for farmers.

Biological pest control, a central tenet of agroecology, involves harnessing natural predators and beneficial organisms to manage pest populations. Recent studies, such as the work of Gurr *et al.* (2022), emphasize the effectiveness of biological pest control strategies in reducing the reliance on chemical pesticides. By promoting natural enemies of pests, agroecological systems mitigate the negative environmental and health impacts associated with synthetic pesticides while maintaining effective pest management. In addition to these principles, agroecology encourages the use of cover cropping, agroforestry, and other sustainable practices that enhance soil health and water retention. Research by Altieri and Nicholls (2020) suggests that agroecological approaches not only contribute to environmental sustainability but also have the potential to enhance the resilience of farming systems in the face of climate change and other stressors. Adopting agroecological principles offers a holistic approach to sustainable agriculture. Diversifying crops, integrating livestock, and promoting biological pest control contribute to resilient and productive farming systems while minimizing the reliance on external inputs.

5.3.3 Precision agriculture

Precision agriculture, driven by technological advancements, stands as a key approach to optimize resource use, reduce waste, and minimize environmental impact. Smart irrigation, a core component of precision agriculture, utilizes sensors and data analytics to tailor irrigation practices based on real-time information about soil moisture levels and crop needs. Recent research by Wang *et al.* (2021b) underscores the effectiveness of smart irrigation systems in conserving water, improving water-use efficiency, and mitigating the environmental consequences associated with excessive water application. Precision nutrient management is another vital aspect of precision agriculture, focusing on the targeted application of fertilizers based on the specific needs of crops. Research by Ogle *et al.* (2020) highlights the potential of precision nutrient management to enhance nutrient use efficiency, reduce nutrient runoff, and mitigate the environmental impacts of nutrient over-application. By precisely matching nutrient inputs to crop requirements, this approach not only optimizes yields but also minimizes the ecological footprint of agricultural activities.

Sensor-based monitoring systems play a crucial role in precision agriculture by providing real-time data on various environmental variables, including soil conditions, weather patterns, and crop health. These systems enable farmers to make informed decisions, optimizing inputs and minimizing waste. The study by Sun *et al.* (2022) emphasizes the importance of sensor-based technologies in improving overall farm management practices, resulting in increased productivity and sustainability. Precision agriculture, with its emphasis on technology-driven optimization, offers a pathway to more sustainable and efficient resource use in farming. Smart irrigation, precision nutrient management, and sensor-based monitoring collectively contribute to reducing environmental impact while enhancing the productivity and resilience of agricultural systems.

5.3.4 Agroforestry

Agroforestry, characterized by the intentional integration of trees into agricultural settings, continues to emerge as a sustainable land-use practice with multifaceted benefits. Recent research by Albrecht *et al.* (2021) reinforces the pivotal role of agroforestry in enhancing carbon sequestration. Trees in agroforestry systems act as carbon sinks, sequestering atmospheric carbon dioxide and storing it in both aboveground biomass and soil organic matter. This carbon sequestration not only mitigates climate change but also enhances overall soil health. The positive impact of trees in agroforestry extends beyond carbon sequestration. The provision of shade by trees offers microclimatic benefits, which can be crucial for the well-being of crops and livestock. Research by Jose (2009) emphasizes that the shade provided by agroforestry systems helps reduce temperature extremes, improve water-use efficiency, and mitigate the impacts of extreme weather events. This microclimatic regulation contributes to increased crop resilience and productivity.

Moreover, agroforestry systems play a crucial role in diversifying income sources for farmers. The integration of tree crops, such as fruits and nuts, alongside traditional crops or livestock, provides additional revenue streams. This diversification not only enhances economic resilience for farmers but also contributes to improved food security. Recent work by van Noordwijk *et al.* (2019) highlights the importance of agroforestry in supporting livelihoods and fostering sustainable land-use practices. In a way, agroforestry emerges as a holistic approach that not only contributes to carbon sequestration but also promotes soil health, microclimatic benefits, and economic diversification. Its adoption holds significant promise for achieving sustainable and resilient agricultural systems.

5.3.5 Renewable energy integration into agriculture

Renewable energy integration into agriculture stands as a pivotal strategy for mitigating the environmental impact of on-farm operations and fostering sustainability. Recent advancements in renewable energy technologies have showcased their potential to revolutionize the energy landscape in agriculture. Research by Zaccaria *et al.* (2021) emphasizes the significance of solar-powered irrigation systems in reducing the carbon footprint of agriculture. These systems harness energy from the sun to power irrigation pumps, offering a clean and sustainable alternative to traditional fossil fuel-based systems. Solar-powered irrigation not only reduces greenhouse gas emissions but also provides a reliable and cost-effective energy source for water management in agriculture. Wind energy represents another promising avenue for renewable energy integration into agriculture. Wind turbines can be strategically deployed on farmlands to generate electricity for various on-farm operations. Studies highlight the potential of wind energy to contribute to a more sustainable energy paradigm in agriculture. The adoption of wind energy not only reduces dependence on non-renewable energy sources but also presents an opportunity for farmers to become energy producers, contributing surplus energy to the grid. Bioenergy derived from agricultural residues offers yet another avenue for sustainable energy in agriculture. The work of Goglio *et al.* (2019) underscores the importance of bioenergy as a renewable resource sourced from agricultural by-products. Agricultural residues, such as crop residues and organic waste, can be converted into bioenergy through processes like anaerobic digestion or combustion. This not only provides an alternative energy source but also addresses waste management issues in agriculture, promoting a circular economy approach. The integration of renewable energy sources, including solar-powered irrigation, wind energy, and bioenergy from agricultural residues, holds substantial promise for reducing the carbon footprint of agriculture. These innovative approaches contribute to a more sustainable and environmentally friendly energy paradigm, aligning with global efforts to transition toward cleaner and renewable energy systems.

5.3.6 Climate-smart crop breeding

Climate-smart crop breeding plays a crucial role in advancing the goals of carbon-neutral agriculture by developing crop varieties that are resilient to the challenges posed by climate change. Recent research by Jarvis *et al.* (2020) emphasizes the importance of climate-smart crop breeding in addressing the impacts of changing environmental conditions on global agriculture. This approach involves the systematic selection and development of crop varieties that exhibit enhanced tolerance to abiotic stresses, such as drought, as well as improved resistance to pests and diseases. Drought tolerance is a primary focus of climate-smart crop breeding, given the increasing frequency and intensity of drought events associated with climate change. Studies, including the work of Voss-Fels *et al.* (2019), highlight the application of advanced breeding techniques such as genomic selection to accelerate the development of drought-tolerant crop varieties. These varieties not only contribute to sustainable water use in agriculture but also enhance the overall resilience of cropping systems to water scarcity. Resistance to pests and diseases is another critical aspect of climate-smart crop breeding. The study by Bisht *et al.* (2019) underscores the importance of incorporating genetic resistance into crop varieties to reduce the reliance on chemical pesticides. By developing crops with built-in resistance mechanisms, climate-smart breeding contributes to environmentally friendly pest management practices, aligning with the principles of sustainable and carbon-neutral agriculture.

Moreover, adaptability to changing environmental conditions is a key focus of climate-smart crop-breeding efforts. The work of Varshney *et al.* (2018) emphasizes the need for crop varieties that can thrive in diverse agroecological zones, ensuring stable yields in the face of shifting climate patterns. By integrating diverse genetic traits, such as heat tolerance and adaptability to different soil types, into crop varieties, climate-smart breeding enhances the ability of agriculture to withstand the uncertainties associated with climate change. Climate-smart crop breeding is instrumental in developing resilient crop varieties essential for carbon-neutral agriculture. By addressing challenges such as drought, pests and diseases, and promoting adaptability to changing environmental conditions, this approach contributes to the sustainability and long-term success of agricultural systems in the face of a changing climate.

5.3.7 Sustainable livestock practices

Sustainable livestock practices are paramount in addressing the environmental impact of livestock production, a sector known for its significant contribution to agricultural emissions. Research by Herrero *et al.* (2020) emphasizes the importance of sustainable livestock practices in mitigating the environmental footprint associated with animal agriculture. Improved feeding strategies stand out as a key component of sustainable livestock management. Recent studies, including work by Hristov *et al.* (2013), highlight the potential of precision feeding and optimized nutritional plans in reducing enteric methane emissions from ruminant animals. By carefully managing the composition of livestock diets, farmers can enhance nutrient utilization and minimize the release of greenhouse gases, contributing to more sustainable and eco-friendly livestock production (Singh *et al.*, 2020, 2021). The incorporation of methane-reducing additives is another innovative approach to sustainable livestock practices. Recent research, such as that by Morgavi *et al.* (2018), explores the use of feed supplements and additives that can modulate microbial activity in the digestive system of livestock, leading to a reduction in methane emissions. These additives aim to improve the efficiency of microbial digestion in the rumen; thereby mitigating one of the major sources of greenhouse gas emissions associated with livestock farming. Agroecological integration in livestock management represents a holistic approach that aligns with sustainable agricultural principles. Studies, including the work of Roschinsky *et al.* (2021), highlight the benefits of integrating livestock into agroecological systems. Agroecological approaches involve the integration of diverse crops, trees, and livestock within a single farming system, promoting resource use efficiency and minimizing environmental impact. Livestock in agroecological systems contribute to nutrient cycling, pest control, and soil fertility, fostering a more sustainable and resilient agricultural landscape. Figure 5.1 gives a clearer picture on how a possible carbon-neutral agriculture action plan detailing the needed actions and its related tasks can be crafted.

5.4 Potential Consequences of Reducing Food Production

As the global population experiences sustained growth, the intricate challenge of simultaneously ensuring food security and mitigating the environmental impact of agriculture becomes increasingly apparent. The pressing need to address environmental concerns associated with current food production practices stems from the recognition that traditional agricultural methods contribute significantly to issues such as deforestation, biodiversity loss, and greenhouse gas emissions. However, the prospect of a substantial reduction in food production raises complex questions and potential consequences that extend beyond immediate environmental benefits. The central theme of this chapter is to delve into the multifaceted ramifications of reducing food production, acknowledging the intricate interplay between environmental, economic, and social factors. The pursuit of sustainable agriculture is essential, but it necessitates a delicate balance to avoid unintended and adverse outcomes. As the global community strives to meet the challenge of

Objective	Strategy	Actions
Carbon-Neutral Agriculture	Sustainable Soil Management	Use of microbes
		Conservation tillage
		Carbon storage
	Agroecological Approaches	Crop diversification
		Inclusion of livestock
		Biological pest control
	Precision Agriculture	Resource optimization
		AI- and sensor-based automation
		Targeted fertilizer use
	Agroforestry	Trees as carbon sinks
		Diversification of income
		Better microclimate and soil health
	Renewable Energy Integration into agriculture	Solar power integration
		Wind energy harvesting
		Bioenergy and biofuel use
	Climate-Smart Crop Breeding	Use of climate-adapted variants
		Genomic selection
		Incorporating better traits
	Sustainable Livestock Practices	Precision feeding and nutrition plans
		Use of methane reducing food additives
		Nutrient cycling and farm integration

Fig. 5.1. A climate-supportive plan of action for carbon-neutral agriculture.

feeding a growing population, it is imperative to critically examine the potential consequences of altering established food production paradigms. The connections between totally disconnected geographical regions with respect to global food production may seem highly non-relational, but one must consider the fact that even in such a context, a challenge exists as food stability is the utmost important parameter for any culture to thrive.

The interconnected web of environmental, economic, and social factors is a key consideration in this exploration. Environmental sustainability in agriculture involves minimizing the negative impact on ecosystems, water resources, and climate while optimizing resource use efficiency (Lal, 2021). Economic aspects encompass the livelihoods of those engaged in agriculture, the global food market, and the economic stability of regions heavily dependent on farming. Social factors encompass issues of food access, nutritional well-being, and the overall socioeconomic fabric of communities. While the need for sustainable agriculture is clear, any significant reduction in food production demands careful evaluation, the consequences are far-reaching and may include alterations in land-use patterns, economic disparities, and challenges to global food security. It is within this intricate context that the chapter unfolds, examining the potential trade-offs and implications associated with decisions aimed at reducing food production.

5.4.1 Environmental impacts: altered land use and biodiversity loss

Reducing food production not only influences land-use patterns but also has significant implications for biodiversity and ecosystem services. Additionally, alterations in agricultural practices may impact the availability of diverse habitats, affecting various species. Research by Newbold *et al.* (2022) underscores the critical role of intact and diverse ecosystems in supporting biodiversity. Habitat loss and fragmentation not only threaten individual species but can also disrupt ecological interactions and lead to imbalances in local ecosystems. Furthermore, biodiversity loss can compromise ecosystem services essential for agriculture, such as natural pest control provided by diverse predator populations. Maintaining a balance between agricultural productivity and biodiversity conservation is crucial for the long-term sustainability of food systems.

5.4.2 Economic implications: rural livelihoods and global food security

The potential reduction in food production holds significant economic implications, particularly for rural communities that heavily rely on agriculture for their livelihoods. Laborde *et al.* (2020) stress the pivotal role of agriculture in providing employment and income, particularly in many developing countries. A decrease in food production could exacerbate poverty levels and food insecurity, particularly in regions where agriculture serves as a primary economic activity. The ripple effect could extend beyond local communities to impact global food security. Research by Robinson *et al.* (2021) highlights the interconnectedness of global food systems and underscores how disruptions in one region can have cascading effects, potentially leading to higher food prices and increased vulnerability to external shocks. Thus, the economic repercussions of reduced food production extend far beyond individual farming communities, with the potential to disrupt the delicate balance of global food markets.

5.4.3 Social consequences: food access and nutritional challenges

The social consequences of reduced food production encompass not only issues of food access and nutritional challenges but also extend to broader aspects of public health and well-being. Hawkes *et al.* (2019) emphasize the intricate link between food production, availability, and nutritional outcomes. A decline in food production may not only limit access to diverse and nutritious foods but also amplify malnutrition and health disparities. Additionally, diminished food production can exacerbate existing vulnerabilities among marginalized populations, such as indigenous communities or those with limited access to healthcare. Research by Fanzo *et al.* (2021) highlights the intricate connections between food systems and public health, emphasizing that disruptions in food production can have long-term consequences on the overall health of communities. Ensuring equitable access to nutritionally rich foods is crucial for fostering resilience against health challenges, especially in the context of a potential reduction in food production.

5.4.4 Climate change and agricultural adaptation: a double-edged sword

Climate change and its impacts on agriculture create a complex interplay between mitigation and adaptation strategies. While reducing food production is considered a measure to mitigate climate change, it simultaneously poses challenges related to the adaptation of agriculture to the changing climate. The altering temperature and precipitation patterns associated with climate change have direct implications for crop yields and overall food production (Wheeler and von Braun, 2013). Studies by Rosenzweig *et al.* (2021) emphasize that variations in temperature and erratic precipitation events can lead to increased risks of crop failure, impacting global food security. Moreover, changes in climate conditions may necessitate shifts in cropping patterns and the introduction of more resilient crop varieties. The delicate balance between mitigating the environmental impact of agriculture and adapting to the evolving climate underscores the need for comprehensive and innovative strategies to ensure the sustainability and resilience of global food systems.

Adapting agriculture to climate change requires a multifaceted approach that considers not only the physical impacts of climate change on crops but also the social and economic dimensions of farming communities. The vulnerability of agriculture to climate change is not uniform across regions or communities. Research by Nelson *et al.* (2014) emphasizes the importance of considering local context and community-specific vulnerabilities in crafting adaptation strategies. Additionally, climate adaptation measures need to be inclusive, addressing the needs of smallholder farmers, who often bear the brunt of climate-related challenges. Collaborative efforts, such as those advocated by Bai *et al.* (2018), stress the significance of involving local communities and stakeholders in the design and implementation of adaptation strategies. Building resilience in agricultural systems requires a nuanced understanding of the socioeconomic landscape, enabling the development of adaptive strategies that not only mitigate climate risks but also enhance the overall sustainability of food production.

5.5 Mitigation of Production Emissions of Greenhouse Gases

As the global community grapples with the urgent need to address climate change, agriculture stands as both a significant contributor to greenhouse gas emissions and a sector vulnerable to its impacts. This chapter explores strategies for mitigating production emissions in agriculture, aiming to pave the way for climate-neutral farming practices. By minimizing the environmental footprint of agricultural activities, we can contribute to global efforts to achieve climate neutrality.

5.5.1 Understanding production emissions in agriculture

To effectively mitigate production emissions, it is crucial to first understand the sources and dynamics of greenhouse gas emissions in agriculture. Recent studies, such as those by Smith *et al.* (2014b) and Ciais *et al.* (2019), provide comprehensive insights into the different components of agricultural emissions, including methane from enteric fermentation, nitrous oxide from fertilizer application, and carbon dioxide from land-use change. By identifying these sources, targeted mitigation strategies can be developed to address specific emission pathways. Advancements in emission quantification and monitoring technologies are enhancing our ability to understand and address greenhouse gas emissions in agriculture. Utilizing satellite and drone-based monitoring, these technologies offer a dynamic and spatially explicit assessment of emissions, allowing for a more nuanced understanding of emission hotspots and patterns. Additionally, research by Wolf *et al.* (2020) emphasizes the importance of integrating machine learning algorithms with emission monitoring tools to predict emission trends and guide timely interventions. The integration of these innovative technologies into emission understanding not only refines our knowledge of agricultural emissions but also provides a foundation for the development of adaptive and responsive mitigation strategies.

5.5.2 Precision agriculture for emission reductions

Precision agriculture, leveraging advanced technologies such as GPS, sensors, and machine learning, offers a promising avenue for reducing production emissions. Pannell and Llewellyn (2014) demonstrate that precision agriculture enables more targeted and efficient use of inputs, minimizing waste and optimizing resource use. By precisely managing fertilizer application, irrigation, and other inputs, farmers can curtail emissions associated with over-application and enhance the overall sustainability of farming practices. Transitioning to renewable energy sources for on-farm operations is a key strategy for reducing greenhouse gas emissions in agriculture. Goglio *et al.* (2019) demonstrate that solar-powered irrigation, wind energy, and bioenergy from agricultural residues contribute to a more sustainable energy paradigm. By replacing fossil fuel-based energy with renewable alternatives, farmers can significantly reduce the carbon footprint of their operations.

5.5.3 Agroecological approaches for sustainable intensification

Agroecological practices, as advocated by Altieri and Nicholls (2020), focus on ecological principles to design farming systems. By diversifying crops, integrating livestock, and fostering biological pest control, agroecology enhances resilience and reduces the need for external inputs. These practices contribute not only to increased productivity but also to emissions reduction, as they minimize reliance on synthetic fertilizers and pesticides associated with high greenhouse gas emissions. Agroecological approaches not only contribute to sustainable intensification but also foster biodiversity and enhance soil health. This biodiversity not only provides essential ecosystem services but also strengthens the resilience of agroecosystems to environmental stressors. Moreover, the promotion of biological pest control within agroecological systems, as highlighted by Letourneau *et al.* (2019), reduces the reliance on chemical pesticides. This not only benefits the environment by minimizing the impact of harmful chemicals but also contributes to emissions reduction associated with pesticide manufacturing and application. Therefore, agroecological practices emerge not only as a means of intensifying agricultural productivity sustainably but also as a holistic approach that positively influences broader ecological dynamics.

Furthermore, the social and economic dimensions of agroecology, as explored by Gliessman (2015), highlight its potential to improve the livelihoods of farmers and contribute to resilient and sustainable food systems. By emphasizing community-based approaches and local knowledge, agroecology becomes not only a set of farming practices but also a transformative movement with the capacity to address global challenges related to emissions, food security, and socioeconomic equity.

5.5.4 Advancing sustainable soil management through innovative techniques with cover cropping and conservation tillage

Cover cropping and conservation tillage practices play a vital role in mitigating production emissions. Paustian *et al.* (2016) highlight that cover crops improve soil health, increase carbon sequestration, and reduce nitrous oxide emissions. Conservation tillage, which minimizes soil disturbance, further enhances these benefits. Together, these practices contribute to climate-neutral agriculture by promoting sustainable soil management.

In addition to cover cropping and conservation tillage, emerging soil management techniques offer promising avenues for further reducing production emissions in agriculture. Recent research by Six *et al.* (2020) underscores the potential of regenerative agriculture practices, such as no-till farming combined with the incorporation of cover crops. This integrated approach not only enhances soil health but also promotes carbon sequestration, mitigating the release of greenhouse gases. By combining traditional conservation practices with innovative technologies, farmers can achieve a synergistic effect, further advancing the goal of climate-neutral agriculture through sustainable soil management.

5.5.5 Enhancing biodiversity for resilient ecosystems

Promoting biodiversity within agricultural systems is not only crucial for ecological practices. Research by Tscharntke *et al.* (2012) emphasizes that fostering biodiversity, through practices like agroecology and the incorporation of diverse cover crops, creates multifunctional landscapes that support a variety of plant and animal species. The presence of diverse vegetation not only provides habitat for beneficial insects but also enhances pollination services, ultimately contributing to increased agricultural productivity. Furthermore, the research of Letourneau *et al.* (2019) highlights the intricate connections between biodiversity and soil health. Diverse plant species contribute to improved soil structure, nutrient cycling, and microbial activity. By incorporating these biodiversity-promoting practices, farmers not only bolster the resilience of their ecosystems but also contribute to the overall health of the soil, laying the foundation for sustainable agriculture.

5.5.6 Soil health as the bedrock of sustainable agriculture

The promotion of soil health is fundamental to sustainable agriculture, and diverse agricultural practices contribute significantly to this objective. Research by Lal (2021) underscores that maintaining soil health involves practices such as cover cropping, crop rotation, and reduced tillage, which enhance organic matter content and microbial diversity. These practices not only sequester carbon in the soil but also improve water retention and nutrient cycling, mitigating the need for synthetic inputs. Additionally, the study by Fanzo *et al.* (2021) emphasizes the vital link between soil health and the nutritional quality of crops. Nutrient-rich soils, enriched through diverse agricultural practices, yield crops with enhanced nutritional content. Thus, promoting biodiversity and soil health within agricultural systems is not only an ecological imperative but also a key strategy for fostering sustainable and nutritious food production.

5.6 Soil Carbon and Bioenergy

5.6.1 Soil carbon

Soil carbon is the overall amount of solid carbon present in the global soil, which is an integral part of soil structure. It defends the properties of soil and safeguards against erosion, desertification, and leaching. This entails Soil Organic Carbon (SOC) and Soil Inorganic Carbon (SIC). The former is present as organic matter and the latter as carbonate minerals (Ferdush and Paul, 2021). It makes up the biggest terrestrial pool of carbon. It stores twice the quantity of carbon in the atmosphere and thrice the quantity in vegetation (Sreenivas *et al.*, 2016). The global stock of SOC varies between 1220 and 1576 petagram (Pg) and soil inorganic carbon varies between 700 and 1700 Pg, in the top 100 cm^3 of the soil. 1 petagram equates to 1 Gigaton (Gt), which is 1 billion metric tons or 10^{15}g (Guo *et al.*, 2016). Soil carbon plays a notable role in soil fertility and productivity as it directly determines the physical, chemical, and biological properties of the soil. Understanding the role of climate change on global carbon stock is of prime importance as it directs toward carbon sequestration and other proactive measures.

5.6.2 Soil organic carbon

Soil organic matter (SOM) depicts the overall organic components present in soil, such as soil microbes, decaying matter of plants and animals, fecal material, and products of decomposition. SOM is crucial for soil stability, structure, nutrients, maintenance and release of plant nutrients, and water-holding capacity. Thus, it portrays the resilient nature of the environment and acts as a key indicator of soil productivity. It contains about 55% of carbon by mass (Food and Agriculture Organization, 2017). Almost all of such carbon constitutes soil organic carbon (SOC), which is resultant of interactions of several ecological processes such as photosynthesis, respiration, and decomposition. Photosynthesis involves the absorption and fixation of atmospheric CO_2 into plant biomass. Soil organic carbon is greatly influenced by the root biomass of plants, as it is responsible for the transfer of carbon compounds from plants to soil. Soil microbes facilitate the process of decomposition of biomass, which results in loss of carbon in the form of CO_2 and replenishment of carbon to the soil in the form of humus. Humus is rich in nutrients, highly resistant to degradation, and is responsible for the moisture retention property of the soil. The deposition of humus in soil is attributed to its characteristic dark color (Liu *et al.*, 2021). An increase in global SOC by 0.01% can tremendously mitigate climate change by reducing atmospheric CO_2 concentration (Lal *et al.*, 1998).

SOC includes dissolved organic carbon (DOC), particulate organic carbon (POC), microbial biomass carbon (MBC), easily oxidizable organic carbon (EOC), and mineral-associated organic carbon (MOC) (Liu *et al.*, 2021). MBC represents the microbial component of SOC, which is involved in microbial-derived processes (Joergensen and Wichern, 2018). DOC represents the energy source and carbon availability to soil microbes. POC is composed of particles whose size ranges from 0.05 mm to 2 mm. EOC represents the oxidizable portion of SOC, while MOC constitutes low molecular compounds mainly derived from plant and microbial activities, typically with a size of less than 0.05 mm (Lavallee *et al.*, 2019).

5.6.3 Soil inorganic carbon (SIC)

SIC is primarily found in arid, semi-arid, and semi-humid regions, which account for one-third of the earth's surface. Lithogenic inorganic carbon (LIC) and pedogenic inorganic carbon (PIC) are the two categories of SIC (Sreenivas *et al.*, 2016). Calcium and magnesium carbonate predominate SIC. Studies have shown that an increase in the fertility of soil increases the carbonate concentration, which in turn affects SIC. Microorganisms play a prominent role in SIC pool in two major ways. First, via photosynthetic CO_2 assimilation and allotment of carbon beneath the ground. The other way is via carbonate recrystallization. Microorganisms break down inorganic carbon, which raises the pH of the soil, subsequently leading to the precipitation of carbonate minerals (Ferdush and Paul, 2021).

In-depth research is necessary to understand the significance and influence of many factors on the entire C pool, soil health and quality, and related ecosystem strategies for management.

5.6.4 Soil carbon sequestration

It is a phenomenon where the atmospheric CO_2 is removed and fixed in the soil carbon pool. This process is primarily driven by plants via photosynthesis, where the carbon is stored in the organic form. In arid and semi-arid regions, the atmospheric CO_2 is converted to an inorganic form of carbon (Lal, 2007). Minimizing soil tillage, application of biochar into soil, and crop rotation are commonly followed techniques that promote carbon sequestration and have become a part of climate-smart agriculture. Conservation tillage, which involves minimization of soil tillage, decreases the decomposition of organic matter and prevents soil disturbances. It also fosters fungal and earthworm biomass, which enhances SOM. Crop rotation substantially increases the nutrient content of soil apart from preventing it from getting leached (Bai *et al.*, 2019). Adoption of soil carbon sequestration practices increases carbon input to the soil, which outweighs the soil output rate, thereby increasing the soil carbon pool. This further improves the resilience and health of the soil.

5.6.5 Bioenergy

The organic biological material, which is derived from living or recently lived organisms, is referred to as biomass. In the context of bioenergy, biomass constitutes agricultural by-products and residues, wastes, woody waste components, food crops, and energy crops. Plants contain energy derived from the sun through photosynthesis. This energy can be converted to usable forms, either directly or indirectly. Every organic compound which is derived either directly or indirectly from photosynthesis is a biomass. Biomass serves as a source for bioenergy generation. Bioenergy is the energy derived from biomass (Rather *et al.*, 2022). Biomass supplies 13–14% of annual energy consumption. The degree to which biomass is utilized determines the progress and generation of bioenergy. Important biomass sources include forestry and agricultural residues, wastes generated in wood processing industries, sewage, aquatic crops, and algae. The chemical composition of biomass sources is distinct. Cellulose, hemicellulose, and lignin are the major constituents of plant-based biomass sources present in varying percentages. Starch, which is a component of cereals, and proteins present in manure are other components. Based on the type of vegetation or ecology, biomass is classified into five groups, namely, wood and woody biomass, herbaceous biomass, aquatic biomass, animal and human waste biomass, and biomass mixtures (Tursi, 2019).

5.6.6 Bioenergy production

The three major conversion processes include thermochemical, physico-chemical, and biochemical conversions.

5.6.6.1 Thermochemical process

This conversion process entails the application of heat and chemicals to generate energy. Combustion, pyrolysis, gasification, and liquefaction are the transformation processes categorized under thermochemical process. Roughly 90% of total bioenergy is generated by combustion. It is primarily an exothermic process which involves burning biomass at a high temperature, which is confined within the combustion chamber. Biomass is burnt in the presence of oxygen to produce CO_2 and energy, which is subsequently released, is transformed into mechanical or electrical form (Alper *et al.*, 2020). Pyrolysis involves partial combustion of certain biomass into charcoal, bio-oil, or combustible gas at a temperature of about 500°C in the absence of oxygen. The products are usually acids, esters, or alcohols (Lebaka, 2013). Gasification entails the conversion of carbonaceous compounds to a synthetic combustible gas at a high temperature in the presence of air. The gases that are usually produced are methane, carbon monoxide, and hydrogen (Sivabalan *et al.*, 2021). Liquefaction encompasses the biomass conversion performed in water at a moderate temperature and high pressure. This process generates oil with a high hydrogen/carbon ratio (López Barreiro *et al.*, 2013).

5.6.6.2 Physico-chemical process

Physico-chemical processes convert animal fat or vegetable oil into biodiesel when it is subjected to a transesterification mechanism. At the end, a high-density fuel is produced. With the application of alcohol, triglycerides are transformed into methyl/ethyl esters leading to biodiesel. Biodiesel is non-toxic and a renewable fuel. High oxygen concentration and excellent lubricating properties are two distinct characteristics that make biodiesel burn optimally in diesel engines (Tursi, 2019).

5.6.6.3 Biochemical processes

Biochemical processes transform into liquid fuels, natural gas, bioproducts, and other chemical compounds. It employs enzymes and microorganisms as catalysts. Anaerobic digestion and fermentation are the two major biochemical conversion processes. Liquid fuels such as alcohols and alkanes, natural gases like hydrogen and methane, bioproducts like carotenoids, omega-3 and omega-6 fatty acids, and chemical compounds such as acetic acid and lactic acid are produced during the process (Liu *et al.*, 2021). Cellulosic bioethanol is one of the significant products of biochemical processes, involving pre-treatment, enzymatic hydrolysis, and fermentation as the major steps in its production. *Saccharomyces cerevisiae* and *Clostridium* sp. are the predominant microbes employed for the fermentation process (Salehi Jouzani and Taherzadeh, 2015).

Anaerobic digestion comprises hydrolysis, acidogenesis, acetogenesis, and methanogenesis. These reactions hydrolyze macromolecules present in the biomass into simpler molecules, with the generation of biogas in anaerobic conditions. Biogas has the potential to be employed directly in gas turbines and ignition gas engines (Liu *et al.*, 2021). Biofuels and biological products are produced from biomass, enabling an environmentally friendly means for the elimination of biodegradable waste. The replacement of fossil fuel-based energy with bioenergy shows a ray of hope toward efforts striving to mitigate climate change. Adoption and incorporation of bioenergy brings a positive outlook across all fronts, ranging from health, environment, and economy. Because biomass is widely available worldwide, it may be transformed into energy under any geographic circumstance and simultaneously assist in the effective management of waste. Owing to their renewability and their large quantity, biomass energy has several advantages compared to fossil fuels. It thus protects the energy security of the world. Efforts should be channeled toward developing more user-friendly and cost-effective technologies in various sectors, which would also attract more investment to the bioenergy industry.

5.7 Overall Potential for Carbon Neutrality

A balance between the amount of carbon released into the atmosphere and the amount of carbon absorbed in carbon sinks is known as carbon neutrality. It can be achieved by carbon offsetting or transition to a circular carbon economy. Since 1997, from its inception in Denmark, this concept has been incorporated across the world.

On December 12, 2015, 197 member countries of the United Nations Framework Convention on Climate Change (UNFCCC) unanimously voted to accept the Paris Agreement (Paris Climate Change Conference), which lays out plans for worldwide action to address climate change after 2020, in response to rising global greenhouse gas concentrations and temperatures (Chen *et al.*, 2022). The Paris Agreement sets a 1.5°C upper limit on global warming over pre-industrial levels. By the middle of the 21st century, the global carbon-neutral target must be reached in order to meet the IPCC's goal (Wu *et al.*, 2022b). Apart from continuously lowering carbon emissions, the pursuit of carbon neutrality additionally has the potential to minimize the concentration of air contaminants, thereby enhancing air quality. Reducing carbon emissions usually necessitates the collaborative impact of several solutions that all aim toward the same goal. It is not dependent solely on a single approach. Multidisciplinary cooperation is essential to achieve carbon neutrality (Broadstock *et al.*, 2021).

For the past 200 years, the world economy has been largely dependent on the unregulated exploitation of natural resources, which has resulted in the disruption of the earth's biogeochemical cycles. In order to meet the rising demand for energy, food, and other goods, there is already an upsurge in the usage of petroleum resources and deforestation. These environmentally harmful practices are the main contributors to the rise in anthropogenic global greenhouse gas (GHG) emissions, which is the main driver of climate change (Lampert, 2019). It is crucial to lower the carbon emissions from fossil fuels and food production systems while fostering the storage of carbon in terrestrial and marine ecosystems in order to attain carbon neutrality and sustainably continue human activities (Cheng, 2020). To truly bridge the gap between the rhetoric and reality of net-zero carbon dioxide emissions, research and the execution of renewable energy from C-free sources (sunlight, tide, wind, water, wave, rain, and geothermal power) and biomass (organic materials from plants or animals) are critically necessary.

5.7.1 Approaches to carbon neutrality

Accelerating the quest for carbon neutrality will help to minimize climate change. Exploring the routes toward a low-carbon society is therefore of the utmost significance. Renewable energy production, food system transformation, carbon-sink preservation, and carbon-negative manufacturing are some of the significant ways in which carbon neutrality can be accomplished.

5.7.2 Renewable energy production

Overconsumption of energy derived from non-renewable resources places humankind in peril by creating energy shortages, emissions of greenhouse gases, climate change, and damage to the environment. Thus, the switch to low or C-free energy and humankind's ecological awareness are more vital today than they have ever been. Coal usage is still the primary component of the energy structure in many countries today. The first step toward a society which aims to be carbon neutral is to change the energy structure and transform coal usage into clean energy (Wang *et al.*, 2021).

5.7.3 Solar energy

Solar energy has the potential to be a significant contributor to the worldwide renewable energy supply due to its clean, renewable, and widespread nature. Photovoltaic technology, which has continued to develop rapidly, is a reliable way to capture solar energy. Owing to their noteworthy operational stability and high-power conversion efficiencies, conventional thin-film solar cells made of inorganic semiconductors like silicon, gallium arsenide (GaAs), copper indium gallium selenide, and cadmium telluride (CdTe) have been widely industrialized. Recent years have witnessed the development of several solar cells as promising photovoltaic technologies, namely organic solar cells, perovskite solar cells, quantum dot solar cells, and other integrated devices. These newly developed types of solar cells can successfully lower CO_2 emissions by providing power generation in many specific regions where normal solar cells fail. It will also serve as an alternative low-cost photovoltaic technology (Kılkış *et al.*, 2020). As opposed to photovoltaic systems, solar thermal technologies rely on photothermal conversion to produce heat, steam, and electricity for C-neutral activities. Energy consumption of fossil fuels and CO_2 emissions are clearly reduced when solar thermal technologies, including concentrated solar power systems, are used in both commercial and residential settings to replace natural gas as a source of energy (Di Leo *et al.*, 2021).

5.7.4 Wind energy

Wind energy is the result of air movement. There are plenty of wind resources on the earth, which are generally located in grasslands, deserts, coastal areas, and islands. The technology, economics, and practicality of wind energy are all greatly influenced by the site location. However, the noise generated by wind turbines is one of the main drawbacks that prevent the adoption of wind energy. There is an urgent need for strategies to further utilize wind sources mindfully and alleviate noise produced by wind turbines. The likelihood of wind turbines damaging birds through collisions, disturbances, or habitat damage if they are erected inappropriately is yet another problem with the production of wind energy (Ren *et al.*, 2021).

The transportation of electrical energy generated by wind turbines is hampered by the uneven presence of wind resources throughout the environment. Another obstacle to the broad adoption of this technology is the significant cost of erecting a wind turbine. To meet the demands of energy consumers, more time and resources must be invested in researching and improvising wind energy technologies.

5.7.5 Ocean energy

The term "ocean energy" describes the clean, renewable energy found in the ocean's water bodies (Wang *et al.*, 2021b). There are five different forms of energy, namely, tidal energy, wave energy, ocean current energy, osmotic energy, and thermal energy. The gravitational interaction of seawater with the moon or the sun creates tides. The overall amount of energy in a tide is known as tidal energy, which also encompasses the kinetic energy of the tide stream and the potential energy associated with the water level. The expected annual production of tidal energy is 1200 TWh, which is relatively low to other ocean energy sources as there are just a few locations where tidal energy can be captured (Nihous, 2006).

The widespread kinetic and potential energy found in water waves is referred to as wave energy. It basically originates from wind, which transfers a portion of its kinetic energy to the surface water of the ocean. The potential of wave energy globally is 29,500 TWh a year. In addition to conventional large-scale systems that use electromagnetic generators, new technologies based on triboelectric nanogenerator networks are being developed to efficiently and affordably obtain wave energy (Wang *et al.*, 2021). Ocean current energy is present in the ocean water circulations. It is the kinetic energy of the motion of water current. Energy extraction employs turbines. As the device needs to be installed in the deep sea, which is very far from the banks, not many efforts have been made in utilizing this energy (Wang *et al.*, 2021b). Sun's irradiation heats up the upper layer of the ocean, which results in a wide temperature difference between the top layer and the deep sea. This temperature difference can be harnessed to extract the thermal energy. The higher the temperature difference, the greater the efficiency to generate the energy. However, the mode of extraction of this energy is still at the developmental stage (Nihous, 2006).

Salinity gradient energy, otherwise called osmotic energy, is the energy that can be harnessed where there is a salinity gradient, for instance in estuaries, where freshwater meets saltwater. Reverse electrodialysis and pressure-retarded osmosis are the two main technologies that are currently being tried to make use of osmotic energy (*Innovation Outlook: Ocean Energy Technologies, 2020*). Ocean energy is of great potential and can substantially reduce our reliance on fossil fuels. Research on efficacious and affordable utilization with further advancements in technologies is quintessential.

5.7.6 Nuclear energy

Nuclear energy is a game changer in achieving a C-neutral world, as it accounts for 40% of low-carbon electricity generation, thwarting about 1.7 Gigatons of CO_2 emissions a year. Nuclear fission is the main technology currently being employed for the generation of nuclear energy, while nuclear fusion is still in the R&D phase. However, the current nuclear fission technology comes with a cost with drawbacks such as maintenance cost, hardships, and emerging challenges to dispose of the nuclear waste, plant safety, and risks of nuclear weaponization. There has been an effort to develop a Gen IV nuclear fission reactor considering the current issues such as reliability, cost-effectiveness, physical safety, and proliferation resistance. Thus, Gen IV reactors serve as a mainstay for sustainable and low-carbon energy (Serp *et al.*, 2014).

5.7.7 Geothermal energy

Geothermal energy is generated from the heat present deep in the interior of the earth. It utilizes the heat from natural geothermal streams to run turbines that produce energy. Flash power, dry steam power, and binary power systems are the present technologies that are involved in geothermal energy generation.

5.7.8 Hydrogen energy

Hydrogen has seen extensive usage in industrial processes over the past two centuries, with the demand continuously rising every year. Fossil fuels have been extensively utilized to produce hydrogen. Hydrocarbon refining and hydrocarbon pyrolysis are the two main technologies from which hydrogen is produced from fossil fuels. Due to greenhouse gas emissions and subsequent climate change, renewable sources have become a choice for hydrogen energy generation. Thermochemical and biological processes are the two technologies through which hydrogen energy is generated from biomass. Apart from biomass, water splitting, where the water serves as a reliable source, can be used to generate hydrogen. Water-splitting processes such as photolysis, electrolysis, and thermo-electrolysis are prominently employed. Despite all these advancements, research gaps regarding hydrogen generation from renewable sources and issues related to hydrogen storage and transport need to be addressed (Nikolaidis and Poullikkas, 2017).

The swift development and widespread adoption of hydropower, solar energy, wind energy, and many other novel sources of energy have drawn increased public interest toward the new energy. In the near future, new energy will be the primary factor toward achieving carbon neutrality. Evolving energy practices will also be a game changer. Nevertheless, there will be a lot of impediments in the way of the carbon-neutral energy transition, which demands proactive steps and cross-sector collaboration at the municipal, state, and even national levels. Additionally, the government's engagement cannot be undermined. To foster integrated pathways to C-neutral energy systems, the next few decades are expected to see an acceleration of the development of revolutionary energy conversion and storage technologies.

5.7.9 Carbon emission reduction in food system

Agriculture has a considerable share in greenhouse emissions, mainly due to crop and animal production. Practices of crop production and animal management, enteric fermentation, manure and pasture management, and applications of synthetic fertilizer and fuels used in livestock production significantly influence greenhouse gas emissions (Poore and Nemecek, 2018).

Crop production and operation: Improved cropping patterns, judicious use of fertilizers, effective ways of irrigation, and implementation of modern digital tools such as multi-sensor drone technology can significantly change the outcome, which helps to mitigate greenhouse gas emissions. New synthetic fertilizers like slow and controlled-release nitrogen fertilizers remarkably improve the food system. Optimization of fertilizer and water use in croplands can considerably reduce emissions. These efforts collectively reduce the footprint of nitrogen-based fertilizers, which ultimately help in achieving a carbon-neutral society (Shang *et al.*, 2021).

Management of animal production: Animal breeding techniques have played a substantial role in the food industry. Selection of genetically superior animals with high productivity and efficient traits can improve agriculture practices and minimize the costs associated with crop production (Knapp *et al.*, 2014). Management of enteric fermentation is one of the prominent strategies for tackling methane emissions in livestock production. This involves the suppression of methanogens and preventing the metabolism that is involved in methanogenesis. Manure management practices, which involve the production and application of organic fertilizer with a low emission factor, play a decisive role (Wallace *et al.*, 2019).

In addition, novel technologies like 3D printing, automatic control technology, biotechnology, and artificial intelligence have been making an impact in agriculture. Production of plant-based and cell-based meat from non-animal sources, stem-cell and large-scale cell technology-based food production, and metabolic engineering of microbes to be a plant growth promoter are some examples of recent progress. However, commercialization and acceptance of non–animal-based meat grown in factory settings are some of the roadblocks at the present moment.

5.7.10 Carbon-sink preservation

Carbon sink refers to the phenomenon where plants absorb CO_2 from the atmosphere and fix it in vegetation, thereby lessening the concentration of greenhouse gases in the atmosphere. Terrestrial and marine carbon sinks are the two major carbon sinks. The forest ecosystem is a prominent terrestrial carbon sink, which accounts for 85–90% of total terrestrial biomass production. It absorbs 45% of total anthropogenic greenhouse gas emissions, thus playing a decisive role in achieving carbon sequestration. Grasslands store around 34% of global terrestrial carbon (Pan *et al.*, 2011). Reforestation and afforestation, biochar inclusions, soil carbon sequestration, and reasonable forest management practices enhance carbon sinks. Forest, which is a major terrestrial C sink, should be substantially increased to promote carbon sequestration by terrestrial ecosystems. In addition to these, technologies that can improve the efficiency of soil carbon sequestration should also be encouraged (Trumbore *et al.*, 2015).

Ocean ecosystems, which occupy approximately 70% of the earth's surface, capture 22.7% of total CO_2 from the atmosphere (Friedlingstein *et al.*, 2020). Atmospheric CO_2 sequestered by marine ecosystems is referred to as blue carbon (Macreadie *et al.*, 2019). The well-known "Solubility pump" and "biological pump" are the two processes that influence oceanic Carbon sink. The former involves mixing atmospheric CO_2 with the upper layer of the ocean. The latter entails photosynthetic absorption by microbes that are prevalent in the ocean and its subsequent transportation deep into the ocean, which is long-term carbon sequestration (Mcleod *et al.*, 2011). Considering the share that ocean ecosystems have in carbon sequestration and as a carbon sink, proper management, conservation and restoration of marine ecosystems is of prime importance. Reducing the application of chemical fertilizers on land preserves the microbial carbon sequestration present in the marine ecosystem. This may also minimize eutrophication and upgrade deep-sea carbon sequestration. Apart from the above-mentioned means, strategies to restore, preserve, and enhance coastal and open-water ecosystems need to be systematically implemented. Eco-engineering techniques and methods are also essential. These include sustainable mariculture, marine artificial upwellings, and land-sea integrated solutions for C sequestration (Jiao *et al.*, 2018).

5.7.11 Technological path to drive carbon neutrality

Carbon capture and storage (CCS), Carbon Capture Utilization and Storage (CCUS) and Bioenergy with Carbon Capture and Storage (BECCS) technologies are notable technological advancements to promote carbon neutrality (Wu *et al.*, 2022b). CCS involves the separation of CO_2 from its emission sources, transportation, and isolating it from the atmosphere for a considerably longer time. Researchers have proposed and attempted CCS but have faced low acceptance and storage issues. However, the long-term benefits cannot be found with short-term practices (Wu *et al.*, 2022b). CCUS seems to be more economical and practically feasible as it purifies and recycles the stored CO_2, which is employed in new production processes. This approach has remarkable potential compared to CCS. CCUS has been popularized across the world in electric power and industrial sectors. Although the CCUS strategy contributes to the objective of being carbon neutral, the high energy consumption and costs associated with the current CCUS technologies make them hard to implement (Wu *et al.*, 2022a).

BECCS integrates bioenergy with carbon capture and storage. This approach greatly minimizes the cost of achieving carbon neutrality. It proficiently removes CO_2 from the atmosphere and helps ease climate change (Wu *et al.*, 2022a). The social behaviors and practices of residents, households, and enterprises have a great deal of potential in the efforts to achieve carbon neutrality. Advocacy activities on the common aim of achieving carbon neutrality and the associated previews critically impact society's behavior and can be a game changer.

5.8 Determining Growth in Food Demand for Estimating Land Area

Estimating the growth in food demand and the associated land area required involves considering various factors, including population growth, dietary changes, economic development, and technological advancements in agriculture (Gao and Bryan, 2017). The world population is increasing every year. This ever-increasing population will lead to food deficiency in the near future. Currently, developed countries' growth is much slower than developing countries (Hertel *et al.*, 2016b). This increasing population and increasing demand for agricultural products turned into new food products, which are based on animal products, fruits, and vegetables. Figure 5.2 gives an overview of factors that are directly affecting the growth in this sector. Research shows that the world population will be 9.7 billion in 2050 and animal-based food will increase instead of agriculture crops (Clements *et al.*, 2019). International

Fig. 5.2. A consolidated view of factors affecting growth in food demand and land area.

Institute for Applied Systems Analysis revealed data that show food demand will increase by 102%, and the growth of developing countries will be 78%. All these circumstances will lead to an increase in the price of staple food, which would include agriculturally produced crops as well as animal-based food products (Fukase and Martin, 2020). World food demand was slower from 1990 to 2010 compared to the last three decades (Mitchell *et al.*, 1997). It was expected that the demand for food grain would increase with the increase in population (Johnson *et al.*, 2015). The cultivation of crops requires land (Hasegawa *et al.*, 2021). To feed this ever-increasing population we need a huge amount of cultivable land, especially we have to check how much crop is yielding per hectare and the consumption pattern of the population (Ishida *et al.*, 2014). The cultivable land is decreasing every year. There are various reasons for the decrease in cropland. Nowadays, food impregnation design is also being changed, which affects total land requirement, and this is directly linked to population growth (Godfray and Garnett, 2014). All the research indicates that global food demand will increase by 35–56% between 2010 and 2050 (Medek *et al.*, 2017). Climate change is also another factor for global food security. When it is considered, then the food scarcity risk becomes 30–62%. This increasing food insecurity becomes −91% to +30% if we take the risk of hunger (Willett *et al.*, 2019).

Here is a general approach to determine the growth in food demand and estimate the corresponding land area:

5.8.1 Population growth

Begin by understanding the expected population growth in the target region. This could be on a global, national, or regional scale. Consider demographic trends and projections to estimate the increase in the number of mouths to feed (Riahi *et al.*, 2017).

5.8.2 Dietary changes

Analyze changes in dietary patterns, especially shifts toward more resource-intensive diets. As economies develop, there tends to be a move toward diets that include more meat and processed foods. We should consider the impact of cultural, social, and economic factors on food choices also (Muhammad *et al.*, 2011).

5.8.3 Economic development

Economic growth often leads to increased food consumption. As people's incomes rise, they tend to demand more diverse and resource-intensive foods. Income distribution and changes in purchasing power also have to be considered (Rask and Rask, 2011).

5.8.4 Technological advancements

Assess the impact of technological advancements in agriculture, including improved farming practices, crop varieties, and precision farming techniques. Consider innovations in aquaculture, vertical farming, and other alternative food production methods (Gerbens-Leenes *et al.*, 2002).

5.8.5 Food loss and waste

Account for food loss and waste in the supply chain. Efficient distribution and storage systems can help reduce the overall demand for production (Hertel *et al.*, 2016b).

5.8.6 Environmental constraints

Recognize the limitations of available land, water, and other natural resources. Factor in environmental sustainability considerations to avoid overestimating the available land for agriculture (Tian *et al.*, 2021).

5.8.7 Climate change

Consider the potential impact of climate change on agricultural productivity. Changes in temperature, precipitation patterns, and extreme weather events can affect crop yields (Smith *et al.*, 2014a).

5.8.8 Policy and trade

Examine government policies, trade agreements, and geopolitical factors that can influence food production and distribution (Timmer *et al.*, 1983).

5.8.9 Determination of land requirements relating to food consumption patterns

Determining land requirements related to food consumption patterns involves assessing the amount of land needed to produce the food consumed by a population (McCalla and Revoredo, 2001). This can be a complex task that requires consideration of various factors, including the type of food, agricultural practices, dietary preferences, and population size (Parody *et al.*, 2018). Here is a general method to estimate land requirements:

Understand Dietary Patterns:

- Analyze the dietary patterns of the population in question.
- Consider the types and quantities of food consumed, including staple crops, fruits, vegetables, meat, and dairy products.

5.8.10 Identify food categories

Categorize the food items into groups based on their source (crops, livestock, etc.) and nutritional content. Different foods have varying land requirements due to differences in cultivation or production methods (Clark and Tilman, 2017).

5.8.11 Determine yields

Research and collect data on average yields per unit of land for each food category. Yields can vary based on factors like crop varieties, farming practices, and climate conditions (Bodirsky *et al.*, 2015).

5.8.12 Calculate total production

Estimate the total production of each food category by multiplying the consumption per capita by the population size (Robinson *et al.*, 2015).

5.8.13 Calculate land requirements

Use the yields per unit of land to calculate the total land required for each food category. The formula is Land Required = Total production/ Yield per unit of land (Lotze-Campen *et al.*, 2008).

5.8.14 Consider sustainability factors

Take into account sustainable farming practices, land-use efficiency, and potential improvements in agricultural productivity. Sustainable practices may include crop rotation, organic farming, or agroforestry (van der Mensbrugghe, 2013).

5.8.15 Account for food losses

Consider post-harvest losses, processing losses, and waste at various stages of the food supply chain. Adjust the land requirements accordingly (Linehan *et al.*, 2013).

5.8.16 Include livestock production

If assessing land requirements for meat and dairy, consider the land needed for livestock farming. This involves not only the land for grazing or feed production but also the infrastructure for animal husbandry (Billen *et al.*, 2015).

5.8.17 Account for import/export

Consider the impact of importing or exporting food on land requirements. If a significant portion of food is imported, the land used in the exporting country should be considered (van Vuuren *et al.*, 2019).

5.8.18 Review and refine

Regularly update the data and refine calculations based on changes in population, dietary habits, agricultural practices, and technology (O'Neill *et al.*, 2020).

5.8.19 GIS

GIS can be employed to analyze spatial data, providing insights into the distribution of land use and potential areas for optimization (Leclère *et al.*, 2020). This method provides an estimate and there may be variations based on regional factors, climate conditions, and technological advancements in agriculture. Additionally, it's important to consider the environmental impact and sustainability of land-use practices (Tomlinson, 2013).

5.8.20 Increase in cropland area and recent trends in world grain production

The expansion or contraction of cropland area depends on various factors, and it can vary across regions due to different economic, environmental, and policy considerations. As of the last knowledge update in January 2022, we do not have the most recent data on global cropland trends. However, some data can be provided which is based on some reliable sources such as Food and Agriculture Organization (FAO) (Shenkut *et al.*, 2013).

5.8.21 Climate change

Weather patterns, including extreme events like droughts, floods, and heat waves, significantly impact grain production. Changes in temperature and precipitation patterns can affect crop yields and lead to fluctuations in global grain production (van Vuuren *et al.*, 2017).

5.8.22 Technological advancement

Advances in agricultural technology, including the use of genetically modified organisms (GMOs), precision farming, and improved crop varieties, can positively influence grain production by enhancing yields and resilience (Pandey and Asif, 2022).

5.8.23 Global demand and population growth

The increasing global population and changing dietary preferences contribute to the rising demand for grains. This demand can influence production trends as farmers seek to meet the needs of a growing population (Hertel *et al.*, 2016a).

5.8.24 Trade policies and market dynamics

International trade policies, market conditions, and geopolitical factors can affect grain production trends. Trade agreements, tariffs, and export/import regulations can impact the flow of grains globally (Diggle, 2002).

5.8.25 Sustainability practices

There is a growing emphasis on sustainable agricultural practices. Farmers and policymakers are increasingly considering environmentally friendly approaches to minimize the ecological impact of grain production (Popp *et al.*, 2017).

5.8.26 Pandemic impact

The COVID-19 pandemic has had various effects on agriculture, including disruptions in supply chains, labor shortages, and shifts in consumer behavior. These factors can influence grain production and distribution (Beckman and Countryman, 2021).

5.8.27 Water scarcity

Water availability is a critical factor in agriculture. Regions facing water scarcity may experience challenges in maintaining or increasing grain production (Dinar *et al.*, 2019).

5.8.28 Government policies and subsidies

Government policies, subsidies, and incentives can play a significant role in shaping grain production trends. Supportive policies may encourage farmers to invest in technology and increase production (Kumar and Sharma, 2020).

5.8.29 Urbanization

The conversion of agricultural land into urban areas or other non-agricultural uses can lead to a reduction in cropland area. Urbanization is a significant factor in many regions, particularly as populations concentrate in urban centers (Tripathi *et al.*, 2019).

5.8.30 Economic factors

Economic considerations, including market demand, commodity prices, and profitability, play a crucial role in decisions related to land use. Economic incentives may drive or limit cropland expansion (Sahoo and Sethi, 2012).

5.9 Carbon Neutrality and Potential Offsets

Carbon neutrality refers to the state in which an individual, organization, or entity has balanced the amount of carbon dioxide (CO_2) emissions it produces with an equivalent amount of emissions removed from the atmosphere or offset through various measures (Dhanda and Hartman, 2011). The goal is to achieve a net-zero carbon footprint, meaning that the entity's overall impact on the climate is neutral, and it is not contributing to the increase of greenhouse gases in the atmosphere (Ramseur, 2007). Climate change is the new enemy of humankind. One of the biggest reasons for climate change is carbon emissions from various sources like vehicles and industrial plants (Barreto *et al.*, 2018). Nowadays, net carbon neutrality is the only option to battle with climate change. In 2020, the governments of 25 nations and the European Union have committed to have zero net carbon emissions by 2050. Also, private entities are also joining to protect environmental policies with the government, voluntarily achieving low- or zero-carbon economy by using different types of innovative techniques (Rayer *et al.*, 2022). Various non-profit and for-profit organizations are now buying renewable electricity, producing carbon neutrality, adopting targets for climate mitigation, and pledging zero carbon emission. There is very little research on carbon neutrality. This research shows there is some work regarding the practical application of carbon neutrality and the realization path of carbon neutrality. However, there are some problems regarding carbon neutrality. We have to study more data to fulfill the quantitative analysis in various sectors. The connection between the carbon neutrality path and industries should be taken care of.

5.9.1 Achieving carbon neutrality involves two main steps

5.9.1.1 Reducing emissions

This involves implementing measures to minimize or eliminate the generation of carbon emissions. This can include increasing energy efficiency, adopting renewable energy sources, and implementing sustainable practices in various aspects of operations (Wilson *et al.*, 2014).

5.9.1.2 Offsetting emissions

For the emissions that cannot be completely eliminated, offsetting comes into play. Carbon offsets are investments in projects or activities that reduce or remove greenhouse gas emissions elsewhere. These projects can include reforestation efforts, renewable energy projects, methane capture, and more. The idea is that the reduction or removal of emissions in one area compensates for the emissions produced in another (Ney and Purman, 2009).

5.9.2 Types of carbon offsets

5.9.2.1 Renewable energy projects

Investing in wind, solar, hydroelectric, or other renewable energy projects to replace or reduce the use of fossil fuels (Sahil *et al.*, 2023).

5.9.2.2 Reforestation and afforestation

Planting trees or restoring forests, as trees absorb CO_2 during photosynthesis (Reilly *et al.*, 2012).

5.9.2.3 Methane capture

Capturing and utilizing methane emissions from landfills, agricultural operations, or other sources (Fennessy *et al.*, 2023).

5.9.2.4 Energy efficiency projects

Implementing measures to reduce energy consumption and associated emissions (Kurramovich *et al.*, 2022).

5.9.2.5 Carbon capture and storage (CCS)

Capturing CO_2 emissions from industrial processes or power plants and storing them underground (Ng'andwe *et al.*, 2015).

5.9.2.6 Challenges and considerations

It is important to ensure that the offset projects are additional, meaning that they would not have occurred without the offset investment (Dong *et al.*, 2022).

5.9.2.7 Permanence

Some offset projects, such as tree planting, face the risk of reversals (e.g. deforestation). Ensuring the long-term permanence of emissions reductions is crucial (Parraga *et al.*, 2019).

5.9.2.8 Verification and certification

Reliable certification standards and third-party verification are essential to ensure the legitimacy and effectiveness of offset projects (Gössling and Buckley, 2016).

5.9.2.9 Local and social impacts

It is important to consider the broader social and environmental impacts of offset projects, such as their effects on local communities, biodiversity, and ecosystems (Mustapha, 2022).

5.9.2.10 Comprehensive approach

Carbon neutrality is most effective when combined with efforts to reduce emissions at the source rather than relying solely on offsets. As the understanding of climate change and sustainable practices evolves, so do discussions around carbon neutrality and the effectiveness of offset mechanisms. It is important for individuals and organizations to stay informed about best practices and continually reassess and update their sustainability strategies (Akimoto *et al.*, 2022).

5.9.3 Carbon neutrality vs climate neutrality

Carbon neutrality, also referred to as "net-zero carbon emissions" or "carbon neutral," means that an entity (individual, organization, country, etc.) has balanced the amount of carbon dioxide (CO_2) emissions it produces with an equivalent amount of emissions removed from the atmosphere or offset through various measures. The term "carbon" specifically refers to CO_2, a major greenhouse gas. Achieving carbon neutrality involves addressing CO_2 emissions through reduction measures and offsetting activities, such as investing in projects that capture or reduce emissions elsewhere (e.g. reforestation, renewable energy projects) (Salvia *et al.*, 2021).

5.9.3.1 Key points of comparison

Carbon neutrality is specific to addressing carbon dioxide emissions, while climate neutrality encompasses all greenhouse gases (Wang *et al.*, 2022).

5.9.3.1.1 GREENHOUSE GASES. Carbon neutrality deals primarily with CO_2, while climate neutrality addresses CO_2 as well as other potent greenhouse gases (Gil-Carrera *et al.*, 2020).

5.9.3.1.2 FOCUS AREA. Carbon neutrality often involves specific initiatives related to energy use, transportation, and industrial processes. Climate neutrality requires a more comprehensive approach; including strategies for agriculture, waste management, and land use (Zhao *et al.*, 2022).

5.9.3.1.3 GLOBAL VS LOCAL. While both terms can be applied at various scales, carbon neutrality is often discussed at the organizational or national level, whereas climate neutrality is more commonly considered on a national or global scale. In summary, while the terms are related and share the common goal of achieving a balance between emissions and removals, carbon neutrality is a subset of climate neutrality that specifically addresses CO_2 emissions. The choice of terminology often depends on the context and the level of detail required in the discussion (Krammer *et al.*, 2013).

5.9.4 Sustainable development and carbon neutrality

Analyzing the integrated assessment of transport transitions in India from the perspective of sustainable development and carbon neutrality involves examining various aspects, including policies, technologies, and behavioral changes (Gupta and Garg, 2020). Here is a comprehensive overview of key considerations:

5.9.5 Policy framework

- Emission Reduction Targets: Evaluate India's commitment to carbon neutrality and sustainable development through its nationally determined contributions (NDCs) and other policy documents (Sudarsan *et al.*, 2023).
- Transport Policy: Analyze existing and proposed policies related to the transport sector, such as fuel efficiency standards, promotion of electric vehicles (EVs), and sustainable urban planning (Chen *et al.*, 2023).

5.9.6 Technological transition

- Promotion of Electric Vehicles (EVs): Assess the progress and effectiveness of initiatives promoting EVs, including subsidies, charging infrastructure development, and incentives for manufacturers (Tiwari *et al.*, 2023).
- Alternative Fuels: Explore the role of alternative fuels such as biofuels, hydrogen, and compressed natural gas (CNG) in reducing carbon emissions from the transport sector (Pradhan *et al.*, 2023).

5.9.7 Infrastructure development

- Public Transportation: Evaluate the expansion and improvement of public transportation systems to reduce reliance on individual vehicles and promote shared mobility (Das and Ghosh, 2023).
- Charging Infrastructure: Examine the growth of charging infrastructure for EVs and its accessibility, especially in urban areas (Jain *et al.*, 2017).

5.9.8 Behavioral changes

- Awareness and Education: Assess government initiatives aimed at creating awareness about sustainable transport practices and educating the public about the environmental impact of their choices (Nassajfar *et al.*, 2023).
- Incentives for Sustainable Choices: Evaluate the effectiveness of incentives

for individuals and businesses adopting sustainable transportation practices (Zhu *et al.*, 2023).

5.9.9 Monitoring and Reporting

- Data Collection: Discuss the importance of robust data collection mechanisms to monitor the impact of transport transitions on carbon emissions and sustainable development indicators (Sen *et al.*, 2021).
- Reporting Mechanisms: Analyze how well India is reporting progress toward its carbon neutrality goals and the transparency of such reporting (Davidson *et al.*, 2021).

5.9.10 Challenges and opportunities

- Barriers to Transition: Identify challenges hindering the transition to sustainable transport, such as economic constraints, technology adoption barriers, and resistance to change (Liu *et al.*, 2022b).
- Opportunities for Innovation: Highlight areas where innovation and collaboration can lead to significant advancements in sustainable transport solutions (Wu *et al.*, 2022a).

5.9.11 International collaboration

- Bilateral and Multilateral Partnerships: Explore collaborations with other countries and international organizations to leverage knowledge, technology, and financial support for sustainable transport initiatives.

5.9.12 Environmental and social impact assessment

- Air Quality and Health: Assess the impact of transport transitions on air quality and public health (Wu *et al.*, 2022b).
- Social Equity: Examine whether the transition is benefiting all segments of the population and not exacerbating existing social disparities (Dwivedi *et al.*, 2023).

5.9.13 Financing mechanisms

- Investment in Sustainable Transport: Evaluate the allocation of financial resources for sustainable transport projects, including public and private investments, international funding, and innovative financing mechanisms (Kong and Wang, 2022).

5.9.14 Adaptation and flexibility

- Adaptive Strategies: Discuss the flexibility of policies and strategies to adapt to changing technological landscapes and unforeseen challenges.
- Application: Carbon neutrality is often discussed in the context of specific entities like companies, cities, or countries. It is a more specific term centered on the carbon component of greenhouse gas emissions (Yang *et al.*, 2023).

5.10 Conclusions

Achieving carbon-neutral agriculture is a complex but imperative goal for global sustainability. Addressing the challenges associated with current agricultural practices requires a comprehensive and integrated approach based on sustainable principles. By adopting practices that prioritize soil health, biodiversity, and resource efficiency, and by integrating innovative technologies, the agricultural sector can transition toward a more sustainable and climate-resilient future. Addressing the environmental impact of food production is crucial, but any strategies to reduce food production must carefully consider the potential consequences. Balancing the need for sustainable agriculture with the socioeconomic implications of reduced food production requires a nuanced approach. Future research and policy initiatives should focus on developing strategies that prioritize environmental sustainability without compromising food security, economic livelihoods, and social well-being. In conclusion, mitigating production emissions in agriculture is a multifaceted endeavor that requires a combination of technological innovation, agroecological principles, and sustainable management practices. The strategies discussed in this chapter offer pathways toward climate-neutral agriculture, aligning with global initiatives to combat climate change and build resilient and sustainable food systems.

References

Akimoto, K., Sano, F. and Nakano, Y. (2022) Assessment of comprehensive energy systems for achieving carbon neutrality in road transport. *Transportation Research Part D: Transport and Environment* 112, 103487. DOI: 10.1016/j.trd.2022.103487.

Albrecht, A., Jucker, T., Albrecht, J. and Lang, F. (2021) Agroforestry reduces emissions from land use change but leads to higher nitrous oxide emissions. *Nature Communications* 12(1), 1–10.

Alper, K., Tekin, K., Karagöz, S. and Ragauskas, A.J. (2020) Sustainable energy and fuels from biomass: a review focusing on hydrothermal biomass processing. *Sustainable Energy and Fuels* 4(9), 4390–4414.

Altieri, M.A. (2018) *Agroecology: The Science of Sustainable Agriculture*, 3rd edn. CRC Press, Boca Raton, Florida.

Altieri, M.A. and Nicholls, C.I. (2020) Agroecology and the reconstruction of a post-COVID-19 agriculture. *The Journal of Peasant Studies* 47(5), 881–898.

Bai, X., Huang, Y., Ren, W., Coyne, M., Jacinthe, P.-A. *et al.* (2019) Responses of soil carbon sequestration to climate-smart agriculture practices: a meta-analysis. *Global Change Biology* 25(8), 2591–2606. DOI: 10.1111/gcb.14658.

Bai, Z.G., Dent, D.L., Olsson, L. and Schaepman, M.E. (2018) Proxy global assessment of land degradation. *Soil Use and Management* 34(1), 1–11.

Barreto, V., Gonzalez, A., Mate, R. and Zuk, E. (2018) *A Study of Carbon Offsets and RECs to Meet Boston's Mandate for Carbon Neutrality by 2050*. BU Institute for Sustainable Energy, Boston, Massachusetts.

Beckman, J. and Countryman, A.M. (2021) The importance of agriculture in the economy: impacts from COVID-19. *American Journal of Agricultural Economics* 103(5), 1595–1611. DOI: 10.1111/ajae.12212.

Billen, G., Lassaletta, L. and Garnier, J. (2015) A vast range of opportunities for feeding the world in 2050: trade-off between diet, N contamination and international trade. *Environmental Research Letters* 10(2), 025001. DOI: 10.1088/1748-9326/10/2/025001.

Bisht, D.S., Bhatia, V. and Bhattacharya, R. (2019) Improving plant-resistance to insect-pests and pathogens: the new opportunities through targeted genome editing. *Seminars in Cell & Developmental Biology* 96, 65–76. DOI: 10.1016/j.semcdb.2019.04.008.

Bodirsky, B.L., Rolinski, S., Biewald, A., Weindl, I., Popp, A. *et al.* (2015) Global food demand scenarios for the 21st century. *PLoS One* 10(11), e0139201. DOI: 10.1371/journal.pone.0139201.

Broadstock, D., Ji, Q., Managi, S. and Zhang, D. (2021) Pathways to carbon neutrality: challenges and opportunities. *Resources, Conservation and Recycling* 169, 105472. DOI: 10.1016/j.resconrec.2021.105472.

Cardinale, B.J., Matulich, K.L., Hooper, D.U., Byrnes, J.E., Duffy, E. *et al.* (2012) The functional role of producer diversity in ecosystems. *American Journal of Botany* 99(5), 1047–1055.

Chen, L., Msigwa, G., Yang, M., Osman, A.I., Fawzy, S. *et al.* (2022) Strategies to achieve a carbon neutral society: a review. *Environmental Chemistry Letters* 20(4), 2277–2310. DOI: 10.1007/s10311-022-01435-8.

Chen, Y.H., Li, Y.Z., Jiang, H. and Huang, Z. (2023) Research on household energy demand patterns, data acquisition and influencing factors: a review. *Sustainable Cities and Society* 99, 104916. DOI: 10.1016/j.scs.2023.104916.

Cheng, H. (2020) Future earth and sustainable developments. *The Innovation* 1(3), 100055. DOI: 10.1016/j.xinn.2020.100055.

Ciais, P., Caldeira, K. and Cox, P. (2019) *IPCC Special Report on Climate Change and Land*. IPCC, Geneva, Switzerland.

Clark, M. and Tilman, D. (2017) Comparative analysis of environmental impacts of agricultural production systems, agricultural input efficiency, and food choice. *Environmental Research Letters* 12(6), 064016. DOI: 10.1088/1748-9326/aa6cd5.

Clements, C.B., Kochanski, A.K., Seto, D., Davis, B., Camacho, C. *et al*. (2019) The FireFlux II experiment: a model-guided field experiment to improve understanding of fire–atmosphere interactions and fire spread. *International Journal of Wildland Fire* 28(4), 308. DOI: 10.1071/WF18089.

Dainese, M., Martin, E.A., Aizen, M.A., Albrecht, M., Bartomeus, I. *et al*. (2021) A global synthesis reveals biodiversity-mediated benefits for crop production. *Science Advances* 7(22), eabg7351.

Das, A. and Ghosh, A. (2023) Vision net zero: a review of decarbonisation strategies to minimise climate risks of developing countries. *Environment, Development and Sustainability* 26(7), 1–37. DOI: 10.1007/s10668-023-03318-6.

Davidson, M., Karplus, V.J., Zhang, D. and Zhang, X. (2021) Policies and institutions to support carbon neutrality in china by 2060. *Economics of Energy & Environmental Policy* 10(2). DOI: 10.5547/2160-5890.10.2.mdav.

Dhanda, K.K. and Hartman, L.P. (2011) The ethics of carbon neutrality: a critical examination of voluntary carbon offset providers. *Journal of Business Ethics* 100, 119–149. DOI: 10.5547/2160-5890.10.2.mdav.

Di Leo, S., Pietrapertosa, F., Salvia, M. and Cosmi, C. (2021) Contribution of the Basilicata region to decarbonisation of the energy system: results of a scenario analysis. *Renewable and Sustainable Energy Reviews* 138, 110544. DOI: 10.5547/2160-5890.10.2.mdav.

Diggle, P. (2002) *Analysis of Longitudinal Data*. Oxford University Press, Oxford, UK. DOI: 10.5547/2160-5890.10.2.mdav.

Dinar, A., Tieu, A. and Huynh, H. (2019) Water scarcity impacts on global food production. *Global Food Security* 23, 212–226. DOI: 10.5547/2160-5890.10.2.mdav.

Dong, Z.Y., Yang, J., Yu, L., Daiyan, R. and Amal, R. (2022) A green hydrogen credit framework for international green hydrogen trading towards A carbon neutral future. *International Journal of Hydrogen Energy* 47(2), 728–734. DOI: 10.1016/j.ijhydene.2021.10.084.

Dwivedi, A., Sassanelli, C., Agrawal, D., Moktadir, Md.A. and D'Adamo, I. (2023) Drivers to mitigate climate change in context of manufacturing industry: an emerging economy study. *Business Strategy and the Environment* 32(7), 4467–4484. DOI: 10.1002/bse.3376.

Fanzo, J., Bellows, A.L., Spiker, M.L., Thorne-Lyman, A.L. and Bloem, M.W. (2021) The importance of food systems and the environment for nutrition. *The American Journal of Clinical Nutrition* 113(1), 7–16. DOI: 10.1093/ajcn/nqaa313.

Fennessy, M.S., Moon, J.B. and Finlayson, C.M. (2023) Linking wetland ecological processes with the delivery of ecosystem services. In: Gell, P.A., Davidson, N.C. and Finlayson, C.M. (eds) *Ramsar Wetlands: Values, Assessment, Management* (1st Edn). Elsevier, pp. 153–178. DOI: 10.1016/B978-0-12-817803-4.00011-5.

Ferdush, J. and Paul, V. (2021) A review on the possible factors influencing soil inorganic carbon under elevated CO_2. *CATENA* 204, 105434. DOI: 10.1016/j.catena.2021.105434.

Food and Agriculture Organization (2017) *Soil Organic Carbon: The Hidden Potential*. Available at: https://openknowledge.fao.org/handle/20.500.14283/i6937en

Friedlingstein, P., O'Sullivan, M., Jones, M.W., Andrew, R.M., Hauck, J. *et al*. (2020) Global carbon budget 2020. *Earth System Science Data* 12(4), 3269–3340. DOI: 10.5194/essd-12-3269-2020.

Fukase, E. and Martin, W. (2020) Economic growth, convergence, and world food demand and supply. *World Development* 132, 104954. DOI: 10.1016/j.worlddev.2020.104954.

Gao, L. and Bryan, B.A. (2017) Finding pathways to national-scale land-sector sustainability. *Nature* 544(7649), 217–222. DOI: 10.1038/nature21694.

Gerbens-Leenes, P.W., Nonhebel, S. and Ivens, W.P.M.F. (2002) A method to determine land requirements relating to food consumption patterns. *Agriculture, Ecosystems & Environment* 90(1), 47–58. DOI: 10.1016/S0167-8809(01)00169-4.

Gil-Carrera, L., Pelaz, G., Mateos, R. and Escapa, A. (2020) Estimating the carbon footprint of microbial electrolysis cells in wastewater treatment plants: case study. *Journal of Sustainable Development of Energy, Water and Environment Systems* 8(3), 537–546. DOI: 10.13044/j.sdewes.d7.0296.

Giller, K.E., Andersson, J.A., Corbeels, M., Kirkegaard, J., Mortensen, D. *et al.* (2020) Beyond conservation agriculture. *Frontiers in Sustainable Food Systems* 4, 1–10. DOI: 10.13044/j.sdewes.d7.0296.

Gliessman, S.R. (2015) *Agroecology: The Ecology of Sustainable Food Systems*. CRC Press, Boca Raton, Florida. DOI: 10.13044/j.sdewes.d7.0296.

Godfray, H.C.J. and Garnett, T. (2014) Food security and sustainable intensification. *Philosophical Transactions of the Royal Society of London. Series B, Biological Sciences* 369(1639), 20120273. DOI: 10.1098/rstb.2012.0273.

Goglio, P., Smith, W.N., Grant, B.B., Desjardins, R.L., Worth, D.E. *et al.* (2019) Assessment of the uncertainty in the emission factors of nitrous oxide from agricultural soils. *Environmental Research Letters* 14(7), 074007. DOI: 10.13044/j.sdewes.d7.0296.

Gössling, S. and Buckley, R. (2016) Carbon labels in tourism: persuasive communication? *Journal of Cleaner Production* 111, 358–369. DOI: 10.1016/j.jclepro.2014.08.067.

Guo, Y., Wang, X., Li, X., Wang, J., Xu, M. *et al.* (2016) Dynamics of soil organic and inorganic carbon in the cropland of upper Yellow River Delta, China. *Scientific Reports* 6(1), 36105. DOI: 10.1038/srep36105.

Gupta, D. and Garg, A. (2020) Sustainable development and carbon neutrality: integrated assessment of transport transitions in India. Transportation Research Part D: Transport and Environment 85, 102474. DOI: 10.1016/j.trd.2020.102474.

Gurr, G.M., Lu, Z., Zheng, X., Xu, H., Zhu, P. *et al.* (2022) Multi-country evidence that crop diversification promotes biological control services through enhanced arthropod predator and parasitoid abundances. *Global Ecology and Biogeography* 31(1), 199–212.

Hasegawa, T., Sakurai, G., Fujimori, S., Takahashi, K., Hijioka, Y. *et al.* (2021) Extreme climate events increase risk of global food insecurity and adaptation needs. *Nature Food* 2(8), 587–595. DOI: 10.1038/s43016-021-00335-4.

Hawkes, C., Ruel, M.T., Salm, L., Sinclair, B. and Branca, F. (2019) Double-duty actions: seizing programme and policy opportunities to address malnutrition in all its forms. *The Lancet* 395(10218), 142–155. DOI: 10.1038/nclimate2925.

Herrero, M., Henderson, B., Havlík, P., Thornton, P.K., Conant, R.T. *et al.* (2016) Greenhouse gas mitigation potentials in the livestock sector. *Nature Climate Change* 6(5), 452–461. DOI: 10.1038/nclimate2925.

Herrero, M., Thornton, P.K., Power, B., Bogard, J.R., Remans, R. *et al.* (2020) Farming and the geography of nutrient production for human use: a transdisciplinary analysis. *The Lancet Planetary Health* 4(6), e244–e255. DOI: 10.1038/nclimate2925.

Hertel, T.W., Baldos, U.L.C., Hertel, T.W. and Baldos, U.L.C. (2016a) Overview of global land use, food security and the environment. *Global Change and the Challenges of Sustainably Feeding a Growing Planet* 1, 1–12. DOI: 10.1038/nclimate2925.

Hertel, T.W., Baldos, U.L.C. and van der Mensbrugghe, D. (2016b) Predicting long-term food demand, cropland use and prices. *Annual Review of Resource Economics* 8, 417–441. DOI: 10.1038/nclimate2925.

Hristov, A.N., Oh, J., Firkins, J.L., Dijkstra, J., Kebreab, E. *et al.* (2013) Mitigation of methane and nitrous oxide emissions from animal operations: I. A review of enteric methane mitigation options. *Journal of Animal Science* 91(11), 5045–5069. DOI: 10.2527/jas.2013-6583.

Innovation Outlook: Ocean Energy Technologies (14 December 2020). Available at: https://www.irena.org/publications/2020/Dec/Innovation-Outlook-Ocean-Energy-Technologies (accessed 2 August 2024). DOI: 10.1016/j.trd.2022.103487.

IPCC (2019) Special Report on Climate Change and Land. Available at: https://www.ipcc.ch/srccl/ (accessed January 2024).

Ishida, H., Kobayashi, S., Kanae, S., Hasegawa, T., Fujimori, S. *et al.* (2014) Global-scale projection and its sensitivity analysis of the health burden attributable to childhood undernutrition under the latest scenario framework for climate change research. *Environmental Research Letters* 9(6), 064014. DOI: 10.1088/1748-9326/9/6/064014.

Jain, S., Agarwal, A., Jani, V., Singhal, S., Sharma, P. *et al.* (2017) Assessment of carbon neutrality and sustainability in educational campuses (CaNSEC): a general framework. *Ecological Indicators* 76, 131–143. DOI: 10.1016/j.ecolind.2017.01.012.

Jarvis, A., Ramirez-Villegas, J. and Campo, B.V.H. (2020) Is cassava the answer to african climate change adaptation? *Tropical Plant Biology* 13(1), 23–39. DOI: 10.1093/nsr/nwy030.

Jiao, N., Wang, H., Xu, G. and Aricò, S. (2018) Blue carbon on the rise: challenges and opportunities. *National Science Review* 5(4), 464–468. DOI: 10.1093/nsr/nwy030.

Jiao, Y., Ma, Q., Liu, Y., Xue, W. and Li, X. (2021) Pesticide residues in soil and water: current issues. *Reviews of Environmental Contamination and Toxicology* 253, 107–140. DOI: 10.1093/nsr/nwy030.

Joergensen, R.G. and Wichern, F. (2018) Alive and kicking: why dormant soil microorganisms matter. *Soil Biology and Biochemistry* 116, 419–430. DOI: 10.1093/nsr/nwy030.

Johnson, M.S., Lathuillière, M.J., Tooke, T.R. and Coops, N.C. (2015) Attenuation of urban agricultural production potential and crop water footprint due to shading from buildings and trees. *Environmental Research Letters* 10(6), 064007. DOI: 10.1093/nsr/nwy030.

Jose, S. (2009) Agroforestry for ecosystem services and environmental benefits: an overview. *Agroforestry Systems* 90(1), 1–10. DOI: 10.1007/s10457-009-9229-7.

Kılkış, Ş., Krajačić, G., Duić, N., Rosen, M.A. and Al-Nimr, M.A. (2020) Advances in integration of energy, water and environment systems towards climate neutrality for sustainable development. *Energy Conversion and Management* 225, 113410. DOI: 10.1016/j.enconman.2020.113410.

Knapp, J.R., Laur, G.L., Vadas, P.A., Weiss, W.P. and Tricarico, J.M. (2014) Invited review: enteric methane in dairy cattle production: quantifying the opportunities and impact of reducing emissions. *Journal of Dairy Science* 97(6), 3231–3261. DOI: 10.3168/jds.2013-7234.

Kong, F. and Wang, Y. (2022) How to understand carbon neutrality in the context of climate change? With special reference to China. *Sustainable Environment* 8(1), 2062824. DOI: 10.1080/27658511.2022.2062824.

Krammer, P., Dray, L. and Köhler, M.O. (2013) Climate-neutrality versus carbon-neutrality for aviation biofuel policy. *Transportation Research Part D: Transport and Environment* 23, 64–72. DOI: 10.1016/j.trd.2013.03.013.

Kumar, P. and Sharma, P.K. (2020) Soil salinity and food security in India. *Frontiers in Sustainable Food Systems* 4, 533781. DOI: 10.3389/fsufs.2020.533781.

Kurramovich, K.K., Abro, A.A., Vaseer, A.I., Khan, S.U., Ali, S.R. *et al.* (2022) Roadmap for carbon neutrality: the mediating role of clean energy development-related investments. *Environmental Science and Pollution Research International* 29(23), 1–20. DOI: 10.1007/s11356-021-17985-3.

Laborde, D., Martin, W., Swinnen, J. and Vos, R. (2020) COVID-19 risks to global food security. *Science* 369(6503), 500–502. DOI: 10.1126/science.abc4765.

Lal, R. (2007) Carbon sequestration. *Philosophical Transactions of the Royal Society B: Biological Sciences* 363(1492), 815–830. DOI: 10.1016/j.enconman.2020.113410.

Lal, R. (2015) Restoring soil quality to mitigate soil degradation. *Sustainability* 7(5), 5875–5895. DOI: 10.3390/su7055875.

Lal, R. (2021) Regenerative Agriculture for Food and Climate. *Frontiers in Sustainable Food Systems* 5, 40.

Lal, R., Kimble, J.M., Follett, R.F. and Cole, C.V. (1998) *The Potential of U.S. Cropland to Sequester Carbon and Mitigate the Greenhouse Effect*. Lewis Publishers, Boca Raton, Florida.

Lampert, A. (2019) Over-exploitation of natural resources is followed by inevitable declines in economic growth and discount rate. *Nature Communications* 10(1), 1419. DOI: 10.1038/s41467-019-09246-2.

Lavallee, J.M., Soong, J.L. and Cotrufo, M.F. (2019) Conceptualizing soil organic matter into particulate and mineral-associated forms to address global change in the 21st century. *Global Change Biology* 26(1), 261–273. DOI: 10.1111/gcb.14859.

Lebaka, V.R. (2013) Potential bioresources as future sources of biofuels production: an overview. In: Gupta, V. and Tuohy, M. (eds) *Biofuel Technologies*. Springer, Berlin, Germany, pp. 223–258. DOI: 10.1007/978-3-642-34519-7_9.

Leclère, D., Obersteiner, M., Barrett, M., Butchart, S.H.M., Chaudhary, A. *et al.* (2020) Bending the curve of terrestrial biodiversity needs an integrated strategy. *Nature* 585(7826), 551–556. DOI: 10.1038/s41586-020-2705-y.

Leip, A., Billen, G., Garnier, J., Grizzetti, B., Lassaletta, L. *et al.* (2015) Impacts of European livestock production: nitrogen, sulphur, phosphorus and greenhouse gas emissions, land-use, water eutrophication and biodiversity. *Environmental Research Letters* 10(11), 115004. DOI: 10.1088/1748-9326/10/11/115004.

Letourneau, D.K., Armbrecht, I., Rivera, B.S., Lerma, J.M., Carmona, E.J. *et al.* (2019) Does plant diversity benefit agroecosystems? A synthetic review. *Ecological Applications* 21(1), 9–21. DOI: 10.1890/09-2026.1.

Linehan, V., Thorpe, S., Gunning-Trant, C., Heyhoe, E., Harle, K, *et al.* (2013) Global food production and prices to 2050. *Australian Bureau of Agricultural and Resource Economics and Sciences*. DOI: 10.1890/09-2026.1.

Liu, J., Hull, V., Godfray, H.C.J., Tilman, D., Gleick, P. *et al.* (2020) Nexus approaches to global sustainable development. *Nature Sustainability* 3(7), 466–476. DOI: 10.1890/09-2026.1.

Liu, J., Geng, X., Ye, J., Wang, Y., Li, H. *et al.* (2022a) Agroecological practices mitigate the negative effects of chemical-intensive agriculture on soil microbiome. *Environmental Pollution* 291, 118274. DOI: 10.1038/s43017-021-00244-x.

Liu, Y., Cruz-Morales, P., Zargar, A., Belcher, M.S., Pang, B. *et al.* (2021) Biofuels for a sustainable future. *Cell* 184(6), 1636–1647. DOI: 10.1016/j.cell.2021.01.052.

Liu, Z., Deng, Z., He, G., Wang, H., Zhang, X. *et al.* (2022b) Challenges and opportunities for carbon neutrality in China. *Nature Reviews Earth & Environment* 3(2), 141–155. DOI: 10.1038/s43017-021-00244-x.

López Barreiro, D., Prins, W., Ronsse, F. and Brilman, W. (2013) Hydrothermal liquefaction (HTL) of microalgae for biofuel production: state of the art review and future prospects. *Biomass and Bioenergy* 53, 113–127. DOI: 10.1016/j.biombioe.2012.12.029.

Lotze-Campen, H., Müller, C., Bondeau, A., Rost, S., Popp, A. *et al.* (2008) Global food demand, productivity growth, and the scarcity of land and water resources: a spatially explicit mathematical programming approach. *Agricultural Economics* 39(3), 325–338. DOI: 10.1111/j.1574-0862.2008.00336.x.

Macreadie, P.I., Anton, A., Raven, J.A., Beaumont, N., Connolly, R.M. *et al.* (2019) The future of blue carbon science. *Nature Communications* 10(1), 3998. DOI: 10.1038/s41467-019-11693-w.

McCalla, A.F. and Revoredo, C.L. (2001) *Prospects for Global Food Security: A Critical Appraisal of Past Projections and Predictions*. Vol. 35. International Food Policy Research Institute, Washington, DC.

Mcleod, E., Chmura, G.L., Bouillon, S., Salm, R., Björk, M. *et al.* (2011) A blueprint for blue carbon: toward an improved understanding of the role of vegetated coastal habitats in sequestering CO_2 . *Frontiers in Ecology and the Environment* 9(10), 552–560. DOI: 10.1890/110004.

Medek, D.E., Schwartz, J. and Myers, S.S. (2017) Estimated effects of future atmospheric CO_2 concentrations on protein intake and the risk of protein deficiency by country and region. *Environmental Health Perspectives* 125(8), 087002. DOI: 10.1289/EHP41.

Minang, P.A., van Noordwijk, M., Freeman, O.E., Mbow, C., de Leeuw, J. *et al.* (2015) *Climate-Smart Landscapes: Multifunctionality in Practice*. World Agroforestry Centre, Nairobi, Kenya.

Mitchell, D.O., Ingco, M.D. and Duncan, R.C. (1997) *The World Food Outlook*. Cambridge University Press, Cambridge, UK.

Molden, D., Sakthivadivel, R., Perry, C.J. and de Fraiture, C. (2007) *Pathways for Increasing Agricultural Water Productivity*. International Water Management Institute (IWMI). Available at: https://ideas.repec.org/p/iwt/bosers/h040200.html (accessed 29 December 2023).

Montgomery, D.R. (2007) Soil erosion and agricultural sustainability. *Proceedings of the National Academy of Sciences* 104(33), 13268–13272. DOI: 10.1073/pnas.0611508104.

Morgavi, D.P., Martin, C. and Martin, C. (2018) Ruminal methane mitigation: a review. *Animal* 12(s2), s224–s235.

Muhammad, A., Seale, J.L., Meade, B. and Regmi, A. (2011) International evidence on food consumption patterns: an update using 2005 international comparison program data. *USDA-ERS Technical Bulletin* (1929). Available at: https://www.ers.usda.gov/webdocs/publications/47579/7636_tb1929summary.pdf?v=0 (accessed 2 August 2024).

Mustapha, C. (2022) The burden of a ton CO_2! Emission trading systems and the air transport business. In: Van de Voorde, E. (ed.) *The Air Transportation Industry*. Elsevier, Amsterdam, pp. 27–53.

Nair, P.K.R., Garrity, D., Michon, G., Dian, N. and Sileshi, G. (2021) Agroforestry, livelihoods and ecosystem services in South and Southeast Asia. *Agroforestry Systems* 95(3), 775–780.

Nassajfar, M.N., Välimäki, M., Hakola, L., Eiroma, K., Immonen, K., Abdulkareem, M. and Horttanainen, M. (2023) The effect of conductive ink alternation on the sustainability and functioning of printed electronics. *Flexible and Printed Electronics* 8(2), 025015.

Nelson, R., Kokic, P., Crimp, S., Martin, P., Meinke, H. *et al.* (2014) The vulnerability of Australian rural communities to climate variability and change: part II—integrating impacts with adaptive capacity. *Environmental Science and Policy* 39, 187–195.

Newbold, T., Hudson, L.N., Arnell, A.P., Contu, S., De Palma, A. *et al.* (2022) Widespread winners and narrow-ranged losers: land use homogenizes biodiversity in local assemblages worldwide. *PLoS Biology* 20(1), e3001010.

Ney, R.A. and Purman, J.R. (2009) Carbon neutrality and the use of offsets. *Facilities Manager* 25(2), 28. DOI: 10.1016/B978-0-12-804090-4.00001-X.

Ng'andwe, P., Mwitwa, J., Muimba-Kankolongo, A. and Ratnasingam, J. (2015) An overview of the forestry sector in Zambia. *Forest Policy, Economics and Markets in Zambia* 1–26. DOI: 10.1016/B978-0-12-804090-4.00001-X.

Nihous, G.C. (2006) A preliminary assessment of ocean thermal energy conversion resources. *Journal of Energy Resources Technology* 129(1), 10–17. DOI: 10.1016/B978-0-12-804090-4.00001-X.

Nikolaidis, P. and Poullikkas, A. (2017) A comparative overview of hydrogen production processes. *Renewable and Sustainable Energy Reviews* 67, 597–611. DOI: 10.1016/j.rser.2016.09.044.

O'Neill, B.C., Carter, T.R., Ebi, K., Harrison, P.A., Kemp-Benedict, E. *et al.* (2020) Achievements and needs for the climate change scenario framework. *Nature Climate Change* 10(12), 1074–1084. DOI: 10.1038/s41558-020-00952-0.

Ogle, S.M., Swan, A., Paustian, K. and Reay, D.S. (2020) The far-reaching scope of the nitrous oxide emissions induced by individual crops: magnitude, emission factors and mitigation strategies. *Global Change Biology* 26(12), 6810–6826. DOI: 10.1016/B978-0-12-804090-4.00001-X.

Pan, Y., Birdsey, R.A., Fang, J., Houghton, R., Kauppi, P.E. *et al.* (2011) A large and persistent carbon sink in the world's forests. *Science* 333(6045), 988–993. DOI: 10.1126/science.1201609.

Pandey, A. and Asif, M. (2022) Assessment of energy and environmental sustainability in South Asia in the perspective of the sustainable development goals. *Renewable and Sustainable Energy Reviews* 165, 112492. DOI: 10.1016/j.rser.2022.112492.

Pannell, D.J. and Llewellyn, R.S. (2014) More reasons why farmers adopt precision agriculture. In: Stafford, J.V. (ed.) *Precision Agriculture '13*. Wageningen Academic Publishers, Wageningen, The Netherlands, pp. 313–326.

Parody, A., Viloria, A., Gaitán, M. and Malagón, L.E. (2018) Application of a central design composed of surface of response for the determination of the flatness in the steel sheets of a colombian steel. In: Tan, Y. and Shi, Y. (eds) *Data Mining and Big Data: Third International Conference, DMBD 2018, Shanghai, China, June 17–22, 2018, Proceedings 3*. Springer International Publishing, Cham, Switzerland, pp. 739–747.

Parraga, J., Khalilpour, K.R. and Vassallo, A. (2019) Polyfeed and polyproduct integrated gasification systems. In: *Polygeneration with Polystorage for Chemical and Energy Hub. Academic Press*. pp. 175–212. DOI: 10.1016/B978-0-12-813306-4.00007-0.

Paustian, K., Lehmann, J., Ogle, S., Reay, D., Robertson, G.P. *et al.* (2016) Climate-smart soils. *Nature* 532(7597), 49–57. DOI: 10.1016/B978-0-12-813306-4.00007-0.

Perfecto, I., Vandermeer, J. and Philpott, S.M. (2021) Ecological complexity and agroecology. *Annual Review of Environment and Resources* 46, 105–130. DOI: 10.1016/B978-0-12-813306-4.00007-0.

Poeplau, C., Kätterer, T., Leblans, N.I., Sigurdsson, B.D., Thomsen, I.K. *et al.* (2019) Carbon sequestration in agricultural soils via cultivation of cover crops—a meta-analysis. *Agriculture, Ecosystems and Environment* 284, 106583. DOI: 10.1016/B978-0-12-813306-4.00007-0.

Poore, J. and Nemecek, T. (2018) Reducing food's environmental impacts through producers and consumers. *Science* 360(6392), 987–992. DOI: 10.1016/B978-0-12-813306-4.00007-0.

Popp, A., Calvin, K., Fujimori, S., Havlik, P., Humpenöder, F. *et al.* (2017) Land-use futures in the shared socio-economic pathways. *Global Environmental Change* 42, 331–345. DOI: 10.1016/j.gloenvcha.2016.10.002.

Pradhan, B.B., Shrestha, R.M. and Limmeechokchai, B. (2023) Attainability of net zero carbon emission targets in Nepal under different effort-sharing approaches. *Global Environmental Research* 26(1–2), 57–64. DOI: 10.1016/j.gloenvcha.2016.10.002.

Ramseur, J.L. (2007) *Voluntary Carbon Offsets: Overview and Assessment*. Congressional Research Service, Washington, DC. DOI: 10.1016/j.gloenvcha.2016.10.002.

Rask, K.J. and Rask, N. (2011) Economic development and food production–consumption balance: a growing global challenge. *Food Policy* 36(2), 186–196. DOI: 10.1016/j.gloenvcha.2016.10.002.

Rather, R.A., Wani, A.W., Mumtaz, S., Padder, S.A., Khan, A.H. *et al.* (2022) Bioenergy: a foundation to environmental sustainability in a changing global climate scenario. *Journal of King Saud University—Science* 34(1), 101734. DOI: 10.1016/j.gloenvcha.2016.10.002.

Rayer, Q., Jenkins, S. and Walton, P. (2022) Defining net-zero and climate recommendations for carbon offsetting. In: *Business and Policy Solutions to Climate Change: From Mitigation to Adaptation*. Springer International Publishing, Cham, Switzerland, pp. 13–35.

Reilly, J., Melillo, J., Cai, Y., Kicklighter, D., Gurgel, A. *et al.* (2012) Using land to mitigate climate change: hitting the target, recognizing the trade-offs. *Environmental Science & Technology* 46(11), 5672–5679. DOI: 10.1021/es2034729.

Ren, K., Tang, X., Wang, P., Willerström, J. and Höök, M. (2021) Bridging energy and metal sustainability: insights from China's wind power development up to 2050. *Energy* 227, 120524. DOI: 10.1016/j.energy.2021.120524.

Riahi, K., van Vuuren, D.P., Kriegler, E., Edmonds, J., O'Neill, B.C. *et al*. (2017) The shared socioeconomic pathways and their energy, land use, and greenhouse gas emissions implications: an overview. *Global Environmental Change* 42, 153–168. DOI: 10.1016/j.gloenvcha.2016.05.009.

Robinson, S., Mason-d'Croz, D., Sulser, T., Islam, S., Robertson, R, *et al*. (2015) The international model for policy analysis of agricultural commodities and trade (IMPACT): model description for version 3. Available at: https://papers.ssrn.com/sol3/Delivery.cfm/SSRN_ID2741234_code1123746.pdf?abstractid=2741234&mirid=1 (accessed 2024).

Robinson, S., Mason d'Croz, D., Islam, S., Sulser, T.B., Robertson, R. *et al*. (2021) Climate change adaptation in global agriculture: insights from country studies. *Agricultural Economics* 52(1), 1–17.

Roschinsky, R., Fernández, M.E. and González, E. (2021) Livestock integration into agroecosystems: challenges and future perspectives. *Journal of Cleaner Production* 287, 125507.

Rosenzweig, C., Antle, J., Elliott, J. and Jones, J.W. (2021) Reanalysis of global warming effects on global food security. *Nature Food* 2(3), 156–160.

Sahil, K., Mehta, P., Bhardwaj, S.K. and Dhaliwal, L.K. (2023) Development of mitigation strategies for the climate change using artificial intelligence to attain sustainability. In: *Visualization Techniques for Climate Change with Machine Learning and Artificial Intelligence*. Elsevier, Oxford, UK, pp. 421–448.

Sahoo, K. and Sethi, N. (2012) Investigating the impact of agriculture and industrial sector on economic growth of India. *OIDA International Journal of Sustainable Development* 5(5), 11–21.

Salehi Jouzani, G. and Taherzadeh, M.J. (2015) Advances in consolidated bioprocessing systems for bioethanol and butanol production from biomass: a comprehensive review. *Biofuel Research Journal* 2, 152–195. DOI: 10.18331/BRJ2015.2.1.4.

Salvia, M., Reckien, D., Pietrapertosa, F., Eckersley, P., Spyridaki, N.A. *et al*. (2021) Will climate mitigation ambitions lead to carbon neutrality? An analysis of the local-level plans of 327 cities in the EU. *Renewable and Sustainable Energy Reviews* 135, 110253. DOI: 10.18331/BRJ2015.2.1.4.

Sen, G., Chau, H.W., Tariq, M.A.U.R., Muttil, N. and Ng, A.W. (2021) Achieving sustainability and carbon neutrality in higher education institutions: a review. *Sustainability* 14(1), 222. DOI: 10.18331/BRJ2015.2.1.4.

Serp, J., Allibert, M., Beneš, O., Delpech, S., Feynberg, O. *et al*. (2014) The molten salt reactor (MSR) in generation IV: overview and perspectives. *Progress in Nuclear Energy* 77, 308–319. DOI: 10.18331/BRJ2015.2.1.4.

Shang, Z., Abdalla, M., Xia, L., Zhou, F., Sun, W. *et al*. (2021) Can cropland management practices lower net greenhouse emissions without compromising yield? *Global Change Biology* 27(19), 4657–4670. DOI: 10.1111/gcb.15796.

Shenkut, A., Tesfaye, K. and Abegaz, F. (2013) Determination of Water Requirement and Crop Coefficient for Sorghum (Sorghum Bicolor L.) at Melkassa, Ethiopia. *Science, Technology and Arts Research Journal* 2(3), 16–24.

Singh, P., Mondal, S. and Singh, R.L. (2020) Introduction. In: Mondal, S. and Singh, R.L. (eds) *Advances in Animal Genomics*. Elsevier, Amsterdam, pp. 1–12.

Singh, P., Singh, P.K. and Singh, R.L. (2021) Transgenic animals. In: Mondal, S. and Singh, R.L. (eds) *Emerging Issues in Climate Smart Livestock Production*. Elsevier, Amsterdam, pp. 171–208.

Sivabalan, K., Hassan, S., Ya, H. and Pasupuleti, J. (2021) A review on the characteristic of biomass and classification of bioenergy through direct combustion and gasification as an alternative power supply. *Journal of Physics* 1831(1), 012033. DOI: 10.1088/1742-6596/1831/1/012033.

Six, J., Calleja-Cervantes, M.E., Bhattarai, H., Carpentier, C.L., Castro-Franco, M. *et al*. (2020) The potential for soil carbon sequestration in the tropics through silvopastoral and agroforestry pathways. *Global Change Biology* 26(2), 1036–1052.

Smith, P., Bustamante, M., Ahammad, H., Clark, H., Dong, H, *et al*. (2014a) Mitigation of climate change. In: Edenhofer, O. (ed.) *Contribution of Working Group III to the Fifth Assessment Report of the Intergovernmental Panel on Climate Change*. Cambridge University Press, Cambridge, UK, pp. 811–922.

Smith, P., Haberl, H., Popp, A., Erb, K.H., Lauk, C. *et al*. (2014b) How much land-based greenhouse gas mitigation can be achieved without compromising food security and environmental goals? *Global Change Biology* 19(8), 2285–2302.

Sreenivas, K., Dadhwal, V., Kumar, S., Harsha, G.S., Mitran, T. *et al.* (2016) Digital mapping of soil organic and inorganic carbon status in India. *Geoderma* 269, 160–173. DOI: 10.1016/j.geoderma.2016.02.002.

Sudarsan, J.S., Vaishampayan, S., Gedam, V.V., Katare, V.D. and Pathak, P.D. (2023) Sustainable resource management through carbon emissions inventorization: a case of the Indian construction industry. *Environmental Quality Management* 33, 467–476.

Sun, Y., Dai, H., Lv, L., Qin, Y. and Yu, B. (2022) Smart agriculture: a review of technologies, applications and benefits. *Journal of Cleaner Production* 341, 130831.

Tian, X., Engel, B.A., Qian, H., Hua, E., Sun, S. *et al.* (2021) Will reaching the maximum achievable yield potential meet future global food demand? *Journal of Cleaner Production* 294, 126285. DOI: 10.1016/j.jclepro.2021.126285.

Timmer, C.P., Falcon, W.P. and Pearson, S.R. (1983) *Food Policy Analysis. Vol. 1983*. Johns Hopkins University Press, Baltimore, Maryland. DOI: 10.1016/j.jclepro.2021.126285.

Tiwari, H., Naresh, R.K., Manisha, R.K., Kumar, L., Singh, S, *et al.* (2023) Ecological significance of tillage systems and precision nutrient management in cereal based systems for carbon-cum-energy efficient and energy-carbon footprint vis a vis system productivity and profitability: a sub-tropical Indian perspectives. Available at: https://www.thepharmajournal.com/archives/?year=2023&vol=12&issue=1&ArticleId=18040 (accessed August 2024).

Tomlinson, I. (2013) Doubling food production to feed the 9 billion: a critical perspective on a key discourse of food security in the UK. *Journal of Rural Studies* 29, 81–90. DOI: 10.1016/j.jclepro.2021.126285.

Tripathi, A.D., Mishra, R., Maurya, K.K., Singh, R.B. and Wilson, D.W. (2019) Estimates for world population and global food availability for global health. In: Takahashi, T. (ed.) *The Role of Functional Food Security in Global Health*. Elsevier Science, London, pp. 3–24. DOI: 10.1016/j.jclepro.2021.126285.

Trumbore, S., Brando, P. and Hartmann, H. (2015) Forest health and global change. *Science* 349(6250), 814–818. DOI: 10.1126/science.aac6759.

Tscharntke, T., Milder, J.C., Schroth, G., Clough, Y., DeClerck, F. *et al.* (2012) Conserving biodiversity through certification of tropical agroforestry crops at local and landscape scales. *Conservation Letters* 5(1), 60–68.

Tursi, A. (2019) A review on biomass: importance, chemistry, classification, and conversion. *Biofuel Research Journal* 6(2), 962–979. DOI: 10.18331/BRJ2019.6.2.3.

UN Water (2021) *Water and Agriculture*. Available at: https://www.unwater.org/publications/un-world-water-development-report-2021 (accessed 30 December 2023).

van der Mensbrugghe, D. (2013) Modeling the global economy–forward-looking scenarios for agriculture. In: *Handbook of Computable General Equilibrium Modelling*, Vol. 1. Elsevier, Amsterdam, the Netherlands, pp. 933–994.

van Noordwijk, M., Hoang, M.H., Neufeldt, H. and Öborn, I. (2019) Tree cover transitions and food security in. *Global Food Security* 23, 35–45.

van Vuuren, D.P., Stehfest, E., Gernaat, D.E.H.J., Doelman, J.C., van den Berg, M. *et al.* (2017) Energy, land-use and greenhouse gas emissions trajectories under a green growth paradigm. *Global Environmental Change* 42, 237–250. DOI: 10.1016/j.gloenvcha.2016.05.008.

van Vuuren, D.P., Bijl, D.L., Bogaart, P., Stehfest, E., Biemans, H. *et al.* (2019) Integrated scenarios to support analysis of the food–energy–water nexus. *Nature Sustainability* 2(12), 1132–1141. DOI: 10.1038/s41893-019-0418-8.

van Zanten, H.H., Herrero, M., Descheemaeker, K. and Peden, D. (2021) Livestock and sustainable development goals. *Nature Sustainability* 4(4), 292–303.

Varshney, R.K., Bansal, K.C., Aggarwal, P.K., Datta, S.K. and Craufurd, P.Q. (2018) Advancing crop productivity for food security under climate change. *Nature* 554(7691), 1–9. DOI: 10.1016/j.resconrec.2022.106576.

Voss-Fels, K.P., Robinson, H., Mudge, S.R., Richard, C., Newman, S. *et al.* (2019) Vernalization modulates root system architecture in wheat and barley. *Molecular Plant* 12(3), 360–373. DOI: 10.1016/j.resconrec.2022.106576.

Wallace, R.J., Sasson, G., Garnsworthy, P.C., Tapio, I., Gregson, E. *et al.* (2019) A heritable subset of the core rumen microbiome dictates dairy cow productivity and emissions. *Science Advances* 5(7), eaav8391. DOI: 10.1126/sciadv.aav8391.

Wang, B., Cai, A., Li, Y., Qin, X., Wilkes, A. *et al.* (2022) Four pathways towards carbon neutrality by controlling net greenhouse gas emissions in chinese cropland. *Resources, Conservation and Recycling* 186, 106576. DOI: 10.1016/j.resconrec.2022.106576.

Wang, F., Harindintwali, J.D., Yuan, Z., Wang, M., Wang, F. *et al.* (2021) Technologies and perspectives for achieving carbon neutrality. *The Innovation* 2(4), 100180. DOI: 10.1016/j.xinn.2021.100180.

Wang, J., Zhang, Y., Wu, X., Huang, X., Zou, Y. *et al.* (2021) Real-time soil moisture monitoring for irrigation decision making: a review. *Journal of Hydrology* 594, 125862.

Wheeler, T. and von Braun, J. (2013) Climate change impacts on global food security. *Science* 341(6145), 508–513. DOI: 10.1126/science.1239402.

Willett, W., Rockström, J., Loken, B., Springmann, M., Lang, T. *et al.* (2019) Our food in the Anthropocene: the EAT-Lancet Commission on healthy diets from sustainable food systems. *Lancet* 393(10170), 447–492. DOI: 10.1016/S0140-6736(18)31788-4.

Wilson, S., Harrison, A.L., Dipple, G.M., Power, I.M., Barker, S.L. *et al.* (2014) Offsetting of CO_2 emissions by air capture in mine tailings at the Mount Keith Nickel Mine, Western Australia: rates, controls and prospects for carbon neutral mining. *International Journal of Greenhouse Gas Control* 25, 121–140. DOI: 10.1016/j.ijggc.2014.04.002.

Wolf, J., Bindraban, P., Dale, V. and Lotze-Campen, H. (2020) Options for keeping the food system within environmental limits. *Nature* 578(7794), 141–148.

Wu, X., Tian, Z. and Guo, J. (2022a) A review of the theoretical research and practical progress of carbon neutrality. *Sustainable Operations and Computers* 3, 54–66. DOI: 10.1016/j.susoc.2021.10.001.

Wu, Z., Huang, X., Chen, R., Mao, X. and Qi, X. (2022b) The United States and China on the paths and policies to carbon neutrality. *Journal of Environmental Management* 320, 115785. DOI: 10.1016/j.jenvman.2022.115785.

Yang, S., Yang, D., Shi, W., Deng, C., Chen, C. *et al.* (2023) Global evaluation of carbon neutrality and peak carbon dioxide emissions: current challenges and future outlook. *Environmental Science and Pollution Research* 30(34), 81725–81744. DOI: 10.1007/s11356-022-19764-0.

Zaccaria, D., Pezzolla, D., De Salvo, M., De Lorenzi, F., Albrizio, R. *et al.* (2021) Economic and environmental benefits of solar energy in agriculture: a case study in southern Italy. *Energy* 217, 119321. DOI: 10.1007/s11356-022-19764-0.

Zhao, C., Ju, S., Xue, Y., Ren, T., Ji, Y. *et al.* (2022) China's energy transitions for carbon neutrality: challenges and opportunities. *Carbon Neutrality* 1(1), 7. DOI: 10.1007/s43979-022-00010-y.

Zhu, Q., Duan, Y. and Sarkis, J. (2024) Supply chain carbon transparency to consumers via blockchain: does the truth hurt? *The International Journal of Logistics Management* 35(3), 833–864. DOI: 10.1108/IJLM-03-2023-0109.

6 Hotspots Associated with Climate Change and Food Security Risks

Himani Chhatwal, Samar Singh and Ashutosh Pandey*
National Institute of Plant Genome Research (NIPGR), New Delhi, India

Abstract
Climate change is an emerging problem, threatening nations at a global stage. Our knowledge about the impact of climate change on agriculture has grown over the past two decades, but recently, it has been found that its effects are even more deep-rooted. Anthropogenic and natural activities are increasing global temperatures at an alarming rate, affecting crop productivity. Extreme weather events will impact food supply, stability, and access. In the current times of increasing population, when wholesome food production is a global concern, climate change becomes an inevitable issue that no nation can escape from. This chapter highlights the major hotspots of the world which face an immediate threat from climate change. It also covers the perils of changing climate on the food production system and the adaptation strategies to overcome it. The need is to focus on developing immediate solutions to develop a sustainable food supply in the future.

Keywords: Agriculture, Crop yield, Food supply, Global warming, Livestock

6.1 Introduction

Food security refers to having consistent, secure access to affordable, wholesome food. Food availability, accessibility, and absorption are the three aspects of food security (FAO, 1996). Climate change is predicted to have a considerable impact on food production and agriculture (Hasegawa *et al.*, 2021). The connections between food security and climate change are complicated, as food security covers food and its production, trade, and nutrition as well as how people and nations sustain access to food in the face of various challenges. Our ability to guarantee global food security, end poverty, and achieve sustainable development is threatened by climate change. The negative effects of climate change are more pronounced among the poor in developing nations, who are heavily dependent on resources that are sensitive to climate change and have the least capacity to withstand climatic shocks (Gitz *et al.*, 2016). The first step in identifying the vulnerable areas is to use the climate change hotspots to pinpoint the regions or places that are affected by the effects of global climate change. A climate change hotspot is defined by Giorgi (2006) as an area where the climate is very sensitive to changes in the global environment. Scholars must address not only changes in global demographics and climatic implications on food security (Lobell and Burke, 2010) but also the extent to which food and agricultural systems can adapt to adverse environments (Lobell and Burke, 2010; Porter *et al.*, 2014). Food access is linked to a stable food supply chain downstream. Climate

*Corresponding author: ashutosh@nipgr.ac.in

DOI: 10.1079/9781800625037.0006

change disrupts the food supply chain and limits physical access to markets in a variety of ways. Heavy precipitation, floods, snow, and storms have an impact on public infrastructure, damaging roads, and bridges, inundating transportation networks, and making it dangerous for people to physically access markets (Nissen and Ulbrich, 2017). The demand for nutritional food is rising at a rate that is even higher than the rate of population expansion, making it difficult to meet global food security needs. These point out the complexities of climate change impacts on food production and the need for a more comprehensive picture of all interacting components. Overall, there is an urgent need to understand and fight climate change, as well as to produce enough food for a growing global population, that not only fulfills its caloric needs but also has an acceptable supply of vitamins and minerals (Singh *et al.*, 2014; Leisner, 2020).

6.2 Major Hotspots

The unprecedented rate of global warming could have significant repercussions on human society and natural ecosystems. Since regional climates are the complex outcome of numerous processes that vary greatly by geographical location, they react differentially to changes occurring at the global scale. To better understand how these might alter at various levels of global warming, there is growing interest in assessing regional responses to climate change. A climate change hotspot is a location whose climate is particularly sensitive to global warming and is more vulnerable to climate change risks than other places (Giorgi, 2006). As a result, identifying hotspots and comprehending the mechanisms that boost responsiveness are essential steps toward comprehending, preparing for, and reducing the dangers associated with future climate change.

The Amazon, central and western Africa, Indonesia, and the Tibetan Plateau are among the locations where the most recent simulations have been identified as strong hotspots of future climatic changes, which will occur at 1.5°C, 2°C, and 3°C warming levels under four scenarios (Fig. 6.1; Fan *et al.*, 2021). The global rise in mean temperatures, the intensification of the extremely hot season in low-latitude areas, and the decline in precipitation over Central Africa are the main causes which are turning this region into a climate hotspot. As global warming advances, the areas of the Arctic, Central America, and southern Africa will become hotspot regions under 4°C warming at the end of the 21st century (Fan *et al.*, 2021). The most populous islands in Indonesia, Java,

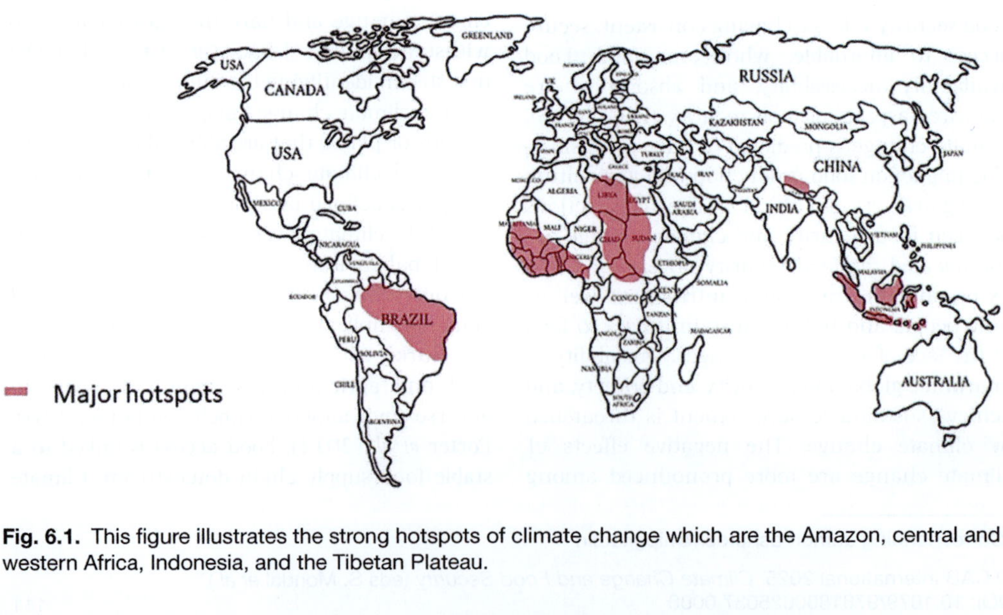

Fig. 6.1. This figure illustrates the strong hotspots of climate change which are the Amazon, central and western Africa, Indonesia, and the Tibetan Plateau.

which is located in the east and west, as well as the coastal sections of Sumatra, parts of western and northern Sulawesi, and the southeast Papua islands, are all extremely sensitive to various climate disasters, such as floods, landslides, and drought (Harris *et al.*, 2020). The hotspots with significant multi-sector risk including West Africa, South and Southeast Asia, the Middle East, and northeastern Brazil where a major part of the population is exposed to high poverty and lack of resources, is where the vulnerable populations will benefit mostly from focused interventions and additional modelling exercises (Johnson, 2017). Climate change would have an impact on Southeast Asia's monsoon rainfall patterns, causing floods in some areas and having major negative impacts on agricultural output. Sub-Saharan Africa is an area of the world which is most vulnerable to climate change, and it is also the region that is most sensitive to changes in weather and other climatic factors (such as temperature, precipitation, and light) (Kotir, 2011). Climate change in sub-Saharan Africa would have severe effects on local agricultural and livestock output due to extreme temperatures and decreased rainfall (Serdeczny *et al.*, 2017). As a result of their vulnerability to climate change, Southeast Asia and sub-Saharan Africa have become research hotspots. Identifying important climate change hotspots and their influencing elements at various levels of global warming gives a starting point for creating climate change mitigation strategies. However, climate change risks are not solely determined by the intensity of climate change; they are also heavily influenced by a number of socioeconomic, demographic, biophysical, and other factors, as well as regional capacity to plan for and manage shifting hazards.

6.3 Impact of Climate Change on Food Security

Food security will be at risk in the coming years because of climate change and rising population demands. High anthropogenic greenhouse gaseous emissions and related climate change are also seen as major threats to the security of food and livelihoods (Richardson *et al.*, 2018; Hasegawa *et al.*, 2021). It will hamper the food production, transport, access, and reach. The future consequences of climate change could be extremely serious if actions are not taken at the right stage (Fig. 6.2). The following effects of climate change have been documented, which will affect a nation's food security:

6.3.1 Effects on agriculture

Climate change has threatened agricultural production in food-insecure regions, particularly in Asian countries. Drought, heat waves, irregular rainfall patterns, storms, floods, and developing insect pests have a negative impact on farmers' livelihoods. Future climatic predictions showed a prominent increase in temperature, as well as irregular rainfall with greater intensity (Gründemann *et al.*, 2022). Climate change damage may jeopardize food production and national economic productivity in the least developed countries (Myers *et al.*, 2017). Due to changes in weather patterns, declining yields in various crops (rice, wheat) varied across locations (Yu *et al.*, 2018). In C_3 plants, CO_2 fertilization can raise crop productivity and balance off the significant consequences of rising temperatures (Obermeier *et al.*, 2017), but it cannot completely counteract such effects (Arunrat *et al.*, 2018). Rising temperatures and unpredictable precipitation have impeded crop growth and development (Rezaei *et al.*, 2018; Asseng *et al.*, 2019). In the Netherlands, extreme weather events have increased in frequency since the turn of the century, which had a significant effect on wheat output there (Powell and Reinhard, 2016). Climate change has been predicted to cause more droughts in the future in the majority of the world's regions, and by 2100, the area impacted by drought is expected to increase from 15.4% to 44.0% of the total land area. The most vulnerable region is said to be in Africa. It is anticipated that main agricultural yields in drought-affected areas will be decreased significantly, almost 90% by 2100 (Li *et al.*, 2009). Crop yield declines can drive up food costs and have a severe impact on agriculture globally, with a 0.3% annual loss in potential global GDP by 2100 (Stevanović *et al.*, 2016). Due to climate change, developing nations will be suffering more than developed

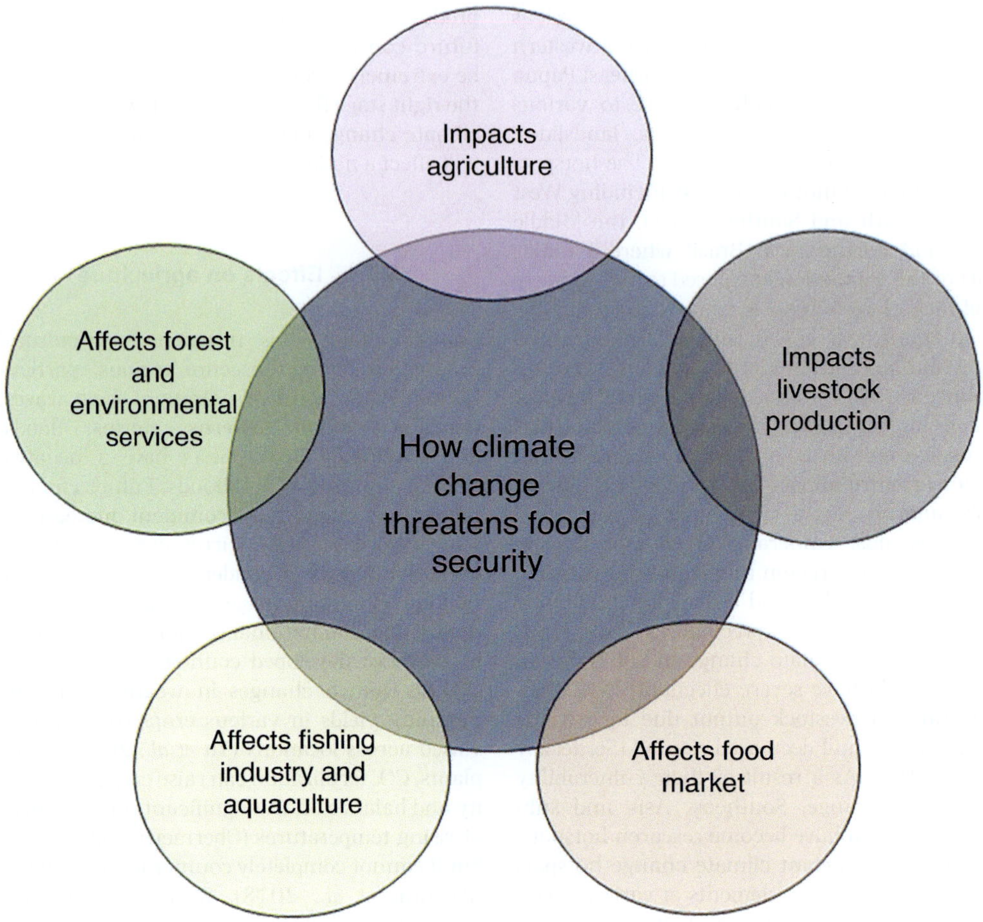

Fig. 6.2. This figure depicts how food security will be impacted in the near future due to increasing levels of global warming and climate change.

nations. The agriculture industry in India may suffer due to the expected rise in temperature in the range of 2.33°C to 4.78°C, doubling of CO_2 concentration, and lengthening of heat waves (Kumar and Raj Gautam, 2014). The average crop yield in sub-Saharan Africa is predicted to decline by 6–24% because of climate change (Waha *et al.*, 2013). Pathogen development and survival are most likely to be impacted by predicted climate changes. As a result of pest infestation, pest management and expenditures have skyrocketed. Data suggests that, in contrast to a decline in wheat in the USA, higher rainfall and temperatures raised the expense of pesticides for crops including corn, potatoes, and soybeans (Chen and McCarl, 2001).

6.3.2 Effects on livestock production

Today, the supply and security of food are significantly impacted by the livestock industry (Singh *et al.*, 2020, 2021). With regional variations, livestock products (meat, milk, and eggs) represent 15% and 31% of the world's per-person calorie and protein supply, respectively. The cattle sector is extremely vulnerable to rising temperatures and decreased precipitation in arid to semi-arid regions (Balamurugan, 2018). Domestic animals prefer temperatures ranging between 10 and 30°C, and each degree increase in temperature results in a 3–5% reduction in the intake of animal feed. Similarly, lowering the temperature would raise the required feed by up to 59%. Furthermore, under climate change scenarios, drought and heat stress might have a significant impact on animal productivity (Habeeb *et al.*, 2018). Change in climate influences the occurrence and spread of various cattle illnesses. Rift Valley Fever (RVF) and tick-borne illnesses (TBDs) have become widespread in sheep, buffalo, goats, cattle, and camels because of increased precipitation and temperature (Bett *et al.*, 2019). Various animal breeds react differently to rising temperatures and water limitations. In addition to heat stress, the increased frequency and severity of storms, fires, floods, and cold waves, can have a considerable negative impact on the cattle industry. Early in 2019, severe floods in northern Australia were followed by cold weather, which resulted in the death of almost 500,000 animals, predominantly cattle but also sheep, goats, and horses (Hall *et al.*, 2020). Similarly, in Georgia, United States, Hurricane Michael killed approximately 2 million hens and destroyed an estimated 84 chicken houses in October 2018 (Georgia Department of Agriculture, 2018).

6.3.3 Impact on forest and environmental services

Forests have been negatively impacted by climate change, including changes in production, carbon dynamics, and shifts in vegetation, as well as depletion of soil resources, heat, and drought stress (Zhu *et al.*, 2021). An increase in disturbance intensity could have a significant impact on the species composition, ecological structure, and carbon sink capacities of forests. The structure, composition, and function of the forest ecosystem, community, or population are being disrupted by disturbances including species invasions, fire, insect attacks, drought and disease outbreaks, and storms like hurricanes. These disturbances also alter the availability of resources or the physical environment. Village and tribal populations obtain many food products from forests like bamboo, dry fruits, spices, as well as medicine. Climate change is threatening their food security by disturbing the ecological balance of the forests. There is proof that over the past centuries, wet soils and flooding have resulted in increased forest damage, erosion, and landslides (Wu *et al.*, 2012). Through its effects on temperature and precipitation, climate change has the potential to directly affect the nutrient cycle. The productivity and health of forest ecosystems depend on the availability of nutrients, and climate change can influence the dynamics of nutrient availability by altering the rate of litter decomposition. The advantages individuals get from the natural processes that maintain ecosystems are called ecosystem services, and they include sustaining, regulating, providing, and cultural services (Fisher *et al.*, 2009). The current state of these ecosystem services, which include both products and processes, is impacted by climate change. Currently, there has been an increase in global temperatures, levels of atmospheric carbon dioxide, and precipitation patterns. The physiology and defenses of the host, connections between pests, their habitat, and other species including natural enemies, mutualists, and competitors are all impacted by climate change. As a result of climate change, invasive species will become more widespread, frequent, and severe. Additionally, previously non-invasive species will become more prone to invasion. It makes it easier for invasive species to colonize the forest ecosystem. When temperature or moisture barriers to dispersal and survival are removed, species' ranges shift, and they can effectively colonize new areas (Hellmann *et al.*, 2008). The abundance and distribution of pests and pathogens in the forest ecosystem are anticipated to vary due to climate change, either directly or indirectly through increasing water stress or wind damage, which will make trees more vulnerable to attack. Climate change reduces plant productivity and the ability of forest ecosystems to survive by limiting the availability of nutrients for plant absorption. The forest ecosystem and net primary productivity are impacted by nutrient availability (Dukes *et al.*, 2014). This will in turn hinder the supply of forest products and services for the food industry.

6.3.4 Impact on fisheries

Changes in fish landings and their economic worth are predicted to increase as climatic changes become more pronounced (Allison *et al.*, 2009; Lam *et al.*, 2020). The fisheries' industry, which is crucial to the social and economic well-being of many coastal communities, faces significant problems due to these changes (Natale *et al.*, 2013; Teh and Sumaila, 2013). According to the reports, extreme weather events can have a direct influence on fishing activities, causing disruptions on land and at sea as well as financial losses (Cheung *et al.*, 2012; Sumaila *et al.*, 2020). Countries that depend heavily on fisheries for their economy are particularly impacted by climate change due to dwindling fish stocks and rising fuel prices (Lam *et al.*, 2020; Huynh *et al.*, 2021). Seas in Europe, especially those in enclosed or partially enclosed basins like the Baltic Sea, Mediterranean Sea, North Sea, and Norwegian Sea, are rapidly warming (Belkin, 2009). Seas in Europe account for more than one-eighth of worldwide fisheries captures, making the EU, behind China, the second-largest trader of aquaculture goods and fisheries in 2019. Many of the species that are captured and eaten across European nations are listed on the IUCN Red List because of the combined effects of climate change and overconsumption (Dureuil *et al.*, 2018). This makes them even more vulnerable. According to Servili *et al.* (2020), rainfall plays a crucial role in fish's accelerated hormonal and sensory responses. Research has demonstrated that excessive temperature and precipitation events have a negative impact on the hatcheries' ability to produce fish (Lebel *et al.*, 2016). In addition, severe weather events are delaying output by harming broodstock and hatchery infrastructure.

6.3.5 Impact on food market

The impact extends from the climate to the environment, to the production sphere, to the social and economic dimensions, posing a variety of additional risks to food availability, access, and consumption, as well as to the stability of the production and consumption chain, for both farm and non-farm households. Climate change impacts the entire food value chain, which has negative effects on humans around the world. The African region is in danger from the effects of climate change (Parry *et al.*, 2005). Two factors will be used to analyze how climate change will affect food systems: (i) food supply networks and (ii) food environments. Food environments include food availability and access, food affordability, food advertising and promotion, and food safety and quality. Food supply chains include food production, storage and distribution, processing and packaging, retailing, and markets. Because of changes in the climate, there is less food produced, which raises demand and drives up market prices (FAO, 2017). China's decreasing wheat production was accompanied by an increase in consumer prices, which led to higher market prices and negatively affected on farmers' livelihoods (Xie *et al.*, 2020). How climate trends, domestic and global market dynamics, and domestic policy changes have been observed to affect the maize industry in Mexico (Eakin *et al.*, 2018). It has also been anticipated that China will see negative food stock returns because of rising global temperatures (Hong *et al.*, 2019).

6.4 Measures to Ensure Food Security

6.4.1 Adaptive measures for increasing crop production

To maintain sustainable production, it is crucial to evaluate how climatic variability affects agricultural output and create adaptation techniques and technology to counteract the negative effects. The negative impact of climate change can be diminished by using resilient and climate-smart practices in agriculture that provide sustainable outputs and food security (Zafar *et al.*, 2018; Ahmad *et al.*, 2019). Crop-specific management techniques like changing sowing times, crop rotation, intercropping, crop diversification, and intensification will prove beneficial adaptation practices (Hisano *et al.*, 2018; Degani *et al.*, 2019). Altering input utilization, adjusting fertilizer rates to increase produce quantity and quality, and introducing cultivars that are resistant to drought are some of the critical adaptation strategies for sustainable agriculture production. Crops with climatic robust genetic features should also be introduced in uncertain environmental conditions to maintain sustainable crop productivity (Bailey-Serres *et al.*, 2019; Raman *et al.*, 2019). Biochar, a porous solid material made from biomass carbonization at relatively low temperatures (400°C) and no oxygen, is regarded as a key tool in the mitigation of climate change due to its role in reducing GHG (greenhouse gases) emissions from soil and sequestering carbon in more stable forms of materials (Joseph *et al.*, 2021). In more recent reports, biostimulants have been listed as one of several significant and potential mitigation strategies that might help plants acquire resilience to a variety of environmental abiotic stresses brought on by rapidly changing climatic conditions (Bhupenchandra *et al.*, 2022). Biostimulants include botanical extracts like seaweed and algae extracts, amino acids, protein hydrolysates, vitamins, antioxidants, and so on (El Boukhari *et al.*, 2020).

6.4.2 Managing genetic resources

Significant communal investments are needed to protect, describe, and value genetic resources. The goals of the breeding program also need to be modified. Breeding programs need to get started well in advance because they require time to succeed. In some areas, it can be required to introduce new breeds and species. Improved ex situ and in situ conservation efforts for domesticated species, their wild relatives, and other wild genetic resources essential for agriculture and food production, as well as policies that support their sustainable use, are therefore imperative (Gitz *et al.*, 2016). The success of plant explant propagation options and cryopreservation is frequently influenced by their developmental stage (Goveia *et al.*, 2004), freezing/chilling (Ballesteros and Pence, 2017), and vigor, and/or tolerance to desiccation (Pammenter and Berjak, 2014). The habitat, in which the donor plants grow, often influences these characteristics (Gutterman, 2000). When adopting modern molecular breeding techniques including genome editing, the preservation of plant genetic resources and the accessibility of their omics data are essential to improve the crop varieties. It may be necessary to protect and use these genetic resources to generate crop types that are climate resilient and to guarantee global food and nutrition security.

6.4.3 Forest protection and renewal

A thorough understanding of the natural carbon cycles in many ecosystems is necessary to ameliorate climate change (Pan *et al.*, 2011). In this regard, forests are of special significance because they provide a significant terrestrial carbon sink and storage even in highly managed areas of the world (Pugh *et al.*, 2019). The degree and longevity of forest carbon stores are significantly influenced by the type and intensity of management (Mikoláš *et al.*, 2021). Forest ecosystems that are healthy and diverse are more adaptable because they can handle stress, repair damage, and change on their own. Ecosystems that have efficient biological processes are more resilient to adverse biotic and abiotic influences than stressed-out ecosystems. Best practices include minimal impact logging in production forests, limiting collection of non-wood forest products or animal grazing at sustainable levels, integrated pest management, disease control, forest fire management, and forest law enforcement. Restoration of degraded forests to a lush green state and reinstating the ecosystem's overall functioning is a crucial strategy for developing resilience (Gitz *et al.*, 2016).

6.4.4 Diversified and integrated fishing systems

Shifting to aquatic species and aquacultural practices, or regions that are less vulnerable or robust to changing environments and resources, may also benefit aquaculture producers (Lim-Camacho *et al.*, 2015). Because it mixes finfish farming with other species, the Integrated Multi-Trophic Aquaculture (IMTA) system is more sustainable, environment friendly, and commercially rewarding, as well as more robust to changing climate, as compared to monoculture (van Osch *et al.*, 2019). Furthermore, one of the most potential avenues for adaptation is developing or shifting to aquaculture species that are more tolerant to climate change (De Young *et al.*, 2012). Sae-Lim *et al.* (2017), for example, stated that species selection based on feed efficiency and establishing a breeding aim that minimizes GHG emissions will be critical in a changing environment. Destructive fishing (e.g. fishing with poisons and toxic compounds) and land-based pollution sources (e.g. contemporary farming and municipal overflow employing pesticides and insecticides) should be restricted. The consequences of climate change on fisheries and aquaculture prospects exhibit significant geographical variability (Merino *et al.*, 2012). Significant political and financial investment in aquaculture in suitable climatic and environmental zones will be necessary if it is to contribute more to food security and meet the growing demand for fish and seafood products.

6.4.5 Increasing investments in research and development

Researchers are deeply concerned with the approaches for adaptation and mitigation in sectoral and geographical contexts. Oil-producing countries may be unable to subsidize food imports in the short term due to the fall in global oil prices and, subsequently, their earnings from oil exports. As a result, these countries must strengthen the agricultural value chain for export. It is possible to enhance income by correcting exchange rate misalignment and adverse trade conditions through R&D and adding value to their food items (Abbass *et al.*, 2022). Many countries have created incentives for research and innovation at the governmental level, including renewable portfolio standards (popular in North America) and feed-in tariffs (popular in Europe) (Alizada, 2018; Zhou *et al.*, 2019). In the food industry, in addition to the expansion of the plant-based burger market, urban food production ("vertical farming") has been researched recently to reduce emissions related to food storage and transportation from fields to metropolitan cities (Wilson, 2018). Marker-assisted selection is crucial for improving quantitative traits and accelerating the breeding process. Crop genomics advancement provides information that can be used to discover DNA markers, which is helpful for marker-assisted breeding programs. Globally, genomics, bioinformatics, and metabolomics resources are crucial for improving crops.

6.4.6 Government laws and policies to reduce the risks

To enable, support, and complement the economic and technical choices for food producers to adapt to climate change, appropriate policies and institutions at the national and international levels are required, especially for small-scale food producers. This necessitates proper institutions and regulations that encourage and promote group development and execution of adaptation actions, whether in a sector (such as along the food chain) or a space (such as a watershed or forest), including the expansion and modification of storage facilities. To achieve this, decision making processes can be made more participatory and transparent, and tools can be provided to encourage activities that will ultimately enhance public and collective adaptation. For the management of natural resources, this is especially crucial (Place *et al.*, 2017). International assistance helps developing nations adapt to a changing climate and lessen its effects, preventing food crises. For instance, USAID is "integrating risk management, climate change adaptation, and mitigation across all planning, programming, and decision-making." The USDA will put into practice a climate-smart agriculture policy that includes techniques for reducing climate change and ensuring sustainable production. Producers will be encouraged to adopt climate-smart and sustainable agricultural practices with financial incentives and technical support. Regional climate hubs concentrate on regional initiatives, such as drought adaptation in western states (Grossman, 2022). Important initiatives which can be taken by the policymakers consist of the promotion of climate-smart technologies and maximizing the synergies between interventions; the strategic design of land-use options based on agroecological analysis and farmer typologies; the provision of value-added weather services to local farmers to manage variability; the promotion of weather-based insurance options for managing climate risk; the facilitation of community partnerships for knowledge sharing; and capacity building for climate change adaptation (Herrero *et al.*, 2015). As a result of climate change, there will likely be an increase in the worldwide movement of genetic resources due to the need for climate-resilient crops, trees, cattle, and aquatic species. For the benefit of plant genetic resources, the International Treaty on Plant Genetic Resources for Food and Agriculture has provisions for genetic resource conservation, information sharing, transfer of technology, capacity development, and exchange of benefits. Additionally, it will be more important than ever to cooperate internationally to control and prevent transnational diseases and pests. An effective tool that can be applied is the International Plant Protection Convention. It promotes actions to stop the spread of pests on plants and plant products and outlines methods for controlling plant pests with the least amount of interference to the movement of people and goods across international borders (Gitz *et al.*, 2016). The Climate Change, Agriculture, and Food Security (CCAFS) Program of the CGIAR (Consultative Group on International Agricultural Research) is fostering climate-resilient and adaptable food and agricultural systems in several nations, including India (Kumari *et al.*, 2020).

6.5 Conclusions

Climate change's consequences on food security, which are anticipated to worsen over time, differ between countries and world areas. Indeed, the effects are projected to be greater in developing countries whose economies rely heavily on the performance of rainfed agriculture. The need is for the authorities to develop stronger laws and policies and generate awareness among farmers and producers. We need to expand genetic, genomic, and gene-editing skills to improve fruit and vegetable crops and to grow plants under more realistic climatic projections to understand their resilience. With the evidence and experience of the adverse effects of climate-related shocks, early preparedness can reduce the uncertainties and potential adverse effects of climate change in vulnerable hotspots on Earth. In view of global climate change, this multimodal strategy that integrates physiology, genetics, and climate modeling is critical to producing a sustainable future food supply.

Acknowledgments

This work was supported by the core grant of the National Institute of Plant Genome Research and Core Research Grant (CRG) from the SERB Department of Science and Technology (CRG/2022/001178) to AP. HC and SS acknowledge the University Grant Commission, the Government of India for Junior and Senior Research Fellowships, respectively. The authors are thankful to DBT-eLibrary Consortium (DeLCON) for providing access to e-resources.

Conflict of Interest

Authors declare no conflict of interest.

References

Abbass, K., Qasim, M.Z., Song, H., Murshed, M., Mahmood, H. *et al.* (2022) A review of the global climate change impacts, adaptation, and sustainable mitigation measures. *Environmental Science and Pollution Research International* 29(28), 42539–42559. DOI: 10.1007/s11356-022-19718-6.

Ahmad, S., Abbas, G., Ahmed, M., Fatima, Z., Anjum, M.A. *et al.* (2019) Climate warming and management impact on the change of phenology of the rice-wheat cropping system in Punjab, Pakistan. *Field Crops Research* 230, 46–61. DOI: 10.1016/j.fcr.2018.10.008.

Alizada, K. (2018) Rethinking the diffusion of renewable energy policies: a global assessment of feed-in tariffs and renewable portfolio standards. *Energy Research & Social Science* 44, 346–361. DOI: 10.1016/j.erss.2018.05.033.

Allison, E.H., Perry, A.L., Badjeck, M., Neil Adger, W., Brown, K. *et al.* (2009) Vulnerability of national economies to the impacts of climate change on fisheries. *Fish and Fisheries* 10(2), 173–196. DOI: 10.1111/j.1467-2979.2008.00310.x.

Arunrat, N., Pumijumnong, N. and Hatano, R. (2018) Predicting local-scale impact of climate change on rice yield and soil organic carbon sequestration: a case study in Roi Et Province, Northeast Thailand. *Agricultural Systems* 164, 58–70. DOI: 10.1016/j.agsy.2018.04.001.

Asseng, S., Martre, P., Maiorano, A., Rötter, R.P., O'Leary, G.J. *et al.* (2019) Climate change impact and adaptation for wheat protein. *Global Change Biology* 25(1), 155–173. DOI: 10.1111/gcb.14481.

Bailey-Serres, J., Parker, J.E., Ainsworth, E.A., Oldroyd, G.E.D. and Schroeder, J.I. (2019) Genetic strategies for improving crop yields. *Nature* 575(7781), 109–118. DOI: 10.1038/s41586-019-1679-0.

Balamurugan, P. (2018) Farmers clubs and livelihood opportunities through livestock development: an analysis in select villages of Nagapattinam district, Tamil nadu. *Productivity* 59(1), 56–61. DOI: 10.32381/PROD.2018.59.01.6.

Ballesteros, D. and Pence, V.C. (2017) Survival and death of seeds during liquid nitrogen storage: a case study on seeds with short lifespans. *Cryo Letters* 38(4), 278–289.

Belkin, I.M. (2009) Rapid warming of large marine ecosystems. *Progress in Oceanography* 81(1–4), 207–213. DOI: 10.1016/j.pocean.2009.04.011.

Bett, B., Lindahl, J. and Delia, G. (2019) Climate change and infectious livestock diseases: the case of rift valley fever and tick-borne diseases. In: Rosenstock, T., Nowak, A. and Girvetz, E. (eds) *The Climate-Smart Agriculture Papers: Investigating the Business of a Productive, Resilient and Low Emission Future*. Springer, Cham, Switzerland, pp. 29–37.

Bhupenchandra, I., Chongtham, S.K., Devi, E.L., Choudhary, A.K, *et al.* (2022) Role of biostimulants in mitigating the effects of climate change on crop performance. *Frontiers in Plant Science* 13, 967665. DOI: 10.3389/fpls.2022.967665.

Chen, C.C. and McCarl, B.A. (2001) An investigation of the relationship between pesticide usage and climate change. *Climatic Change* 50(4), 475–487. DOI: 10.1023/A:1010655503471.

Cheung, W.W.L., Pinnegar, J., Merino, G., Jones, M.C. and Barange, M. (2012) Review of climate change impacts on marine fisheries in the UK and Ireland. *Aquatic Conservation* 22(3), 368–388. DOI: 10.1002/aqc.2248.

De Young, C., Soto, D., Bahri, T., Brown, D., De Young, C. *et al.* (2012) Building resilience for adaptation to climate change in the fisheries and aquaculture sector. *Building Resilience for Adaptation to Climate Change in the Agriculture Sector* 23(103).

Degani, E., Leigh, S.G., Barber, H.M., Jones, H.E., Lukac, M. *et al.* (2019) Crop rotations in a climate change scenario: short-term effects of crop diversity on resilience and ecosystem service provision under drought. *Agriculture, Ecosystems & Environment* 285, 106625. DOI: 10.1016/j.agee.2019.106625.

Dukes, J.S., Classen, A.T., Wan, S. and Langley, J.A. (2014) Using results from global change experiments to inform land model development and calibration. *The New Phytologist* 204(4), 744–746. DOI: 10.1111/nph.13083.

Dureuil, M., Boerder, K., Burnett, K.A., Froese, R. and Worm, B. (2018) Elevated trawling inside protected areas undermines conservation outcomes in a global fishing hot spot. *Science* 362(6421), 1403–1407. DOI: 10.1126/science.aau0561.

Eakin, H., Sweeney, S., Lerner, A.M., Appendini, K., Perales, H. *et al.* (2018) Agricultural change and resilience: agricultural policy, climate trends and market integration in the Mexican maize system. *Anthropocene* 23, 43–52. DOI: 10.1016/j.ancene.2018.08.002.

El Boukhari, M.E.M., Barakate, M., Bouhia, Y. and Lyamlouli, K. (2020) Trends in seaweed extract based biostimulants: manufacturing process and beneficial effect on soil-plant systems. *Plants* 9(3), 359. DOI: 10.3390/plants9030359.

Fan, X., Miao, C., Duan, Q., Shen, C. and Wu, Y. (2021) Future climate change hotspots under different 21st century warming scenarios. *Earth's Future* 9(6), e2021EF002027. DOI: 10.1029/2021EF002027.

FAO (1996) *World Food Summit Plan of Action*. FAO Corporate Document Repository, Rome, Italy.

FAO (2017) Climate change: implications for food safety. *African Crop Science Journal* 12, 1–30.

Fisher, B., Turner, R.K. and Morling, P. (2009) Defining and classifying ecosystem services for decision making. *Ecological Economics* 68(3), 643–653. DOI: 10.1016/j.ecolecon.2008.09.014.

Georgia Department of Agriculture (2018) Press release—Hurricane Michael Devastates Georgia's Agricultural Industry. WWW Document. Available at: http://agr.georgia.gov/hurricane-michael-devastates-georgias-agricultural-industry.aspx (accessed 28 July 2024).
Giorgi, F. (2006) Climate change hot-spots. *Geophysical Research Letters* 33(8), L08707. DOI: 10.1029/2006GL025734.
Gitz, V., Meybeck, A., Lipper, L., Young, C.D. and Braatz, S. (2016) Climate change and food security: risks and responses. *Food and Agriculture Organization of the United Nations (FAO) Report* 110(2), 3–36.
Goveia, M., Kioko, J.I. and Berjak, P. (2004) Developmental status is a critical factor in the selection of excised recalcitrant axes as explants for cryopreservation: a study on *Trichilia dregeana* Sond. *Seed Science Research* 14(2), 241–248. DOI: 10.1079/SSR2004173.
Grossman, M.R. (2022) Climate change impact on agrarian law: legal aspects of food security in the united states. *The American Journal of Comparative Law* 70(Supplement_1), i118–i154. DOI: 10.1093/ajcl/avac014.
Gründemann, G.J., van de Giesen, N., Brunner, L. and van der Ent, R. (2022) Rarest rainfall events will see the greatest relative increase in magnitude under future climate change. *Communications Earth & Environment* 3(1), 235. DOI: 10.1038/s43247-022-00558-8.
Gutterman, Y. (2000) Environmental factors and survival strategies of annual plant species in the Negev Desert, Israel. *Plant Species Biology* 15(2), 113–125. DOI: 10.1046/j.1442-1984.2000.00032.x.
Habeeb, A.A., Gad, A.E. and Atta, M.A. (2018) Temperature-humidity indices as indicators to heat stress of climatic conditions with relation to production and reproduction of farm animals. *International Journal of Biotechnology and Recent Advances* 1(1), 35–50. DOI: 10.18689/ijbr-1000107.
Hall, T.J., Milson, J. and Hall, C. (2020) Pasture recovery, land condition and some other observations after the monsoon flooding, chill event in north-west Queensland in January–March 2019. Available at: https://era.daf.qld.gov.au/id/eprint/7443/ (accessed 3 August 2024).
Harris, I., Osborn, T.J., Jones, P. and Lister, D. (2020) Version 4 of the CRU TS monthly high-resolution gridded multivariate climate dataset. *Scientific Data* 7(1), 109. DOI: 10.1038/s41597-020-0453-3.
Hasegawa, T., Sakurai, G., Fujimori, S., Takahashi, K., Hijioka, Y. *et al.* (2021) Extreme climate events increase risk of global food insecurity and adaptation needs. *Nature Food* 2(8), 587–595. DOI: 10.1038/s43016-021-00335-4.
Hellmann, J.J., Byers, J.E., Bierwagen, B.G. and Dukes, J.S. (2008) Five potential consequences of climate change for invasive species. *Conservation Biology* 22(3), 534–543. DOI: 10.1111/j.1523-1739.2008.00951.x.
Herrero, M., Wirsenius, S., Henderson, B., Rigolot, C., Thornton, P. *et al.* (2015) Livestock and the environment: what have we learned in the past decade? *Annual Review of Environment and Resources* 40(1), 177–202. DOI: 10.1146/annurev-environ-031113-093503.
Hisano, M., Searle, E.B. and Chen, H.Y.H. (2018) Biodiversity as a solution to mitigate climate change impacts on the functioning of forest ecosystems. *Biological Reviews of the Cambridge Philosophical Society* 93(1), 439–456. DOI: 10.1111/brv.12351.
Hong, H., Li, F.W. and Xu, J. (2019) Climate risks and market efficiency. *Journal of Econometrics* 208(1), 265–281. DOI: 10.1016/j.jeconom.2018.09.015.
Huynh, P.T., Le, N.D., Le, S.T. and Nguyen, H.X. (2021) Vulnerability of fishery-based livelihoods to climate change in coastal communities in central Vietnam. *Coastal Management* 49(3), 275–292. DOI: 10.1080/08920753.2021.1899927.
Johnson, N. (2017) Global models and the land-energy water nexus: status, challenges and opportunities. *Global Environmental Change* 11(11), 2223.
Joseph, S., Cowie, A.L., Van Zwieten, L., Bolan, N., Budai, A. *et al.* (2021) How biochar works, and when it doesn't: a review of mechanisms controlling soil and plant responses to biochar. *GCB Bioenergy* 13(11), 1731–1764. DOI: 10.1111/gcbb.12885.
Kotir, J.H. (2011) Climate change and variability in Sub-Saharan Africa: a review of current and future trends and impacts on agriculture and food security. *Environment, Development and Sustainability* 13(3), 587–605. DOI: 10.1007/s10668-010-9278-0.
Kumari, S., Yadav, B.S. and Yadav, R.B. (2020) Synthesis and modification approaches for starch nanoparticles for their emerging food industrial applications: a review. *Food Research International* 128, 108765. DOI: 10.1016/j.foodres.2019.108765.
Kumar, R. and Raj Gautam, H. (2014) Climate change and its impact on agricultural productivity in India. *Journal of Climatology & Weather Forecasting* 2(1), 1–3. DOI: 10.4172/2332-2594.1000109.

Lam, V.W., Allison, E.H., Bell, J.D., Blythe, J., Cheung, W.W. *et al.* (2020) Climate change, tropical fisheries and prospects for sustainable development. *Nature Reviews Earth & Environment* 1(9), 440–454. DOI: 10.1038/s43017-020-0071-9.

Lebel, P., Sriyasak, P., Kallayanamitra, C., Duangsuwan, C. and Lebel, L. (2016) Learning about climate-related risks: decisions of Northern Thailand fish farmers in a role-playing simulation game. *Regional Environmental Change* 16(5), 1481–1494. DOI: 10.1007/s10113-015-0880-4.

Leisner, C.P. (2020) Climate change impacts on food security-focus on perennial cropping systems and nutritional value. *Plant Science* 293, 110412. DOI: 10.1016/j.plantsci.2020.110412.

Li, Y., Ye, W., Wang, M. and Yan, X. (2009) Climate change and drought: a risk assessment of crop-yield impacts. *Climate Research* 39(1), 31–46. DOI: 10.3354/cr00797.

Lim-Camacho, L., Hobday, A.J., Bustamante, R.H., Farmery, A., Fleming, A. *et al.* (2015) Facing the wave of change: stakeholder perspectives on climate adaptation for Australian seafood supply chains. *Regional Environmental Change* 15(4), 595–606. DOI: 10.1007/s10113-014-0670-4.

Lobell, D.B. and Burke, M.B. (2010) On the use of statistical models to predict crop yield responses to climate change. *Agricultural and Forest Meteorology* 150(11), 1443–1452. DOI: 10.1016/j.agrformet.2010.07.008.

Merino, G., Barange, M., Blanchard, J.L., Harle, J., Holmes, R. *et al.* (2012) Can marine fisheries and aquaculture meet fish demand from a growing human population in a changing climate? *Global Environmental Change* 22(4), 795–806. DOI: 10.1016/j.gloenvcha.2012.03.003.

Mikoláš, M., Svitok, M., Bače, R., Meigs, G.W., Keeton, W.S. *et al.* (2021) Natural disturbance impacts on trade-offs and co-benefits of forest biodiversity and carbon. *Proceedings. Biological Sciences* 288(1961), 20211631. DOI: 10.1098/rspb.2021.1631.

Myers, S.S., Smith, M.R., Guth, S., Golden, C.D., Vaitla, B. *et al.* (2017) Climate change and global food systems: potential impacts on food security and undernutrition. *Annual Review of Public Health* 38, 259–277. DOI: 10.1146/annurev-publhealth-031816-044356.

Natale, F., Carvalho, N., Harrop, M., Guillen, J. and Frangoudes, K. (2013) Identifying fisheries dependent communities in EU coastal areas. *Marine Policy* 42, 245–252. DOI: 10.1016/j.marpol.2013.03.018.

Nissen, K.M. and Ulbrich, U. (2017) Increasing frequencies and changing characteristics of heavy precipitation events threatening infrastructure in Europe under climate change. *Natural Hazards and Earth System Sciences* 17(7), 1177–1190. DOI: 10.5194/nhess-17-1177-2017.

Obermeier, W.A., Lehnert, L.W., Kammann, C.I., Müller, C., Grünhage, L. *et al.* (2017) Reduced CO_2 fertilization effect in temperate C_3 grasslands under more extreme weather conditions. *Nature Climate Change* 7(2), 137–141. DOI: 10.1038/nclimate3191.

Pammenter, N.W. and Berjak, P. (2014) Physiology of desiccation-sensitive (recalcitrant) seeds and the implications for cryopreservation. *International Journal of Plant Sciences* 175(1), 21–28. DOI: 10.1086/673302.

Pan, Y., Birdsey, R.A., Fang, J., Houghton, R., Kauppi, P.E. *et al.* (2011) A large and persistent carbon sink in the world's forests. *Science (New York, N.Y.)* 333(6045), 988–993. DOI: 10.1126/science.1201609.

Parry, M., Rosenzweig, C. and Livermore, M. (2005) Climate change, global food supply and risk of hunger. *Philosophical Transactions of the Royal Society B* 360(1463), 2125–2138. DOI: 10.1098/rstb.2005.1751.

Place, F., Meybeck, A., Colette, L., de, C., Gitz, V. *et al.* (2017) Food security and sustainable resource use? what are the resource challenges to food security. *Pharmacology* 23, 43–52.

Porter, J.R., Xie, L., Challinor, A.J., Cochrane, K., Howden, S.M. *et al.* (2014) Food security and food production systems. In: Mastrandrea, M.D. (ed.) *Climate Change 2014: Impacts, Adaptation, and Vulnerability. Part A: Global and Sectoral Aspects. Contribution of Working Group II to the Fifth Assessment Report of the Intergovernmental Panel on Climate Change*. Cambridge University Press, Cambridge, UK, and New York, pp. 485–533.

Powell, J.P. and Reinhard, S. (2016) Measuring the effects of extreme weather events on yields. *Weather and Climate Extremes* 12, 69–79. DOI: 10.1016/j.wace.2016.02.003.

Pugh, T.A., Lindeskog, M., Smith, B., Poulter, B., Arneth, A. *et al.* (2019) Role of forest regrowth in global carbon sink dynamics. *Proceedings of the National Academy of Sciences* 116(10), 4382–4387. DOI: 10.1073/pnas.1810512116.

Raman, H., Uppal, R.K. and Raman, R. (2019) Genetic solutions to improve resilience of canola to climate change. In: Kole, C. (ed.) *Genomic Designing of Climate-Smart Oilseed Crops*. Springer, Cham, Switzerland, pp. 75–131.

Rezaei, E.E., Siebert, S., Hüging, H. and Ewert, F. (2018) Climate change effect on wheat phenology depends on cultivar change. *Scientific Reports* 8(1), 4891. DOI: 10.1038/s41598-018-23101-2.

Richardson, K.J., Lewis, K.H., Krishnamurthy, P.K., Kent, C., Wiltshire, A.J. *et al.* (2018) Food security outcomes under a changing climate: impacts of mitigation and adaptation on vulnerability to food insecurity. *Climatic Change* 147(1–2), 327–341. DOI: 10.1007/s10584-018-2137-y.

Sae-Lim, P., Kause, A., Mulder, H.A. and Olesen, I. (2017) Breeding and genetics symposium: climate change and selective breeding in aquaculture. *Journal of Animal Science* 95(4), 1801–1812. DOI: 10.2527/jas.2016.1066.

Serdeczny, O., Adams, S., Baarsch, F., Coumou, D., Robinson, A. *et al.* (2017) Climate change impacts in Sub-Saharan Africa: from physical changes to their social repercussions. *Regional Environmental Change* 17(6), 1585–1600. DOI: 10.1007/s10113-015-0910-2.

Servili, A., Canario, A.V.M., Mouchel, O. and Muñoz-Cueto, J.A. (2020) Climate change impacts on fish reproduction are mediated at multiple levels of the brain-pituitary-gonad axis. *General and Comparative Endocrinology* 291, 113439. DOI: 10.1016/j.ygcen.2020.113439.

Singh, P., Mondal, S. and Singh, R.L. (2020) Introduction. In: Mondal, S. and Singh, R.L. (eds) *Advances in Animal Genomics*. Elsevier, London, pp. 1–12. DOI: 10.1016/B978-0-12-820595-2.00001-1.

Singh, P., Singh, P.K. and Singh, R.L. (2021) Transgenic animals. In: Mondal, S. and Singh, R.L. (eds) *Emerging Issues in Climate Smart Livestock Production*. Elsevier, London, pp. 171–208. DOI: 10.1016/B978-0-12-822265-2.00001-6.

Singh, R.L., Vishwakarma, S.P. and Singh, P. (2014) Vitamins and minerals: roles and plant sources. In: Prakash, D. (ed.) *Phytochemicals of Nutraceutical Importance and Their Role in Human Health*. CAB International, Wallingford, UK, pp. 248–265.

Stevanović, M., Popp, A., Lotze-Campen, H., Dietrich, J.P., Müller, C. *et al.* (2016) The impact of high-end climate change on agricultural welfare. *Science Advances* 2(8), e1501452. DOI: 10.1126/sciadv.1501452.

Sumaila, U.R., Palacios-Abrantes, J. and Cheung, W. (2020) Climate change, shifting threat points, and the management of transboundary fish stocks. *Ecology and Society* 25(4), 40. DOI: 10.5751/ES-11660-250440.

Teh, L.C. and Sumaila, U.R. (2013) Contribution of marine fisheries to worldwide employment. *Fish and Fisheries* 14(1), 77–88. DOI: 10.1111/j.1467-2979.2011.00450.x.

van Osch, S., Hynes, S., Freeman, S. and O'Higgins, T. (2019) Estimating the public's preferences for sustainable aquaculture: a country comparison. *Sustainability* 11(3), 569. DOI: 10.3390/su11030569.

Waha, K., Müller, C., Bondeau, A., Dietrich, J.P., Kurukulasuriya, P. *et al.* (2013) Adaptation to climate change through the choice of cropping system and sowing date in Sub-Saharan Africa. *Global Environmental Change* 23(1), 130–143. DOI: 10.1016/j.gloenvcha.2012.11.001.

Wilson, C. (2018) Disruptive low-carbon innovations. *Energy Research & Social Science* 37, 216–223. DOI: 10.1016/j.erss.2017.10.053.

Wu, J., van der Linden, L., Lasslop, G., Carvalhais, N., Pilegaard, K. *et al.* (2012) Effects of climate variability and functional changes on the interannual variation of the carbon balance in a temperate deciduous forest. *Biogeosciences* 9(1), 13–28. DOI: 10.5194/bg-9-13-2012.

Xie, W., Huang, J., Wang, J., Cui, Q., Robertson, R. *et al.* (2020) Climate change impacts on china's agriculture: the responses from market and trade. *China Economic Review* 62, 101256. DOI: 10.1016/j.chieco.2018.11.007.

Yu, H., Zhang, Q., Sun, P. and Song, C. (2018) Impact of droughts on winter wheat yield in different growth stages during 2001–2016 in Eastern China. *International Journal of Disaster Risk Science* 9(3), 376–391. DOI: 10.1007/s13753-018-0187-4.

Zafar, S.A., Hameed, A., Nawaz, M.A., Ma, W., Noor, M.A. *et al.* (2018) Mechanisms and molecular approaches for heat tolerance in rice (Oryza Sativa L.) under climate change scenario. *Journal of Integrative Agriculture* 17(4), 726–738. DOI: 10.1016/S2095-3119(17)61718-0.

Zhou, Q., Leng, G., Su, J. and Ren, Y. (2019) Comparison of urbanization and climate change impacts on urban flood volumes: importance of urban planning and drainage adaptation. *Science of The Total Environment* 658, 24–33. DOI: 10.1016/j.scitotenv.2018.12.184.

Zhu, X., Zhang, S., Liu, T. and Liu, Y. (2021) Impacts of heat and drought on gross primary productivity in China. *Remote Sensing* 13(3), 378. DOI: 10.3390/rs13030378.

7 Food Security Through Sustainable Livestock Production

Bikash Kanti Biswas[1]* and Sudip Kumar Das[2]
[1]DREF, W.B. University of Animal & Fishery Sciences, Kolkata, India; [2]W.B. University of Animal & Fishery Sciences, Kolkata, India

Abstract
Food security does not only mean the availability of food but also accessibility to food. Food availability may be secured through the rise of food production, and accessibility to food can be achieved through employment generation rather than income generation. This can be assured through livestock production. To meet the demands for food, with the increasing human population, the production of food items (i.e. grains, vegetables, and fruits) mainly animal produce, must be amplified. Farmers face many hurdles in escalating the production of milk, meat, and eggs. The animal husbandry production system can be managed better than the cereal grain production systems. So, for the food security of the human population, by all means, there is no alternative to livestock production. With the changing world, the livestock production system is to be developed sustainably for food security and to keep hunger away.

Keywords: Global warming, Sustainable production, Livestock production Arthropod-borne Disease, Heat resistance breed

7.1 Introduction

The growth of the world's human population is around 1.1% annually, or about 83 million per year. The global population was 1 billion in 1800 and increased to 7.9 billion in 2020. The UN has projected that the population will reach 8.6 billion by mid-2030, 9.8 billion by mid-2050, and 11.2 billion by 2100. With the increase in human population, the production of food items (grains, vegetables, fruits, animal produce) must be augmented to meet food and nutrition demands. At the same time, it needs to maintain the purchasing capacity for food security. Farmers face many hurdles in escalating the production of milk, meat, and eggs. To meet the demand, it needs to produce an additional 70% of food for the projected 2.1 billion extra people by 2050. The global meat demand will be around 0.2 billion tons per year by 2050, which is about double the present need. It will happen due to population growth and dietary changes (Tilman *et al.*, 2011).

The livestock sector is vital because it accounts for a minimum of one-third of the global agricultural GDP. Globally, the livestock sector is the largest user of land resources. Compared with other economic sectors, livestock is the fastest growing, particularly in the third world. The livestock sector gives employment to 8- to 80-year-old people and equally to women. Food security and nutrition (FSN) are very

*Corresponding author: bkbiswasmicro@rediffmail.com

DOI: 10.1079/9781800625037.0007

important in underdeveloped and developing countries. This sector can create a hunger-free world even if it reduces poverty by generating employment opportunities. Herbivore animals have the unique potency of converting low-value, inedible, less palatable, and less digestible protein foods into high-value nutrients and very palatable to human protein foods (Smith *et al.*, 2013). These high-value animal-source proteins are very important for children's and infants' growth as well as the development of the brain compared with plant-source proteins (Willett *et al.*, 2019). Livestock produce can meet the need for micronutrient deficiencies in women and children and provide a balanced diet to all.

The High-Level Panel of Experts of Food Security and Nutrition (HLPE) has published the report "*Sustainable Agricultural Development for Food Security and Nutrition: What Roles for Livestock*" at the Committee on World Food Security (CFS). The concept of sustainability is not only limited to environmental sustainability, but it also ensures nutrition and food security for all, now and in the future, which is key to long-lasting human development and well-being. There are links between livestock production and plant-based production systems and dependency on each other. To augment or sustain livestock production, it needs to have sustainable feed production (i.e. cereal and fodder), which may hamper adverse climatic conditions. Animal-source food includes livestock, aquatic animals, wild animals, and insects. The production of all types of animal sources of food may change with the change of environment. There may need to be more than the existing farming systems to sustain livestock production in the long run (Singh *et al.*, 2020, 2021). The shift in farming systems will vary a lot between and within countries and regions in the world. The farmers are facing the constraints and challenges for sustaining animal-source food production, which may vary with respect to climatic zones, type of animals, and type of farming system adopted between and within countries.

To overcome the constraints, some steps may be adopted. In the farming system, genetically modified animals can be reared. Those animals can withstand climate change (i.e. high temperatures, changes in feed ingredients, and patterns of farming systems). Disease-resistant animals can also be raised in suitable areas (Eisler *et al.*, 2014). Similarly, high temperature, high precipitation, and even drought-resistant varieties of cereal and fodder should be chosen for cultivation in different climatic zones to keep sustainable production. Research and development are needed to keep sustainable livestock production through better farming practices that maintain natural resources and protect existing ecosystems (Dumont *et al.*, 2019; Twine, 2021).

7.1.1 Population

The United Nations estimated that the world's population had already exceeded eight billion in mid-November 2022 (Kight and Lysik, 2022). The human population's growth has continued over time, and it has experienced the Great Famine of 1315–1317. At the end of the Black Death in 1350, the population reached nearly 0.37 billion (Biraben, 1980). The projected growth rate of the human population will reach 141 million per year during 2040–2045, where it is currently 140 million per year (2015–2020). Again, the growth will decline to 126 million annually by 2100.

The UN Department of Economics and Social Affairs foresees that the world population will attain between 9 and 10 billion by 2050, and they confidently forecast that the population will reach 10–12 billion at the end of the 21st century when the growth rate will reach zero. The human population will start to decline in the second half of the 21st century, as predicted by another group of demographers. FAO has predicted that the world's population will reach 7.6–10 billion between 2017 and 2067. At the same time, the demand for milk is likely to increase globally by 48% between 2005 and 2050 (Alexandratos and Bruinsma, 2012). With the growth of people's incomes, the choice of diets typically will shift from those containing mostly grains to those comprising a greater proportion of animal products (i.e. meat, dairy, and eggs) (Tilman *et al.*, 2011; Kastner *et al.*, 2012).

Meeting the food demands from animal sources for such a growing global population is a great challenge. Sustainable animal-sourced food production, compared with the need, will be a challenging task with climate change and other constraints.

The constraints are:

- Change of climate/global warming
- Shortage of grazing land
- Scarcity of good quality fodder
- Competition between man and animal for grains
- Animal health problems
- Lack of adaptive animal breeds in adverse weather

7.1.2 Change of climate/global warming

Global warming does not only mean the rise of temperature, but it refers to climate change. Other components of the atmosphere (i.e. air pressure, humidity, wind speed and direction, evapotranspiration, precipitation, condensation, cloud cover, and radiation) can also be changed—even flood, storms, flood and monsoon duration and pattern change. The effect of climate change on the production of crops and livestock and its influence on farming and rearing systems is more negative in nature (Porter *et al.*, 2014). The effect of climate change is not only changes in temperature, rain, or other components, but it also results in erratic weather. The date and duration of monsoon, rainfall, snowfall, and so on, have changed. There is no rhythm in precipitation; instead, sudden heavy rain causes floods in unexpected times and geographical areas. Heat waves and drought are widespread nowadays. The result of this phenomenon is that the production of grass is at stake. Crops and vegetables are cultivated in this situation through irrigation from underground water. It causes environmental mainly imbalance because this loss cannot be reversed. High temperature or heat affects the critical factors of livestock production. The critical factors are the availability of drinking water and also water for irrigation for the cultivation of good quality feed and fodder. Heat stress affects animal production and reproduction performances. Stressed animals become immunocompromised and easily become infected by common commensal organisms. In high temperatures, the pests and arthropods are in habitation areas and spread in new domains. The arthropod-borne diseases also invade new areas. Water stagnation places increase with the increase in precipitation, and the breeding places of arthropods increase (Rojas-Downing *et al.*, 2017).

Climate change has a great impact on livestock productivity, particularly of mixed and extensive systems, which are strongly linked to impacts on rangelands and pastures. The quality of pasture and the quantity of grass and fodder production were mainly affected by the increase in temperature, amount, and duration of the rainfall and carbon dioxide. It also affects their biomass and nutritional quality (Morton, 2007; Steinfeld, 2010). Pastoral systems are vulnerable to climate change among all the livestock systems adopted (Dasgupta *et al.*, 2014). The impacts of climate change led to a rise in the cost of feeding, transport, and housing and scarcity of quality water and feedstuff. The infrastructures are destroyed by extreme events like floods, storms, and cyclones. Due to these reasons, animal production and profitability from animal husbandry become uncertain (Lopez-i-Gelats, 2014; Rivera-Ferre *et al.*, 2016). In the lack of grazing land, industrial or landless livestock farming systems face several risk factors because of the cost of grains and smooth availability. Similarly, mixed farming systems are also facing challenges. There are fewer adaptive varieties of grasses and animals in the changed climate (Nardone *et al.*, 2010). The impact of climate change on livestock production is negative. There has been a reduction in livestock production due to contaminated water, malnutrition due to poor-quality fodder, and increased susceptibility to disease (Rojas-Downing *et al.*, 2017).

7.1.3 Shortage of grazing land

In a grass-based dairy farming system, the production cost of milk per unit is less than in any other system. Because grass is used as a significant feed source, which is relatively cheaper, better-quality grasses need to be cultivated and the pastureland maintained properly. Conversion of agricultural land to human habitation is unavoidable because of population explosion. In present days as per modern lifestyle, common people like to live in modern cities and like to have all contemporary amenities. There is a continuous loss of agricultural land to establish high-rise dwelling houses, marketplaces, amusement parks, hobby farms, and so on. Competition between land users leads to negative pressure on farmers (Du *et al.*, 2014). The loss of fertile agricultural land during the last decades in western European areas has occurred due to the expansion of the built-up area, which has caused important environmental and socioeconomic impacts (Lambin *et al.*, 2001; Frumkin, 2016). United Nations Convention to Combat Desertification (UNCCD) has said that urbanization will cause the loss of about 1.6–3.3 million ha of prime agricultural land annually between 2000 and 2030. This loss of land has a great impact because the expansion takes place on prime agricultural lands.

One of the results of global warming is the rise of average global sea levels by 15–25 cm (6–10 in), or an average of 1–2 mm per year between 1901 and 2018 (IPCC, 2019). In the 2013–2022 decade, this rate accelerated to 4.62 mm per year (Farrell *et al.*, 2002). The effects of the rise of the sea level include high tide and more frequent storm-surge flooding, inhibition of primary production processes, increased coastal erosion, and more extensive coastal inundation, along with changes in surface water quality and groundwater. These can lead to a more significant loss of coastal habitats, agricultural land, and production, including human and animal life loss (TAR Climate Change, 2001). The impacts of coastal flooding are worsened by land-use changes such as deforestation or urbanization of low-lying coastal zones. Low-lying coastal regions are vulnerable to rising sea levels and struggle with coastal flooding, washing away land and altering the landscape (Wu, 2021).

As the availability of pastureland is questionable, it is required to maximize production efficiencies of grass and fodders, minimizing any adverse environmental impacts using a lesser area of land. Studies have shown that pasture-based dairy farming systems of milk production have a distinct advantage over high input-intensive systems (Shalloo *et al.*, 2004). It is also reported that improved animal welfare, increased product quality, and increased labor efficiency associated with greater global sustainability are far better in free-range grazing systems (O'Brien *et al.*, 2012). The profitability of the farm business will sustain and increase the efficiency of animal production in pasture-based systems (Shalloo *et al.*, 2011). Loss of pastureland occurs due to soil erosion. Heavy rainfalls, floods, earthquakes, landslides in hilly regions, changes in river flow lines, and sandstorms in desert land can lead to more soil erosion. The result of soil erosion is a major environmental threat that leads to a decrease in crop production, animal feed, and fodder production. It has a direct impact on animal husbandry (Gowda, 2018).

Drought also reduces total cultivable land. Very few pasturelands are under irrigation. Because the cultivation of fodder or grass by irrigation is not profitable, water sources for irrigation are not available everywhere. For example, the southwest US has been facing a long-term drought since 2020, which is a record for the last 1200 years. It happened due to very low precipitation (Mankin, 2021). With climate change, farmers are losing their interest in the cultivation of crops, fodder, and vegetables because the yield of crops and grains (i.e. corn, rice, and oats) is decreasing. This is happening due to the rise in temperature and the increase in carbon dioxide concentration (Gowda, 2018).

Wildfires are also a cause of loss of grassland. Climate change is the main cause of increasing wildfires. The woodfires become a challenge to farmlands, grasslands, and rangelands. This happens due to the rise in temperature and less precipitation. Sylvian pastureland can be a substitute for the loss of grassland.

7.1.4 Scarcity of good quality and quantity fodder

Out of the total diversity of vegetation on the earth's surface, about 40% is covered with grass (excluding Greenland and Antarctica) (White *et al.*, 2000). Lands are used for different types of cultivation, such as grains, oilseeds, vegetables, and orchards. The single most extensive use of land is for grazing purposes (Erb *et al.*, 2016). In need of production of cereal grains, the pasture or grazing lands are converted to agricultural land.

In developed countries like Europe, for the past 20 years, the production systems have been moved into more high-cost and intensive systems from grass-based systems. However, there is competition between animal feed, human food, and renewable energy systems (Taube *et al.*, 2014). However, the potential of the grazing system is robust and economically and environmentally friendly. The pasture-based dairy farming systems are desirable for keeping up the efficiency and sustainability of milk production, where the pasture-based system is currently a production role model. As the climate is changing rapidly, grassland systems will be role models due to their flexibility and adaptive management. Grass-based systems generally have lower production costs per unit product than indoor systems and rely more on high volume and lower margin per unit.

Some popular fodders are cultivated for animal feeding. They are accustomed to global warming. The following fodders will be the source of animal nutrition.

7.1.5 Common fodders cultivated for intensive dairy animal farming

7.1.5.1 Cowpea/karamani (Vigna unguiculata)

It is legume-type fodder and is used as a protein source for animals. It is an annual crop grown in warm temperature regions (i.e. tropics and subtropic regions). In tropic and subtropic regions, cowpeas can be grown in kharif, rabi, and summer seasons and in temperate regions only in summer seasons (Padulosil and Ng, 1997). This fodder can be used in green form and stored as hay. This hay can be used as a protein source in a mixture with maize. In dry periods, it needs irrigated water for good harvest (Langyintuo *et al.*, 2003).

7.1.5.2 Desmanthus-hedge/lucerne/velimasal (Desmanthus virgatus)

This legume is a perennial plant. It can be cultivated at any time of the year in tropical and subtropical regions. But it can be sown in the temperate zone in the summer. It needs irrigation in dry seasons/times. This fodder can be used as protein sources in green conditions and hay for livestock. This plant is cultivated in many countries as a chief forage crop and used for grazing. The plant is also used for the preparation of hay and silage and as a green manure. This legume is used as fodder for high-yielding dairy cows in some countries because it contains high protein and digestible fiber. It is also used as fodder for horses, sheep, and goats (Heuzé *et al.*, 2013).

7.1.5.3 Lucerne (Medicago sativa)

This legume is an excellent source of crude protein and is palatable to animals. That is why they are called the "Queen of Forages." It is also a perennial plant and lives about 4–8 years. Sometimes, it lives up to 20 years, depending upon variety and climate. It is a deep-rooted plant that can grow under various climatic zones from tropical to alpine. This plant increases the fertility of land by binding nitrogen in the soil. Good silage can be made from this fodder (Heuzé *et al.*, 2013). Lucerne plants are used for grazing and preparing hay and silage. Hay will be more nutritious if it is cut before the seed head fully matures, and the stem of the plant becomes tough and thick.

7.1.5.4 Stylo (Stylosanthes guianensis var. guianensis)

It is also a perennial but short-lived legume that can be cultivated in pasture. The plant is 0.6–1.8 m tall. This plant is resistant to drought, adapted to tropical climates, and can tolerate soils with low fertility, acidic soils, and soils with stagnated water. This plant is mainly suited for forage in subhumid tropical and subtropical areas in the dry season. Stylo is used for hay, cut-and-carry systems, and pasture (Cook *et al.*, 2005). Stylo can be cultivated at any time with irrigation facilities.

7.1.5.5 Maize-makkacholam (Zea mays L.)

It is an annual crop that can be grown in the tropics, subtropics, and warm temperature regions. This fodder can be grown throughout the year in a warm region under irrigation facilities. This fodder is used in green form for animal consumption and stored in hay for future uses.

This forage in the green stage, particularly when it contains the stalks, leaves, and ears, is a source of high energy and feed for ruminant livestock. Year-round maize production in the tropics may allow the continuous harvesting of green forage stored for future use.

7.1.5.6 Sorghum-cholam/jowar (Sorghum bicolor (L.) Moench)

It is a drought-resistant annual crop cultivated mainly for grains and fodder. It can grow suitably in a temperature range of 25–35°C and annual rainfall of 300–350 mm. For better yielding, it needs irrigation. Jowar is one of the best drought and heat-resistant kharif forage crops that are grown for silage and hay. This rapidly growing fodder gives heavy yields on a small amount of land. Jowar can be grown in well-drained heavy soil, though sandy loam soil is better for its cultivation. This crop contains such nutrients that alone are enough for maintaining ration for good health of all livestock (Matthys, 2018).

7.1.5.7 Napier grass (Cenchrus purpureus)

It is a monocot perennial grass. This plant is tall and forms robust bamboo-like clumps. It generally reproduces vegetatively through stolons that are horizontal shoots above the soil. The shoots extend from the parental plant to the offspring. Though it is a heterogeneous plant, the seeds are seldom developed to complete form (Khan *et al.*, 2007).

This plant needs less amount of nutrients and water for growth. It can be cultivated in semi-arid and even in arid regions. It is an essential fodder for dairy cattle in different parts of the world. This plant can prevent soil erosion and serve as a windbreak and firebreak. It also increases soil fertility (Farrell *et al.*, 2002).

7.1.5.8 Hybrid Napier-Cumbu Napier Ottupull/Bazra Napier hybrid

This grass grows in hot and humid climates. Loam soil is very good for the cultivation of hybrid Napier. This grass is very popular with dairy farmers due to its prolific yield. This perennial grass is widely distributed in tropical and subtropical regions. It is adopted in a wide range of agroecological conditions, and it has vigorous growth. Within 3 months of sowing, it becomes ready for harvesting and successive cutting can be done within 2 months. This grass should be mixed with legumes to make proper ration (Bhakar and Ram, 2019).

7.1.5.9 Guinea grass (Megathyrsus maximus)

Guinea grass is a bunch of grass that grows naturally in open grassland. It can grow under or near trees, shrubs, ponds, and riverbanks. This plant cannot be destroyed by wildfire and can withstand drought. They can also reproduce through apomixis and effectively clone themselves through their seeds. This grass is treated as foraging grass if the grassland irrigated and fertilized. This grass is suitable for cut-and-carry for ruminants. It can also be used for silage. This grass has a high protein value. In some countries, this grass is treated as a weed (Dhanesh, 2007).

7.1.5.10 Para grass (Brachiaria mutica)

Para grass is a rapidly growing plant in the summer season. It is a vigorous, semi-prostrate perennial grass with creeping stolons that can grow up to 5 m long and grow perennially. It prefers waterlogging areas like riverbanks, canal sides, pond sides, sides of sewages, and heavy rainfall areas for its growth. This plant reproduces by roots and stem cutting. Their seeds are not viable for propagation.

These grasses are used as green fodder and dry hay. This grass is very palatable and highly nutritious. It contains a good amount of crude protein (10.2%) and crude fiber (23.6%) (Diop *et al.*, 2020).

7.1.5.11 Blue buffel grass (Cenchrus ciliaris)/dhaman grass, anjan grass, koluk katai

This grass grows in hot and dry zones. It is a deep-rooted plant and a drought-tolerant grass. It has widely adapted to desert areas. It can survive in very low rainfall (100 mm annually). It can resist soil erosion. This grass is used for grazing, as dry hay, and for silage. Nitrogen content improves with frequent grazing. For the preparation of hay, the plant should be cut in the early flowering stage to maintain nutritive value (Truter *et al.*, 2014). Fire cannot destroy the grass; the old, vegetative, woody leaves burn without damaging crowns. After burning, the plant recovers, and fresh, young leaves with higher nutritive value appear.

7.1.5.12 Subabul/kubabool (*Leucaena leucocephala*)

It is a wonder fodder tree. The forage is highly digestible and a good source of energy. It is a protein-rich (29%) plant that can be planted in the boundaries and borders of any orchard as a hedge. It is a fast-growing fodder plant. It is a deep-rooted tree that helps to grow in drought-prone areas. This tree produces enormous seeds, which allow for rapid propagation of the plant. Though the plant is very nutritious to animals, the leaves contain the toxin mimosine. During use for young animals, one should be careful. This legume has been introduced in many countries because of its high nutritive value for animals. The residual parts of the plant are used for firewood and the preparation of charcoal.

7.1.5.13 Gliricidia (*Gliricidia sepium*)

Gliricidia is a small, fast-growing, and thornless tree with bright pink spring flowers that appear entirely outcrossing. It has been spread internationally as a green manure, fodder, firewood tree, and ornamental plant. It is a protein-rich fodder that can complement the low quality of fodder available in tropical regions. This plant is used as a cut-and-carry forage for cattle, goats, and sheep. Depending on precipitation and temperature, this plant can tolerate repeated cutting every 2–4 months. This cutting helps to retain the leaves in the dry season, and the animal can forage on that alone (Stuttle, 2015).

7.1.5.14 Sesbania-agathi (*Sesbania grandiflora*)

They are fast-growing, soft, wooden trees that can thrive under extreme exposure to sunlight and are frost sensitive. The plant prefers to grow in good soil and needs a hot and humid climate. The taste of the leaves is bitter, but it is perfect for health. It is a small leguminous tree with edible leaves and flowers.

7.2 Competition Between Man and Animal for Grains

It is a common thought that livestock share the food grains grown for human consumption, and that grain can reduce the hunger of poor people. It is estimated that in India, 58 million tons of protein is consumed by livestock for the production of 77 million tons of protein yearly. It is, at most, a 25% loss of protein. This happens because more concentrate-based feed is used in the in-house farming system. It can be reduced in the free-range system farming, which means livestock is provided with a good pasture. On the contrary, the fact is that animal feedstuff's protein source is primarily unsuitable for human consumption. Animal-produced protein carries better nutrients and is palatable (Mallikarjuna *et al.*, 2018). Comparing nutritive value, the protein efficiency of feedstuff of dairy cattle to milk is 96% to 276%, and in the case of meat animals, feedstuff to meat is 52% to 104% (Oltjen and Beckett, 1996).

The demand for animal protein is increasing with the growth of the human population. Rapid changes in human food habits occurred due to urbanization, modernization, and changes in living standards. Due to climate changes, existing farming systems may not be enough to meet the same. In this situation, cultured meat may be the alternate source of animal protein for human nutrition. Cultured meat is meat that is grown in the laboratory rather than in the animal body. Initially, stem cells for muscle are collected from the animal and cultured in artificial scaffolds, which are used to obtain the muscle texture. Nutrition for the growth of cell culture is supplied through the synthetic medium. Once the cell lines are established, no further slaughter of animals is needed. Then, no question of inhumanity or cruelty will arise. The environment or ecosystem will not be disturbed in this system because it requires less water and land, and no greenhouse gas emissions are produced compared with the traditional animal-rearing system.

7.3 Animal Health Problems

For better production, livestock needs to be healthy. Good health depends on the ambient environment, nutritious feed, and good management, including hygiene maintenance and biosecurity adopted. In a changed climate or high environmental temperature, the livestock are under stress, which will aggravate different diseases. The average global temperature could increase by 1.4–5.8°C by 2100. This increased temperature will be unbearable for the livestock in their present habitat. This temperature, along with high moisture, will be excruciating for the animals. Heat stress causes metabolic disorders in animals. This change has direct and indirect effects on the health and production levels of livestock. In the case of health, the animal will suffer from metabolic disorders, poor response to vaccination, common commensal organism-causing diseases, and animals becoming fatigued, among others. The growth performance, milk and meat production, egg production, and the quality of milk, meat, and eggs will go down. Even reproductive failure or poor reproductive performance will be observed (Nardone *et al.*, 2010; Henry *et al.*, 2012). Due to heat stress (HS) in summer, post-partum anestrus in cattle is frequently observed (Talukder *et al.*, 2018). Pigs show decreased reproductive performance in late summer and early autumn compared with winter and spring. This happens due to changes in temperature and the length of the daytime (Auvigne *et al.*, 2010). Heat stress also has effects on semen quality (Nichi *et al.*, 2006), oocyst and embryo development (Stewart *et al.*, 2011), and estrus cycle (Wolfenson *et al.*, 2000).

Animals in an uncomfortable zone due to heat stress will take less food and more water. During their comfort period, animals take more feed at a time, which causes ruminal acidosis that causes lameness. During uncomfortable periods, the animal used to stand comfortably. Due to heat stress, when animals keep standing for an extended period, it causes structural changes in their claws. The respiration rate and perspiration will increase. The animal's normal physiology will change. Other health markers, like cell-mediated immunity, will decrease. Immunoglobulin in the colostrum will be reduced, which will cause the death of neonates by commensal organisms (Nardone *et al.*, 2010). Vaccination programs will also fail. Heat stress also causes depression in lymphocyte function (Lacetera *et al.*, 2005). As the function of neutrophils is reduced, the protection of the mammary gland becomes weak, and the incidence of mastitis increases. Due to this immune suppression, common organisms, even commensals, become pathogenic and cause illness. The resulting use of antimicrobial agents increased, which led to antimicrobial resistance (O'Neill, 2016). In these circumstances, the lactating animals become infested with mastitis. There are many diseases called vector-borne diseases (VBD), which are diseases transmitted by arthropods or small animal vectors from one animal to another. These diseases cause enormous suffering to the animals. These diseases cause not only production loss, but sometimes, they cost animal lives. Vectors act as transmitting medium of pathogens, and in some cases, the essential stages of pathogens take place in the vector. Many vectors are mounting their range to transfer newer diseases.

The habitats of vectors are expanding to the newer zones with the change in climate. The effects of global warming are irregular rainfall and a rise in temperature. Water stagnation areas are also increasing due to unexpected precipitation. These temperatures and stagnated water are raising the breeding opportunity of vectors/arthropods. Climate changes have an effect on the way pathogens interact with both the arthropod vector and the animal host. The life cycle, the behaviors, and the interaction between vector and host animals of arthropods depend upon the components of climate (Lindgren, 1998).

Some tick-borne animal diseases:

- Babesiosis
- Bluetongue
- African horse sickness (AHS)
- Lyme disease
- Bovine ephemeral fever
- Equine infectious anemia
- Rift Valley fever (RVF)

7.3.1 Babesiosis

This disease is caused by infection with the eukaryotic blood parasite *Babesia*. It is the second-most common blood parasite of mammals after trypanosomes. This disease mainly occurs in the worm or moderate worm areas and has a major effect on the health of domestic animals (Berger *et al.*, 2006). This disease is transmitted by tick (Ixodes) bite. This disease is common in the area of the Northeastern and Midwestern United States and parts of Europe and is now spreading throughout the rest of the world, where the environment became suitable for the breeding and survival of ticks. Animals can get infected with Babesia parasites by the bite of ticks who are infected by getting a blood transfusion from an infected donor of blood products. This parasite also has a congenital transmission through ticks (an infected mother to her baby). The disease is known as Texas cattle fever, red water, or piroplasmosis in cattle (Hunfeld *et al.*, 2008). *Babesia canis rossi*, *Babesia bigemina*, and *Babesia bovis* cause babesiosis in animals. The disease causes "hemoglobinuria red-water," hemolytic anemia, "cerebral babesiosis," and disseminated intravascular coagulation. In the case of cattle, initially, the mucous membrane becomes pale. In advanced cases, the red blood cells are lysed, bilirubin is produced as a by-product, and the visible mucous membranes become yellow in color. This icterus condition happens due to the failure of the liver to metabolize the excess bilirubin. Hemoglobinuria is observed due to lysed red-blood-cell products excreted through urine. Temperatures can reach as high as 105°F. The disease can be controlled by controlling the ticks.

7.3.2 Bluetongue

Bluetongue disease is an insect-borne viral disease. This disease mainly affects sheep and is less common in goats, cattle, buffalo, deer, and antelopes. Orbivirus is the cause of the disease transmitted by bites from midges, and this occurs during summer and fall. Morbidity and mortality of this disease is high in sheep. Other ruminants are affected, causing a loss of productivity (Roy, 2008). Prime symptoms of the disease are swelling of the face, cyanosis, swelling of the tongue, and excessive salivation with high fever. Cyanosis of the tongue gives its typical blue appearance and the name of the disease was given. Some infected animals become carriers and do not show any symptoms. Sometimes, the diseased animals show foot lesions beginning with coronets, followed by lameness. If an infected animal shows cardinal clinical symptoms, it rapidly loses condition, becomes sicker, and may die within a week. If it survives, it shows a very slow recovery. Other ruminants like cattle, goats, and wild ruminants remain asymptomatic in spite of high virus levels in the blood (Jensen and Swift, 1982).

The vector *Culicoides* cannot survive in cold winter temperatures (i.e. frost temperature). The bluetongue virus infections typically do not last for more than 60 days. In the temperate zone, the disease does not last till the next summer. The disease's existence depends upon the vectors' survival (i.e. the midge, Culicoides spp. adult) to the next spring. The disease is seasonal in the temperate zone, but it may exist around the year in tropical weather (Baylis *et al.*, 2017). The disease can be controlled through practical midge control measures. Control may be done by keeping the vector away from the animal during maximum biting time from dusk to dawn, through simple husbandry practice. At that time, livestock may be housed and kept enclosed under fine mesh netting or coarser material impregnated with insecticide. Proper management or treatment of the breeding grounds, that is, animal dung and moist soils, either bare or covered in short grass to break the breeding cycle will significantly reduce the local midge population as well as midge bites (Gairdner, 2007; Abel, 2008).

7.3.3 African Horse Sickness (AHS)

It is a non-contagious viral disease of horses, mules, donkeys, and zebra. The horse is the most susceptible and the mortality rate is about 90%, followed by mules (50%) and donkeys (10%). It is a vector-borne disease caused by *Orbivirus* under the family *Reoviridae*. The disease is transmitted by the biting of arthropods. The main biological vector is Culicoides (midges). Mosquitoes like *Culex*, *Anopheles*, and *Aedes* and ticks such as *Hyalomma* and *Rhipicephalus* can spread the viruses (Du Toit, 1944; Bouayoune *et al.*, 1998; Meiswinkel, 1998). Cardinal symptoms of the disease are high fever, depression, and respiratory troubles. The affected animals suffer from trouble breathing, cough, and frothy fluid from the nostrils and mouth, followed by pulmonary edema within 4 days. The animal dies within 24 hour of the onset of the clinical symptoms due to serious lung congestion, which causes respiratory failure (Erasmus, 1972). The occurrence and spread of the disease depend on the presence of high numbers of competent vectors and host-vector interaction. The important vector for transmitting the virus in endemic areas is the biting midge *Culicoides imicola* (Du Toit, 1944). The disease can be controlled through quarantine of diseased animal, vector control, and vaccination. The uninfected horses are to be vaccinated against the virus, and affected horses are to be slaughtered.

7.3.4 Lyme disease

Lyme disease is a tick-borne disease caused by spirochetal bacteria belonging to the *Borrelia burgdorferi*. This disease affects cattle, horses, dogs, and cats. Tick vectors of *B. burgdorferi* sensu lato are hard-shelled *Ixodes* ticks. The common clinical signs of Lyme disease in horses are chronic weight loss, low-grade fever, swollen joints, muscle tenderness, lameness, and anterior uveitis. Neurological signs like depression, head tilt, dysphagia, behavioral changes, and encephalitis are seen. In the case of cattle, it is a major problem. In acute cases of the disease, cattle show fever, swollen joints, stiffness, and decreased milk yield. In chronic cases, cattle show weight loss, laminitis, and often abortion. The incidences of Lyme disease are increasing due to the expanding geographic range of tick vectors globally. Protection of the animals from ticks is the prime method of controlling the disease.

7.3.5 Bovine ephemeral fever

It is an arthropod-borne disease of cattle caused by rhabdovirus. Occasionally, buffaloes suffer from this disease. Common symptoms of the disease are a short fever, shivering, muscular stiffness, and lameness. The fever continues for 3 days and is called a 3-day fever. Mosquitoes and biting midges (sandflies) are thought to be vectors. Mortality of this disease is not significant, but economic loss due to the disease is major. The sudden drop in milk yield (70%) cannot be recovered after the cure. Milk yield is reduced by a minimum of 15% after recovery. Further, fertility problems arise in both males and females. Abortion may occur, and the milking cow becomes dry. Though the mortality rate is low, morbidity is too high. All the animals in herd will be affected. The bulls may be infertile temporarily (Walker and Klement, 2015).

7.3.6 Equine Infectious Anaemia (EIA)

EIA is a non-contagious disease of horses and other Equidae. The disease is caused by *A. Lentivirus*. The virus can be spread between horses by biting insects, grooming implements, multiple use of surgical instruments and other equipment such as stomach tubes, and from a dam to her foal via the milk. Common clinical symptoms are deep depression, rapid loss of condition and generalized weakness, fever where the temperature reaches up to 41°C, and a clear nasal discharge. With the progression of the disease, anemia develops. The mucous membrane becomes pale, and petechial hemorrhages and jaundice symptoms are seen. The brisket, ventral abdomen, and limbs become edematous. In some cases, diarrhea dysentery is seen. Abortion is also not uncommon (Craigo *et al.*, 2005).

7.3.7 Rift Valley Fever (RVF)

RVF is an emerging acute viral hemorrhagic fever commonly seen in domesticated animals like cattle, buffalo, goats, sheep, and camels. The causative virus is the Rift Valley Fever virus, a member of the genus Phlebovirus. This virus can cause human illness. The disease is a mosquito-borne disease. Here, mosquitoes (Aedes and Culex) act as reservoirs and vectors of RVF. The virus has vertical transmission in the mosquito, meaning the virus can be transmitted to the offspring through eggs. The number of mosquitoes increases during heavy rainfall, flooding, and high temperatures. This condition is favorable for the breeding of mosquitoes and the hatching of eggs (Murphy *et al.*, 1999). The effect of the disease on livestock is 100% abortion of fetuses. Cardinal symptoms of the disease are diarrhea, vomiting, respiratory problems, fever, anorexia, lethargy, and sudden death of animals.

7.3.8 Akabane disease

Akabane disease is caused by arbovirus. It leads to reproductive failure and fetal deformities in cattle, sheep, and goats. This disease mainly affects ruminants, but pigs and horses are also affected.

Immune-competent pregnant animals are generally affected. Generally, abortion occurs during late-term pregnancy, or it leads to stillbirth. Congenital deformities like arthrogryposis (fixed contracted joints), hydranencephaly (part of the brain being absent), torticollis, scoliosis, kyphosis, and spina bifida of joints are common. These lead to difficult labor and death of the fetus. Clinical manifestations of hydranencephaly are ataxia, blindness, and a poor or absent sucking reflex (Konno and Nakagawa, 1982).

7.4 Lack of Adaptive Animal Breeds in Adverse Weather

Milk and milk products are the capital source of essential protein and fat for most of the people of the world. Heat stress reduces milk production and lowers normal health status, affecting breeding performance, growth rate, and quality of milk production and profitability. The core body temperature of the animals increases due to heat stress, which affects the fat production in the mammary gland. A result of heat stress is a reduction of milk yield and altered milk composition such as protein, fat, lactose, and solid-not-fat percentages (Dunn *et al.*, 2014). Some physiological parameters of cattle that change due to heat stress are an increase in rectal temperature, core body temperature, respiration rate, panting score, pulse rate, sweating rate, and heart rate. In tropical countries where the climate is always hot and humid, the fertility and health status of cattle go down (Boonkum *et al.*, 2011) Performance of dairy cattle breeds in tropical countries is different. Heat-adapted Bos indicus cattle and less resilient Bos taurus breeds respond differently to the effect of heat street due to their genetic makeup (Cheruiyot *et al.*, 2021). Heat-tolerant dairy breeds need to be identified depending on the climatic zone and the foreseeable temperature for the long-term security of nutrition from dairy products. Heat-tolerant breeds may have evolved through crossbreeding between cattle of the temperate zone and tropical zone. It is found that crossbreeding of high-yielding Holstein cows with heat-tolerant like Gir and Sahiwal improved milk yield and reproductive performance under heat stress conditions (Sigdel *et al.*, 2019). Indicine breeds of cattle are heat tolerant, ectoparasite resistant, less susceptible to gastrointestinal parasites, and have low maintenance cost, traits that are far better than taurine breed cattle. All these physical traits help Bos indicus grow under hot and humid tropical and subtropical climates (Iqbal *et al.*, 2019; Sajjanar *et al.*, 2023).

The ambient temperature for cows is 41–77°F, which means that livestock around us are struggling to cope with high temperatures and longer summers. The cows with shorter coats and more active sweat glands can maintain a healthy body temperature, even in hot summers. The high-yielding cows generate more heat during the production period, which makes it harder to resist the heat stress in summer. In this situation, the cow grazes less (eating about 3–5% less per additional degree of ambient temperature), causing a decline in milk production. It becomes difficult to get pregnant. The cow becomes more susceptible to diseases due to immune suppression. Though the traditional Holsteins originated in the temperate zone (i.e. the Netherlands) and are the top milk-producing cows in terms of volume, they are the least comfortable with heat and humidity. The Holstein breeds with the slick mutation are able to keep their body temperature about 1°F cooler. That is why their milk production and fertility performance do not drop as much as non-slick cows during the hottest months (Bole-Feysot *et al.*, 1998). Some good dairy cattle rearing worldwide for milk production.

7.4.1 Gir/Gyr

The Gir or Gyr is an Indian breed of zebuine cattle that originated in the Kathiawar peninsula in Gujarat. The name of the breed derives from the Gir Hills in that region. The average milk yield of the Gir breed in the Indian climate is about 2110 kg per lactation, with approximately 4.5% fat. Brazil imported this breed from India in the early 20th century, and they developed separate dairy (Gir Leiteiro) and beef strains through selective breeding in the 1960s. This breed was also used in the development of the Brahman breed in North America. This breed is well known for its milk-producing qualities. In different countries, Gir is used frequently because this Bos indicus breed is resistant to hot temperatures and tropical diseases.

7.4.2 Sahiwal

This breed originated from the Sahiwal area in Montgomery district, presently in Punjab, Pakistan. Sahiwal is considered one of the best milk-producing breeds in India. In organized farm conditions, the breed produces as much as 6000 l of milk in a lactation period. Observing the performance, Australia imported this breed, and it was used to develop a synthetic crossbred breed of Australian Milking Zebu (AMZ) cattle. This breed has also adapted to the hot and humid environment of the tropics (Mason, 1996).

7.4.3 Holstein Friesian

Now, Holstein Friesian is known as the international dairy cattle breed of the world. The breed originated in North Holland and Northern Germany. This breed dominates industrial dairy farming in about 160 countries (Transboundary breed, 2023). Today, this breed is used for both milk and meat production. In some countries, the breed is further developed and used purely for dairy purposes. The average milk production is about 10,220 kg per lactation, but the fat content is low (3.1–3.7%). This breed is suitable in the temperate zone and cannot tolerate high temperatures.

7.4.4 Jersey

The Jersey is a small dairy cattle breed that originated in the British Channel Island. The breed produces 5000–8000 kg of milk per year with high fat (4.84%) (Valerie *et al.*, 2016). These cows are very popular for dairy farming in various countries and almost all climatic zones. Other good characteristics of the cows are small size, calm and docile, high fertility, low rate of dystocia, and the ability to thrive on locally produced feeds. They are very popular for crossbreeding. They can give their average production in hot and humid climates. They are more heat resistant compared with many other dairy breeds.

7.5 Mitigation Strategies for Heat Stress in Dairy Cattle in the Hot and Humid Climate Due to Global Warming

A single strategy cannot ease the negative impacts of heat stress in dairy cattle farming. The following strategies can reduce the effects of heat stress in dairy cattle (Bernabucci *et al.*, 2014): (i) genetic development of heat-tolerant animals (Corazzin *et al.*, 2020), (ii) physical modification of the environment (shading, cooling) (Marumo *et al.*, 2022), and (iii) improved feeding management strategies (Hammami *et al.*, 2015). These three practices in hot and humid tropical and subtropical areas enhance the milk yield. The cooling technology has an exclusively positive impact on reducing the adverse effects of the high temperature-humidity index in dairy cows (Bernabucci *et al.*, 2014). However, those stratagems cannot completely remove the adverse effects of heat stress on animal-related factors such as genetics, hair coats, and metabolic heat generated (Wang *et al.*, 2022). Thus, highly identifying heat-resistant cattle will be suitable to mitigate the problems and improve animal welfare and production (Tresoldi *et al.*, 2016).

7.5.1 Selection of genetic heat-tolerant dairy cattle

For selecting or developing a thermotolerant dairy cattle breed through breeding policy, the following strategies are to be considered: (i) a more heat-tolerant animal compared with the others should be selected for breed or cross for raising more heat-tolerant cows, (ii) introgression of thermotolerance genes into the high-yielding cows, and (iii) upgradation through selective breeding connected with heat tolerance trait (Pryce *et al.*, 2022). In tropical countries, Bos indicus, which is relatively highly adapted to heat stress, should be chosen for breeding with high-yielding cattle breeds of the temperate zone (Cuellar *et al.*, 2023). The *SLICK* haplotype cows can be chosen for rearing in hot and humid areas, as the animals have short, fine hair and greater body temperature control (Camargo and Pereira, 2022).

7.5.2 Physical modification of the environment

Change in rearing and management systems can bring comfort to the animals. Dry matter intake reduces as the environmental temperature increases. During a hot period, watering, sprinklers, providing sheds for animals, tunnels and cross ventilation can reduce temperature in the microenvironment. This helps in body heat loss, resulting in reduced body temperature, and the animal will take more dry matter (West, 2003). Providing a shed to protect the animals from direct sunlight during hot periods is a method to give comfort. It has been proved that providing sheds during hot sunny days can improve milk yield and composition (Van Laer *et al.*, 2015). A shed may be given by trees, hay, straw shades, or a roofing with ridged sheets decorated with a deliberate coating to throw back sunlight in the pasture. It will reduce the water intake (Palacio *et al.*, 2015). Cooling through evaporative measures (i.e. watering, sprinkler) as a heat abatement system has been used successfully. This system can be used around the year, but this process is not practical in cooler and highly humid areas (Bohmanova *et al.*, 2007). It is reported that dairy cattle under both shade and cooled by fans and spray produced 2 kg of milk per day compared with dairy cattle under shade alone (Anderson *et al.*, 2013). Providing natural shade (trees) or artificially constructed shade within the pastureland production system saves livestock from extreme solar radiation and changes their radiation balance (Reis *et al.*, 2021).

7.5.3 Feeding and nutritional strategies for minimizing heat stress in animals

Animals eat less feed in stress conditions. Milk production reduces due to lower intake of feeds. To provide sufficient nutrients to lactating cattle, they need to consume high-quality feed, including a lot of grains and supplementary fat (Garner *et al.*, 2022). To meet the nutrient demand for dairy cattle under heat stress, in ration, the use of concentrate will be more compared with forage. Feeding of excessive concentrate causes ruminal disorder, which produces acidosis and causes cows to cease feeding (Shwartz *et al.*, 2009). The use of sodium bicarbonate in feed can accommodate a greater amount of concentrate feeding. To maintain ruminal health from dilatory changes, yeast, fungal culture, and niacin can be used. Where yeast enhances fiber degradation, niacin, and fungal culture enhance energy utilization (Zimbelman *et al.*, 2010). Fresh, palatable, and high-quality feed should be used to combat the adverse effects of heat stress. Feed ingredients should be easily digestible to reduce the heat increment through nutrient utilization in the animal (Godara *et al.*, 2016).

7.6 Conclusions

Food security for the future is a most uttered subject. Food security does not mean only the availability of food but also access to food for all people. It is predicted that the production of food, particularly from animal sources, will be hampered due to climate change. The global population is increasing rapidly, and to meet the demand for food, the production of food needs to be enhanced. At the same time, to have access to food, employment needs to be generated, which may be through animal husbandry. The animal production system needs to be changed to cope with this situation. More attention is to be given to the erosion of land, production of good quality feed (i.e. grasses), control of arthropod-borne diseases, and use of high-yield, heat-resistant breed animals.

Climate change is an unavoidable phenomenon. Research should be done to evolve heat-tolerant varieties of fodders. Similarly, heat-tolerant breeds of animals should be identified, or genetically heat-tolerant breeds need to be bred. This research is needed to keep the balance between demand and supply of food to keep away hunger in the future.

References

Abel, C. (2008) Bluetongue vaccine BTV8 questions answered. *Farmers Weekly,* Archived 2008-06-11 at the Wayback Machine. Available at: https://www.fwi.co.uk/livestock/health-welfare/bluetongue-vaccine-btv8-questions-answered

Alexandratos, N. and Bruinsma, J. (2012) *World agriculture towards 2030/2050: The 2012 revision*. Rome, Italy.

Anderson, S.D., Bradford, B.J., Harner, J.P., Tucker, C.B., Choi, C.Y. *et al.* (2013) Effects of adjustable and stationary fans with misters on core body temperature and lying behavior of lactating dairy cows in a semiarid climate. *Journal of Dairy Science* 96(7), 4738–4750. DOI: 10.3168/jds.2012-6401.

Auvigne, V., Leneveu, P., Jehannin, C., Peltoniemi, O. and Sallé, E. (2010) Seasonal infertility in sows: A five year field study to analyze the relative roles of heat stress and photoperiod. *Theriogenology* 74(1), 60–66. DOI: 10.1016/j.theriogenology.2009.12.019.

Baylis, M., Caminade, C., Turner, J. and Jones, A.E. (2017) The role of climate change in a developing threat: The case of bluetongue in Europe. *Revue Scientifique et Technique-Office International des Épizooties* 36(2), 467–478. DOI: 10.20506/rst.36.2.2667.

Berger, S., Marr, A. and John, S. (2006) *Human Parasitic Diseases Sourcebook*. Jones and Bartlett Learning, Burlington, MA.

Bernabucci, U., Biffani, S., Buggiotti, L., Vitali, A., Lacetera, N. *et al.* (2014) The effects of heat stress in Italian holstein dairy cattle. *Journal of Dairy Science* 97(1), 471–486. DOI: 10.3168/jds.2013-6611.

Bhakar, A. and Ram, H. (2019) Hybrid Napier grass a boon to forage production. *Indian Farming* 69(2), 23–25.

Biraben, J.N. (1980) An essay concerning mankind's evolution. *Population, Selected Papers* 4, 1–13.

Bohmanova, J., Misztal, I. and Cole, J.B. (2007) Temperature-humidity indices as indicators of milk production losses due to heat stress. *Journal of Dairy Science* 90, 1947–1956. DOI: 10.3168/jds.2006-513.

Bole-Feysot, C., Goffin, V., Edery, M., Binart, N. and Kelly, P.A. (1998) Prolactin (PRL) and its receptor: Actions, signal transduction pathways and phenotypes observed in PRL receptor knockout mice. *Endocrine Reviews* 19(3), 225–268. DOI: 10.1210/edrv.19.3.0334.

Boonkum, W., Misztal, I., Duangjinda, M., Pattarajinda, V., Tumwasorn, S. *et al.* (2011) Genetic effects of heat stress on milk yield of Thai holstein crossbreds. *Journal of Dairy Science* 94, 487–492. DOI: 10.3168/jds.2010-3421.

Bouayoune, H., Touti, J., El Hasnaoui, H., Baylis, M. and Mellor, P.S. (1998) The culicoides vectors of African horse sickness virus in Morocco: Distribution and epidemiological implications. *Archives of Virology Supplements* 14, 113–125.

Camargo, D.A.S.L. and Pereira, J.F. (2022) Genome-editing opportunities to enhance cattle productivity in the tropics. *CABI Agriculture and Bioscience* 3, 1–11. DOI: 10.1186/s43170-022-00075-w.

Cheruiyot, E.K., Haile-Mariam, M., Cocks, B.G., MacLeod, I.M., Xiang, R. *et al.* (2021) New loci and neuronal pathways for resilience to heat stress in cattle. *Scientific Reports* 11, 1–16. DOI: 10.1038/s41598-021-95816-8.

Cook, B.G., Pengelly, B.C., Brown, S.D., Donnelly, J.L., Eagles, D.A. *et al.* (2005) *Tropical forages*. CSIRO, DPI&F(Qld), CIAT and ILRI, Brisbane, Australia.

Corazzin, M., Saccà, E., Lippe, G., Romanzin, A., Foletto, V. *et al.* (2020) Effect of heat stress on dairy cow performance and on expression of protein metabolism genes in mammary cells. *Animals* 10, 1–13. DOI: 10.3390/ani10112124.

Craigo, J.K., Li, F., Steckbeck, J.D., Durkin, S., Howe, L. *et al.* (2005) Discerning an effective balance between equine infectious anemia virus attenuation and vaccine efficacy. *Journal of Virology* 79(5), 2666–2677. DOI: 10.1128/JVI.79.5.2666-2677.2005.

Cuellar, C.J., Saleem, M. and Jensen, L.M. (2023) Differences in body temperature regulation during heat stress and seasonal depression in milk yield between Holstein, Brown Swiss, and crossbred cows. *Journal of Dairy Science* 106, 22725. DOI: 10.3168/jds.2022-22725.

Dasgupta, P., Morton, J., Dodman, D., Karapinar, B., Meza, F. *et al.* (2014) Rural areas. Climate change 2014: impacts, adaptation and vulnerability. Part A: Global and sectoral aspects. In: Field, C.B. (ed.) *Contribution of Working Group II to the Fifth Assessment Report of the Intergovernmental Panel on Climate*. Cambridge University Press, Cambridge, New York, pp. 613–657.

Dhanesh, W. (2007) First known record of guinea grass cultivation in Sri Lanka, 1801–1802. *Journal of the Royal Asiatic Society of Sri Lanka* 53, 219–222.

Diop, F.N., Lansdown, R.V. and Gupta, A.K. (2020) Brachiaria mutica. *IUCN Red List of Threatened Species* 2020:e.T164272A67773905.

Du Toit, R.M. (1944) The transmission of blue tongue and horse sickness by Culicoides. *Onderstepoort Journal of Veterinary Science and Animal Industry* 19(1944), 7–16.

Du, S., Shi, P. and Van Rompaey, A. (2014) The relationship between urban sprawl and farmland displacement in the pearl river delta, China. *Land* 3(1), 34–51. DOI: 10.3390/land3010034.

Dumont, B., Ryschawy, J., Duru, M., Benoit, M., Chatellier, V. *et al.* (2019) Review: Associations among goods, impacts and ecosystem services provided by livestock farming. *Animal* 13(8), 1773–1784. DOI: 10.1017/S1751731118002586.

Dunn, R.J.H., Mead, N.E., Willett, K.M. and Parker, D.E. (2014) Analysis of heat stress in UK dairy cattle and impact on milk yields. *Environmental Research Letters* 9(6), 064006. DOI: 10.1088/1748-9326/9/6/064006.

Eisler, M.C., Lee, M.R., Tarlton, J.F., Martin, G.B., Beddington, J. *et al.* (2014) Agriculture: Steps to sustainable livestock. *Nature* 507(7490), 32–34. DOI: 10.1038/507032a.

Erasmus, B.J. (1972) The pathogenesis of African horse sickness. In: Bryans, J.T. and Gerber, H. (eds) *Equine Infectious Diseases III, Proc. 3rd Int. Conf. Equine Inf. Dis*. Paris, France, 1972. Karger, Basel, Switzerland, 1973, pp. 1–11. DOI: 10.1159/000393512.

Erb, K., Lauk, C., Kastner, T., Mayer, A. and Theurl, M.C. (2016) Habits and weeds of Napier grass, *Pennisetum Purpureum*: A review. *International Journal of Pest Management* 48(1), 39–48. DOI: 10.1080/09670870110065578.

Farrell, G., Simons, S.A. and Hillocks, R.J. (2002) Pests, disease. WMO annual report highlights continuous advance of climate change. World Meteorological Organization, 21 april 2023. Archived from the original on 20 june 2023. Press release Number: 21042023.

Frumkin, H. (2016) *Urban sprawl and public health*. Public Health Reports, Sage, Los Angeles, California.

Gairdner, J. (2007) Bluetongue (blue tongue) outbreak in the UK-fwi's special report. *Farmers Weekly, 24 September 2007. Offers Practical Advice for Preventing Bluetongue in Livestock. Archived 2007-10-12 at the Wayback Machine*. Available at: https://www.fwi.co.uk/business/bluetongue-blue-tongue-outbreak-in-the-uk-fwi-s-special-report

Garner, J.B., Williams, S.R.O., Moate, P.J., Jacobs, J.L., Hannah, M.C. *et al.* (2022) Effects of heat stress in dairy cows offered diets containing either wheat or corn grain during late lactation. *Animals* 12(16), 2031. DOI: 10.3390/ani12162031.

Godara, A.S., Bhat, S.A., Yogi, R.K., Devi, S. and Sahoo, S.P. (2016) Feeding strategies to ameliorating the impact of heat stress in bovine. *International Journal of Science and Nature* 7, 25–29.

Gowda, P. (2018) Agriculture and rural communities. In: *Impacts, Risks and Adaptation in the United States: Fourth National Climate Assessment*, Vol. II. US Global Change Research Program, Washington, DC, p. 415.

Hammami, H., Vandenplas, J., Vanrobays, M., Rekik, B., Bastin, C. *et al.* (2015) Genetic analysis of heat stress effects on yield traits, udder health, and fatty acids of Walloon Holstein cows. *Journal of Dairy Science* 98(7), 4956–4968. DOI: 10.3168/jds.2014-9148.

Henry, B., Charmley, E., Eckard, R., Gaughan, J.B. and Hegarty, R. (2012) Livestock production in a changing climate: Adaptation and mitigation research in Australia. *Crop and Pasture Science* 63, 191–202. DOI: 10.1071/CP11169.

Heuzé, V., Tran, G., Boval, M., Lebas, F., Lessire, M. *et al.* (2013) *Alfalfa (Medicago sativa)*. Feedipedia.org. A programme by INRA, CIRAD, AFZ and FAO.

Hunfeld, K.P., Hildebrandt, A. and Gray, J.S. (2008) Babesiosis: Recent insights into an ancient disease. *International Journal for Parasitology* 38(11), 1219–1237. DOI: 10.1016/j.ijpara.2008.03.001.

IPCC Pörtner, H.-O., Roberts, D.C., Masson-Delmotte, V., Zhai, P. and Tignor, M. (eds). (2019) *Summary for Policymakers. In: IPCC Special Report on the Ocean and Cryosphere in a Changing Climate*. Cambridge University Press, Cambridge, UK, and New York. DOI: 10.1017/9781009157964.001.

Iqbal, N., Liu, X., Yang, T., Huang, Z., Hanif, Q. *et al.* (2019) Genomic variants identified from whole-genome resequencing of indicine cattle breeds from Pakistan. *PloS One* 14, 1–21. DOI: 10.1371/journal.pone.0215065.

Jensen, R. and Swift, B.L. (1982) *Diseases of Sheep*. Lea and Febiger, Philadelphia, Pennsylvania.

Kastner, T., Rivas, M.J.I., Koch, W. and Nonhebel, S. (2012) Global changes in diets and the consequences for land requirements for food. *Proceedings of the National Academy of Sciences of the United States of America* 109, 6868–6872. DOI: 10.1073/pnas.1117054109.

Khan, Z.R., Midega, C.A.O., Wadhams, L.J., Pickett, J.A. and Mumuni, A. (2007) Evaluation of Napier grass (*Pennisetum purpureum*) varieties for use as trap plants for the management of African stemborer (*Busseola fusca*) in a push–pull strategy . *Entomologia Experimentalis et Applicata* 124(2), 201–211. DOI: 10.1111/j.1570-7458.2007.00569.x.

Kight, S.W. and Lysik, T. (2022) The human race at 8 billion. *Axios,* 14 November 2022.

Konno, S. and Nakagawa, M. (1982) Akabane disease in cattle: Congenital abnormalities caused by viral infection. *Experimental Disease Veterinary Pathology* 19(3), 267–279. DOI: 10.1177/030098588201900305.

Lacetera, N., Bernabucci, U., Scalia, D., Ronchi, B., Kuzminsky, G. *et al.* (2005) Lymphocyte functions in dairy cows in hot environment. *International Journal of Biometeorology* 50(2), 105–110. DOI: 10.1007/s00484-005-0273-3.

Lambin, E.F., Turner, B.L., Geist, H.J., Agbola, S.B., Angelsen, A. *et al.* (2001) The causes of land-use and land-cover change: Moving beyond the myths. *Global Environmental Change* 11(4), 261–269. DOI: 10.1016/S0959-3780(01)00007-3.

Langyintuo, A.S., Lowenberg-DeBoer, J., Faye, M., Lambert, D., Ibro, G. *et al.* (2003) Cowpea supply and demand in West and Central Africa. *Field Crops Research* 82(2–3), 215–231. DOI: 10.1016/S0378-4290(03)00039-X.

Lindgren, E. (1998) Climate change, tick-borne encephalitis and vaccination needs in sweden—a prediction model. *Ecological Modelling* 110(1), 55–63. DOI: 10.1016/S0304-3800(98)00041-6.

Lopez-i-Gelats, F. (2014) Impacts of climate change on food availability: livestock. In: *Global Environmental Change*. Springer, Dordrecht, Netherlands, pp. 689–694. Available at: http://link.springer.com/10.1007/978-94-007-5784-4_118 (accessed 29 July 2024).

Mallikarjuna, S.N., Gururaj, B. and Ramesh, (2018) Livestock: A driving force of food security and sustainable development in India. *International Journal of Agriculture Sciences* 10(9), 5937–5939.

Mankin, J.S. (2021) *NOAA Drought Task Force Report on the 2020–2021 Southwestern US Drought*. National Oceanic and Atmospheric Administration (NOAA) Drought Task Force; NOAA Modeling, Analysis, Predictions and Projections Programs; and National Integrated Drought Information System, Silver Spring, Maryland, p. 4.

Marumo, J.L., Lusseau, D., Speakman, J.R., Mackie, M. and Hambly, C. (2022) Influence of environmental factors and parity on milk yield dynamics in barn-housed dairy cattle. *Journal of Dairy Science* 105(2), 1225–1241. DOI: 10.3168/jds.2021-20698.

Mason, I.L. (1996) *A World Dictionary of Livestock Breeds, Types and Varieties*, 5th edn. CAB International, Wallingford, UK.

Matthys, D. (2018) Manual on how to produce animal feed from local encroached bush launched. *Namibia Economist*.

Meiswinkel, R. (1998) The 1996 outbreak of african horse sickness in south africa—the entomological perspective. *Archives of Virology Supplement* 14, 69–83. DOI: 10.1007/978-3-7091-6823-3_8.

Morton, J.F. (2007) The impact of climate change on smallholder and subsistence agriculture. *Proceedings of the National Academy of Sciences of the United States of America* 104, 19680–19685. DOI: 10.1073/pnas.0701855104.

Murphy, F.A., Gibbs, E.P.J., Horzinek, M.C. and Studdert, M.J. (1999) *Veterinary Virology*, 3rd edn. Academic Press, San Diego, California.

Nardone, A., Ronchi, B., Lacetera, N., Ranieri, M.S. and Bernabucci, U. (2010) Effects of climate changes on animal production and sustainability of livestock systems. *Livestock Science* 130(1–3), 57–69. DOI: 10.1016/j.livsci.2010.02.011.

Nichi, M., Bols, P.E.J., Züge, R.M., Barnabe, V.H., Goovaerts, I.G.F. *et al.* (2006) Seasonal variation in semen quality in Bos indicus and Bos taurus bulls raised under tropical conditions. *Theriogenology* 66(4), 822–828. DOI: 10.1016/j.theriogenology.2006.01.056.

O'Brien, D., Shalloo, L., Patton, J., Buckley, F., Grainger, C. *et al.* (2012) A life cycle assessment of seasonal grass-based and confinement dairy farms. *Agricultural Systems* 107, 33–46. DOI: 10.1016/j.agsy.2011.11.004.

Oltjen, J.W. and Beckett, J.L. (1996) Role of ruminant livestock in sustainable agricultural systems. *Journal of Animal Science* 74(6), 1406–1409. DOI: 10.2527/1996.7461406x.

O'Neill, J. (2016) *Tackling Drug-Resistant Infections Globally: Final Report and Recommendations.* HM Government and Welcome Trust, London. Available at: https://amr-review.org/sites/default/files/160518_Final%20paper_with%20cover.pdf(accessed 23 September 2019).

Padulosil, S. and Ng, N.Q. (1997) Origin, taxonomy and morphology of *Vigna unguiculata* (L.) Walp. In: Singh, B.B., Mohan, D.R., Dashiell, K.E. and Jackai, L.E.N. (eds) *Advances in Cowpea Research*. International Institute of Tropical Agriculture and Japan International Research Center for Agricultural Sciences, Ibadan, Nigeria, pp. 1–19.

Palacio, S., Bergeron, R., Lachance, S. and Vasseur, E. (2015) The effects of providing portable shade at pasture on dairy cow behavior and physiology. *Journal of Dairy Science* 98(9), 6085–6093. DOI: 10.3168/jds.2014-8932.

Porter, J.R., Xie, L., Challinor, A.J., Cochrane, K., Howden, S.M. *et al.* (2014) Food security and food production. In: *Climate Change: Impacts, Adaptation and Vulnerability*. Cambridge University Press, Cambridge, UK, pp. 201–222.

Pryce, J.E., Nguyen, T.T.T., Cheruiyot, E.K., Marett, L., Garner, J.B. *et al.* (2022) Impact of hot weather on animal performance and genetic strategies to minimise the effect. *Animal Production Science* 62(8), 726–735. DOI: 10.1071/AN21259.

Reis, N.S., Ferreira, I.C., Mazocco, L.A., Souza, A.C.B., Pinho, G.A.S. *et al.* (2021) Shade modifies behavioral and physiological responses of low to medium production dairy cows at pasture in an integrated crop-livestock-forest system. *Animals* 11(8), 2411. DOI: 10.3390/ani11082411.

Rivera-Ferre, M.G., Di Masso, M., Vara, I., Cuellar, M., Calle, A. *et al.* (2016) Local agriculture traditional knowledge to ensure food availability in a changing climate: Revisiting water management practices in the Indo-Gangetic plains. *Agroecology and Sustainable Food Systems* 40(9), 965–987. DOI: 10.1080/21683565.2016.1215368.

Rojas-Downing, M.M., Nejadhashemi, A.P., Harrigan, T. and Woznicki, S.A. (2017) Climate change and livestock: Impacts, adaptation, and mitigation. *Climate Risk Management* 16, 145–163. DOI: 10.1016/j.crm.2017.02.001.

Roy, P. (2008) Molecular dissection of bluetongue virus. In: Mettenleiter, T.C. and Sobrino, F. (eds) *Animal Viruses: Molecular Biology*. Caister Academic Press, Norfolk, UK, pp. 305–354.

Sajjanar, B., Aalam, M.T., Khan, O., Tanuj, G.N., Sahoo, A.P. *et al.* (2023) Genome-wide expression analysis reveals different heat shock responses in indigenous (Bos indicus) and crossbred (Bos indicus X Bos taurus) cattle. *Genes and Environment* 45, 1–15. DOI: 10.1186/s41021-023-00271-8.

Shalloo, L., Dillon, P., Rath, M. and Wallace, M. (2004) Description and validation of the moorepark dairy system model. *Journal of Dairy Science* 87(6), 1945–1959. DOI: 10.3168/jds.S0022-0302(04)73353-6.

Shalloo, L., Creighton, P. and O'Donovan, M. (2011) The economics of reseeding on a dairy farm. *Irish Journal of Agricultural and Food Research* 50, 113–122.

Shwartz, G., Rhoads, M.L., VanBaale, M.J., Rhoads, R.P. and Baumgard, L.H. (2009) Effects of a supplemental yeast culture on heat-stressed lactating holstein cows. *Journal of Dairy Science* 92(3), 935–942. DOI: 10.3168/jds.2008-1496.

Sigdel, A., Abdollahi-Arpanahi, R., Aguilar, I. and Peñagaricano, F. (2019) Whole genome mapping reveals novel genes and pathways involved in milk production under heat stress in US Holstein cows. *Frontiers in Genetics* 10, 1–10. DOI: 10.3389/fgene.2019.00928.

Singh, P., Mondal, S. and Singh, R.L. (2020) Introduction. In: Mondal, S. and Singh, R.L. (eds) *Advances in Animal Genomics*. Elsevier, Amsterdam, the Netherlands, pp. 1–12. DOI: 10.1016/B978-0-12-820595-2.00001-1.

Singh, P., Singh, P.K. and Singh, R.L. (2021) Transgenic animals. In: Mondal, S. and Singh, R.L. (eds) *Emerging Issues in Climate Smart Livestock Production*. Elsevier, Amsterdam, the Netherlands, pp. 171–208. DOI: 10.1016/B978-0-12-822265-2.00001-6.

Smith, J., Sones, K., Grace, D., MacMillan, S., Tarawali, S. *et al.* (2013) Beyond milk, meat, and eggs: Role of livestock in food and nutrition security. *Animal Frontiers* 3(1), 6–13. DOI: 10.2527/af.2013-0002.

Steinfeld, H. (2010) Livestock in a changing landscape. In: Steinfeld, H. (ed.) *Drivers, Consequences and Responses*, Vol. 1. Island Press, Washington, DC, p. 396.
Stewart, B.M., Block, J., Morelli, P., Navarette, A.E., Amstalden, M. *et al.* (2011) Efficacy of embryo transfer in lactating dairy cows during summer using fresh or vitrified embryos produced *in vitro* with sex-sorted semen. *Journal of Dairy Science* 94(7), 3437–3445. DOI: 10.3168/jds.2010-4008.
Stuttle, J.M. (2015) *Gliricidia sepium (Jacq.)*. Food and Agriculture Organization of the United Nations, Rome, Italy.
Talukder, M.A.I., Rahman, S.M., Miah, M.A.H., Shahjahan, M., Ali, M.Y. *et al.* (2018) Prevalence of different dairy cattle diseases in selected dairy areas and farms of Bangladesh. *Asian-Australasian Journal of Bioscience and Biotechnology* 3(3), 331–336. DOI: 10.3329/aajbb.v3i3.64834.
TAR Climate Change (2001) *The Scientific Basis (PDF) (Report)*. International Panel on Climate Change, Cambridge University Press, Cambridge, UK.
Taube, F., Gierus, M., Hermann, A., Loges, R. and Schönbach, P. (2014) Grassland and globalization – challenges for North-West European grass and forage research. *Grass and Forage Science* 69(1), 2–16. DOI: 10.1111/gfs.12043.
Tilman, D., Balzer, C., Hill, J. and Befort, B.L. (2011) Global food demand and the sustainable intensification of agriculture. *Proceedings of the National Academy of Sciences of the United States of America* 108(50), 20260–20264. DOI: 10.1073/pnas.1116437108.
Transboundary breed: Holstein (black and white) (2023) *Domestic Animal Diversity Information System*. Food and Agriculture Organization of the United Nations, Rome, Italy.
Tresoldi, G., Schütz, K.E. and Tucker, C.B. (2016) Assessing heat load in drylot dairy cattle: Refining on-farm sampling methodology. *Journal of Dairy Science* 99, 8970–8980. DOI: 10.3168/jds.2016-11353.
Truter, W.F., Dannhauser, C.S., Smith, H. and Trytsman, G. (2014) *Cenchrus ciliaris* (Foxtail/Blue buffalo grass). In: *Integrated Crop and Pasture-Based Livestock Production Systems*. Conservation Agriculture—Part 6, SA Grain, pp. 80–82. South Africa.
Twine, R. (2021) Emissions from animal agriculture—16.5% is the new minimum figure. *Sustainability* 13(11), 6276. DOI: 10.3390/su13116276.
Valerie, P., Alderson, L., Hall, S.J.G. and Sponenberg, D.P. (2016) *Mason's World Encyclopedia of Livestock Breeds and Breeding*, 6th edn. CAB International, Wallingford, UK.
Van Laer, E., Tuyttens, F.A.M., Ampe, B., Sonck, B., Moons, C.P.H. *et al.* (2015) Effect of summer conditions and shade on the production and metabolism of Holstein dairy cows on pasture in temperate climate. *Animal* 9, 1547–1558. DOI: 10.1017/S1751731115000816.
Walker, P.J. and Klement, E. (2015) Epidemiology and control of bovine ephemeral fever. *Veterinary Research* 46, 124. DOI: 10.1186/s13567-015-0262-4.
Wang, Z., Niu, K., Rushdi, H.E., Zhang, M., Fu, T. *et al.* (2022) Heat stress induces shifts in the rumen bacteria and metabolome of buffalo. *Animals* 12(10), 1300. DOI: 10.3390/ani12101300.
West, J.W. (2003) Effects of heat-stress on production in dairy cattle. *Journal of Dairy Science* 86, 2131–2144. DOI: 10.3168/jds.S0022-0302(03)73803-X.
White, R., Murray, S. and Rohweder, M. (2000) *Pilot Analysis of Global Systems: Grassland Ecosystems*. World Resources Institute. Available at: https://wriorg.s3.amazonaws.com/s3fs-public/pdf/page_grasslands.pdf (accessed July 2024).
Willett, W., Rockström, J., Loken, B., Springmann, M., Lang, T. *et al.* (2019) Food in the Anthropocene: The EAT–lancet commission on healthy diets from sustainable food systems. *The Lancet* 393(10170), 447–492. DOI: 10.1016/S0140-6736(18)31788-4.
Wolfenson, D., Roth, Z. and Meidan, R. (2000) Impaired reproduction in heat-stressed cattle: Basic and applied aspects. *Animal Reproduction Science* 60, 535–547. DOI: 10.1016/s0378-4320(00)00102-0.
Wu, T. (2021) Quantifying coastal flood vulnerability for climate adaptation policy using principal component analysis. *Ecological Indicators* 129, 108006. DOI: 10.1016/j.ecolind.2021.108006.
Zimbelman, R.B., Baumgard, L.H. and Collier, R.J. (2010) Effects of encapsulated niacin on evaporative heat loss and body temperature in moderately heat-stressed lactating holstein cows. *Journal of Dairy Science* 93, 2387–2394. DOI: 10.3168/jds.2009-2557.

8 Aquatic Food Security: Challenges and Opportunities

Bimal Prasanna Mohanty[1]*, Himanshu Sekhar Swain[2] and Yasmeen Basade[1]
[1]Indian Council of Agricultural Research, Fisheries Science Division, KAB II, New Delhi, India; [2]ICAR—Central Institute of Freshwater Aquaculture, Bhubaneswar, India

Abstract
Fisheries and aquaculture have increasingly contributed to global food security and nutrition in the past few decades. Aquaculture is one of the fastest-growing food production sectors with great promise for food and nutrition security and achieving Sustainable Development Goals (SDGs), especially Goal #2, Zero hunger by 2030, besides other goals (# 1, 3, 12, 14, and 17). Degraded ecosystems, severe climate crisis, biodiversity loss, biological hazards, pathogens of human concern, marine biotoxins, chemical hazards in wild-capture production, aquaculture and supply chains, social environmental welfare, and ethics are the key challenges faced by the aquatic food industry around the globe. This chapter explores the potential effects of different challenges, especially those of climate change on aquatic food security. Their implications on the sector's sustainability, as well as suitable mitigation measures against the challenges for the aquatic food industry and opportunities, have been discussed.

Keywords: Aquatic foods, Aquaculture, Climate change, Fisheries, Food security

8.1 Introduction

Globally, the demand for aquatic food is increasing, with a rise in income as well as awareness among the population about its nutritional facts and health benefits. Food from the aquatic habitats comes from diverse sources comprising finfish, shellfish, mollusks, seaweeds, and many other aquatic plants and animals. Aquatic foods are rich sources of proteins, essential amino acids, omega-3 polyunsaturated fatty acids (PUFAs)—namely, EPA (eicosapentaenoic acid) and DHA (docosahexaenoic acid)—and micronutrients such as vitamins and minerals. The nutrient-rich aquatic foods support the overall development of human health, strengthen the immune system, as well as prevent cardiovascular diseases (Gupta *et al.*, 2015; Singh *et al.*, 2016; Gupta *et al.*, 2023). The nutritive value of many of the commercially available aquatic foods has been documented. Nutritional information on amino acid composition, macronutrients (Na, K, Ca, Mg, P), micronutrients, trace elements (Fe, Cu, Zn, Mn, Se), and fat-soluble vitamins (A, D, E, K) content of important food fishes has been documented (Mohanty *et al.*, 2014, Mohanty *et al.*, 2016). This gives an understanding of their utility in human nutrition and in assessing their potential contribution (daily value %) to the recommended daily allowance (RDA).

*Corresponding author: bimal.mohanty@icar.gov.in

DOI: 10.1079/9781800625037.0008

On the other hand, seaweeds have also been explored for their novel biomolecules, and various nutraceutical products have been developed. It is observed that the aquatic food demand varies from one region of the world to another, based on the people's culture, food preferences, and availability. Furthermore, to attract consumers, various value-added aquatic food products offering enhanced health benefits are available in the market.

8.2 Climate Change and Aquatic Food Security

The challenges of population growth, emerging economies, and growing instability associated with land and water resources, compounded by the impact of climate change, are seriously affecting aquatic food resources. Under this scenario, there is a growing need to promote efficient and sustainable use of natural resources and develop climate-resilient fisheries and aquaculture for the betterment of the population that is dependent on them for food, nutrition, and livelihood. Effective addressing of climate change impacts is crucial to foster the resilience of this sector for sustained and greater contribution to sustainable development goals (SDGs) related to no poverty, zero hunger, good health and well-being, and rational use of natural resources—life below water and life on land.

Climate change has become a global issue. Climate change refers to the long-term changes in the statistical distribution of weather patterns around the globe, which take decades to millions of years (Yazdi and Shakouri, 2010; IPCC, 2014), and the change is too slow for humans to sense. After the Industrial Revolution, climate change was contributed at a rapid rate by humans through the use of fossil fuels (coal and oil). In other words, the emission of CO_2 was lower before the industrial era. The global average atmospheric concentration of CO_2 was 280 parts per million (ppm) or less before the industrial revolution started in the mid-1700s and has increased up to 419.3 ppm in 2023, setting a new record (Lindsey, 2024).

Global food production is currently seen to be at risk from climate change, which also poses a serious threat to the amount and quality of production (Hamdan *et al.*, 2015; Myers *et al.*, 2017). It has also posed a serious threat to the sustainability of the agricultural food production sectors. Fisheries and aquaculture cannot keep them away from the impact of climate change. The effect of climate change can put much pressure on the fisheries sector by destroying their marine and freshwater ecosystems. If action is not taken in time to minimize the situation, this may jeopardize not only people's livelihoods but also the economies of entire nations. In a high-emissions scenario, the countries of Southeast Asia are at "high" to "very high" risk for one or more food system outcomes by the middle of this century. The nations with the biggest emissions of greenhouse gases (GHGs) are not always the ones most at risk from climate change, but everyone will be affected equally and severely.

The anticipated effects of climate change are posing a danger to food security, specifically to the availability of dietary proteins (Kandu, 2017). All four dimensions (availability, accessibility, utilization and stability) could be impacted, along with human health, livelihood assets, food production and distribution, shifting purchasing power, and market dynamics. More than 500 million people in developing nations rely on aquaculture and fisheries for livelihood support. With 7% yearly growth, aquaculture is one of the fastest-growing food production systems in the world.

As aquaculture contributes significantly to global food security, nutrition, and livelihoods, the consequences of climate change on sustainability have attracted a lot of attention recently (Blanchard *et al.*, 2017; Dabbadie *et al.*, 2018; FAO, IFAD, UNICEF, WFP and WHO, 2020). Numerous projections indicate that the aquaculture value chain as a whole is susceptible to climate change (Cochrane *et al.*, 2009; Fleming *et al.*, 2014; Bueno and Soto, 2017; Barange *et al.*, 2018; Dabbadie *et al.*, 2018). However, with unsustainable harvests from capture fisheries, the aquaculture sector is seen as the only solution to meeting the rising demand for aquatic products globally (FAO, IFAD, UNICEF, WFP and WHO, 2020). Aquaculture needs to grow sustainably during the coming years to meet the growing demand for aquatic foods worldwide, particularly in areas with food shortages, and to generate new jobs and income channels.

The crucial contention is whether the sector is expanding quickly and responsibly enough to satisfy the need that is anticipated, which will be again made worse by the changing climate and a rapidly expanding human population.

Most of the literature that is accessible is heavily skewed toward demonstrating how aquaculture production systems will be adversely affected by climate change, leaving other stages of the value chain uncertain (Fleming *et al.*, 2014). Furthermore, quantifying the effects of climate change on aquaculture is challenging due to the lack of global models that can account for both the direct and indirect effects. However, there has been a trend in most of these studies to focus more on the expected negative consequences of climate change on aquaculture, with significantly less attention paid to the beneficial effects that are crucial for developing adaptation measures. A more thorough examination of climate change's benefits and drawbacks will increase producer readiness and reduce risks.

8.3 Causes and Elements of Climate Change

There are various human-made causes for the rapid climate change. The immediate cause of climate change is global warming. Global warming is the rise in average global temperature, and it is caused by the increase in GHGs such as CO_2, CO, methane (CH_4), and nitrous oxide (N_2O), which traps sunlight; in turn, it traps heat in the atmosphere. Deforestation and forest degradation for construction and agriculture are also some of the main reasons for climate change since they release GHGs such as CO_2. Other causes of climate change include the use of transportation (mainly relying upon fossil fuels) and the manufacturing of goods such as plastics, metals, clothes, and electronics, which emit GHG in enormous amounts.

Apart from that, we are aware that the sun is Earth's main energy source; nonetheless, minor changes made over time can result in a shift in the planet's climate. Numerous natural phenomena like the earth's orbit around the sun, volcanic eruptions, and atmospheric concentrations of GHGs all contribute to changes in the planet's temperature. The main factors associated with climate change are expected to be increased temperatures, ocean acidification, toxic algal blooms, altered rainfall and precipitation patterns, sea level rise, altered sea surface salinity and extreme weather events (Brander, 2007; Ficke *et al.*, 2007; Barange *et al.*, 2018). The element-wise impact especially on aquatic ecosystems and fisheries is discussed below.

8.3.1 Temperature

For aquatic species to grow and thrive, temperature is essential (Ngoan, 2018). The average world temperature is expected to rise by 1.5°C this century, which means that most fish species—especially those in cold water species like Atlantic halibut, salmon, and cod—will likely suffer at greater rates due to thermal stress (Gubbins *et al.*, 2013). As a result, extended heat stress may have many kinds of effects on aquaculture productivity, the main one being decreased output. Further, chronic stress alters immunological responses, neuroendocrine, and osmoregulatory systems in economically significant species (Gazeau *et al.*, 2014; Paukert *et al.*, 2016; Stévant *et al.*, 2017; Stewart *et al.*, 2019; Zhang *et al.*, 2019). Furthermore, the metabolism, physiology, feeding behavior, and growth performance of most fish species are being affected (Marcogliese, 2008; Akegbejo-Samsons, 2009; Lemasson *et al.*, 2018). Additionally, the ocean's ability to absorb carbon dioxide is gradually diminished by rising ocean temperatures and the ensuing ocean acidification, which results in changes to water systems' hydrology and hydrography as well as the occurrence of red tides (Cochrane *et al.*, 2009). The physiological responses of the most commercially significant species to temperature rise, however, are still poorly understood and skewed toward adult stages, leaving early ontogenetic stages like eggs, larvae, and fingerlings in the dark.

8.3.2 Ocean acidification

Ocean acidification is the outcome of atmospheric CO_2 uptake causing a drop in the pH values of ocean water. The carbon dioxide dissolves in the ocean producing carbonic acid, thus lowering its pH and raising its acidity. Since the start of the industrial revolution, the pH of the ocean's surface waters has dropped from 8.21 to 8.10, causing ocean acidification (Lindsey, 2024). According to Seggel and De Young (2016), the ocean is thought to be able to retain roughly 50 times as much CO_2 as the atmosphere. A number of aquatic species' growth, development, calcification, survival, and abundance will be negatively impacted by the anticipated increase in CO_2 uptake by seas at 1.5°C or higher global warming (IPCC, 2018).

Increased CO_2 buildup in water causes pH levels to drop, making the water acidic. This could endanger the environmental sustainability of aquaculture production systems by deteriorating the water quality and resulting in low productivity. Furthermore, increasing ocean acidity decreases the availability of carbonate needed by shell-forming organisms such as prawns, mussels, oysters, and corals (Hoegh-Guldberg *et al.*, 2007; Weatherdon *et al.*, 2016; Kibria *et al.*, 2017). This could pose a threat to the production of marine aquaculture (Yazdi and Shakouri, 2010; Kroeker *et al.*, 2014). According to Cooley *et al.* (2012), Mangan *et al.* (2017), and Thomsen *et al.* (2017) with species' adaptive capability, rate of change, and complex biophysical feedback between the organism and environment, the species will be able to adjust to changes in ocean acidity (Silbiger and Sorte, 2018). Finfish, especially marine species, are susceptible to ocean acidification due to the presence of calcified otoliths (Clements and Chopin, 2017). It may have an impact on their growth and development (Frommel *et al.*, 2014), RNA viability (Franke and Clemmesen, 2011), and damage to tissues and respiration (Munday *et al.*, 2009; Frommel *et al.*, 2012). Despite this, the effects of ocean acidification on finfish are still not completely understood (Wittmann and Pörtner, 2013; Clements and Chopin, 2017).

8.3.3 Diseases and harmful algal blooms

The aquatic organism becomes more prone to illness when subjected to thermal stress conditions and rising temperatures may encourage the emergence of more exotic diseases (Collins *et al.*, 2020). Therefore, under a changing climate scenario, it is expected that warm water disease outbreaks will occur more frequently in addition to the possibility of occurrence of new diseases (Sae-Lim *et al.*, 2017). Increased water temperature has the potential to quicken pathogen replication, virulence spread across various finfish and shellfish, and lengthened life cycles. In salmon culture, diseases like argulus and sea lice could always be a problem.

8.3.4 Rainfall

Increased rainfall (flooding) and periods of low or no rainfall (drought) are the two directly opposing effects of changes in the pattern of precipitation, or rainfall. Rising rainfall will increase the hazards in lowland areas. These hazards include fish loss from ponds during floods, undesired species invading the pond and embankment wall erosion, and infilling causing damage to the pond (Rutkayová *et al.*, 2018). Water stress from droughts can result in shortages and degradation, which can harm aquaculture productivity. The anticipated scarcity of water resulting from climate change is expected to escalate disputes over water among many user groups, including aquaculture, agriculture, households, and industries (Barange *et al.*, 2018). This will have an impact on diversification and sustainability across the board in aquaculture. To understand how various fish species and life stages—especially those with significant economic implications—respond to variations in precipitation and rainfall patterns, however, in-depth research is required.

8.3.5 Sea level rise

According to IPCC (2018), sea level rise under 1.5°C global warming will be about 0.1 m, and the global warming temperature rise will be 2°C by 2100. It is anticipated that the rise will continue after 2100, with the rate and amount of increase most likely to be determined by the future GHG pathways (IPCC, 2018). Sea level rise poses a threat to several coastal habitats, including salt marshes and mangroves, which are vital for preserving wild fish populations and providing seed for aquaculture (Kibria *et al.*, 2017). Saltwater intrusion from rising sea levels is expected to have an impact on aquaculture production infrastructure, including ponds, cages, tanks, and pens, especially in lowland areas (Kibria *et al.*, 2017). Freshwater fisheries, aquaculture, and agricultural productivity are all thought to be negatively impacted by groundwater salinization (Kibria *et al.*, 2017). Inland and marine aquaculture output may be threatened by changes in species composition, organism abundance and distribution, ecosystem productivity, and phenological shifts brought on by sea level rise (Doney *et al.*, 2012). The fishmeal and fish oil supplies, as well as the sustainability of fish breeding programs due to the increased scarcity of wild seeds, will be significantly affected by the impact of climate change on capture fisheries through changes in the abundance and distribution of fish species (Bradley *et al.*, 2015; Blanchard *et al.*, 2017; Bueno and Soto, 2017).

8.3.6 Variations in sea surface salinity

Sea salinity fluctuations can be caused by greater evaporation brought on by rising temperatures, and the higher salinity effect has a direct correlation with coastal aquaculture production systems. Higher salinity had detrimental effects on freshwater prawn production (Nguyen *et al.*, 2018); meanwhile, Baker *et al.* (2005) found that juvenile clam mortality had increased with an increase in sea surface salinity. In contrast, oysters at lower salinity were found to be more susceptible to bacterial invasion (Rodrick, 2008). Additionally, Rodrick (2008) noted that changes in salinity may have an impact on oysters' immune systems, specifically on hemocytes' (blood cells') potential to stave off bacterial invasion.

8.3.7 Extreme weather events

Extreme weather phenomena like storms, waves, and cyclones are predicted to have an impact on aquaculture growth, particularly for items that are marine ornamental and located near coasts (Toussaint *et al.*, 2018). The most significant risks to aquaculture production and other associated coastal industries are thought to be storm surges, waves, and coastal erosion. For example, the effects of waves and cyclones will probably pose a greater risk to open-water cage aquaculture. The social and economic sustainability of aquaculture in these places may be significantly impacted by equipment failure, site relocation, or modifications to production methods brought on by extreme weather events (Gubbins *et al.*, 2013).

8.4 Effects of Climate Change

Climate change has a definite role in the environment and the organism. The rise in temperature and change in the precipitation pattern are the main consequences of climate change. The scarcities of freshwater, ocean acidification, and sea level rise are the series of effects caused by climate change. It is challenging to link certain weather occurrences to global warming because rising temperatures might trigger more significant changes, such as the retreat of glaciers, the shrinking of the Arctic, and an increase in sea level globally. Drought and flooding can be caused by variations in the quantity and timing of precipitation. The majority of global climate change models predict that many mountain ranges would see a decrease in snowpack. Although faster sea level rise on century time intervals cannot be ruled out, such changes are anticipated to occur over millennial timescales. Sea surface temperatures are expected to rise by 2–5°C by 2100, according to current climate change models (IPCC, 2001).

Climate change, directly and indirectly, affects the food production sectors and the food safety of the global population. Also, climate change would be a serious hindrance to the sustainability of agriculture and other allied food-producing sectors like aquaculture. Agriculture needs to change significantly in order to feed the world's expanding population, support economic expansion, and reduce poverty. Climate change will make this endeavor more challenging. The agricultural and food security sectors have already been impacted by an increase in extreme weather events and the unpredictability of weather patterns, which has resulted in lower incomes and decreased production in vulnerable areas. Agriculture may reduce the availability of aquatic food and restrict the intake of vital minerals including iron, zinc, vitamin B12, and omega-3 fatty acids in populations where shortage thresholds have already been reached.

8.4.1 Food security

The availability, use, accessibility, stability across time, and sustainability of food are all aspects of food security (HLPE, 2020). In particular, for populations already vulnerable to food insecurity due to lower income, lower physical access to nutritious food, social discrimination, or other factors, the climate change risk to food security refers to the breakdown of food systems, including crops, livestock, and fisheries, as well as disruptions in food distribution linked to global warming, drought, flooding, and precipitation variability and extremes (O'Neill *et al.*, 2022).

The world currently has a high level of food insecurity, 720–811 million people are undernourished, and roughly 2.3 billion from malnourished. SDG 2 calls for achieving zero hunger in the world by 2030 (FAO, IFAD, UNICEF, WFP and WHO, 2021). In order to distinguish between acute and chronic food insecurity, the Integrated Food Security Phase Classification (IPC) states that 200,000 people experience catastrophic food insecurity, 32.3 million experience emergency food insecurity, 112.3 million experience crisis food insecurity, and 210 million experience stressed food insecurity.

When the food supply is adequate, secure, sustainable, resilient, and sound, the aquatic food supply helps to ensure food security. Consumer demands and preferences are satisfied in terms of both quantity and nutritional value when there is an adequate supply of food. However, a sustainable food supply provides food now, today, and tomorrow.

8.4.2 Effect of climate change on aquaculture sector and its sustainability

When the food supply is adequate, secure, sustainable, resilient, and sound, the aquatic food supply helps to ensure food security. Consumer demands and preferences are satisfied in terms of both quantity and nutritional value when there is an adequate supply of food. Since aquaculture is a risky sector that needs more precautions and is vulnerable to environmental conditions, climate change will influence more on the same (Fig. 8.1). The direct way of affecting aquaculture production will be through change in physiology and the aquatic animals' response to the environmental parameters. For example, the temperature rise will affect the fish and shellfish's homeostasis, and so, there will be an increase in the difficulties to culture those animals, as the farmers would face severe mortalities and disease outbreaks during the culture. Climate change indirectly affects aquaculture sustainability, disturbing the primary and secondary productivity in the aquaculture system. In turn, it affects the supply of raw materials to produce animal feed such as fish meal and fish oil. Therefore, there will be an increase in the price of production, and so the animal protein produced through aquaculture will not be able to reach the people who cannot afford the protein-rich food.

Climate change makes the seasons more intense and extreme. The summer would become hotter, and the winter would become colder. The extremity and the seasonal changes bring difficulties into agriculture and allied sectors for the rearing of plants and animals. It also creates pressure for the organisms to adapt to climate change. The organisms that have lower adaptability toward acute environmental changes tend to fail to survive. The survival loss diminishes the sustainability of the culture of those plants and animals. The ocean acidification of the water results in a decrease in the ability of skeletal formation of shellfish species such as shrimps, crabs, and mollusks. In addition to that, the acidification leads to a collapse in the intercellular transport mechanism, which leads to disintegration in their physiology and metabolism. The rise in the water temperature will affect the aquatic species to a larger extent by inducing thermal stress. Besides the animals with thermal stress being prone to the pathogens, replication rate, virulence, and longevity of the disease-causing pathogens would become greater in higher temperatures. The elevated degree of disease outbreaks forces the farmers

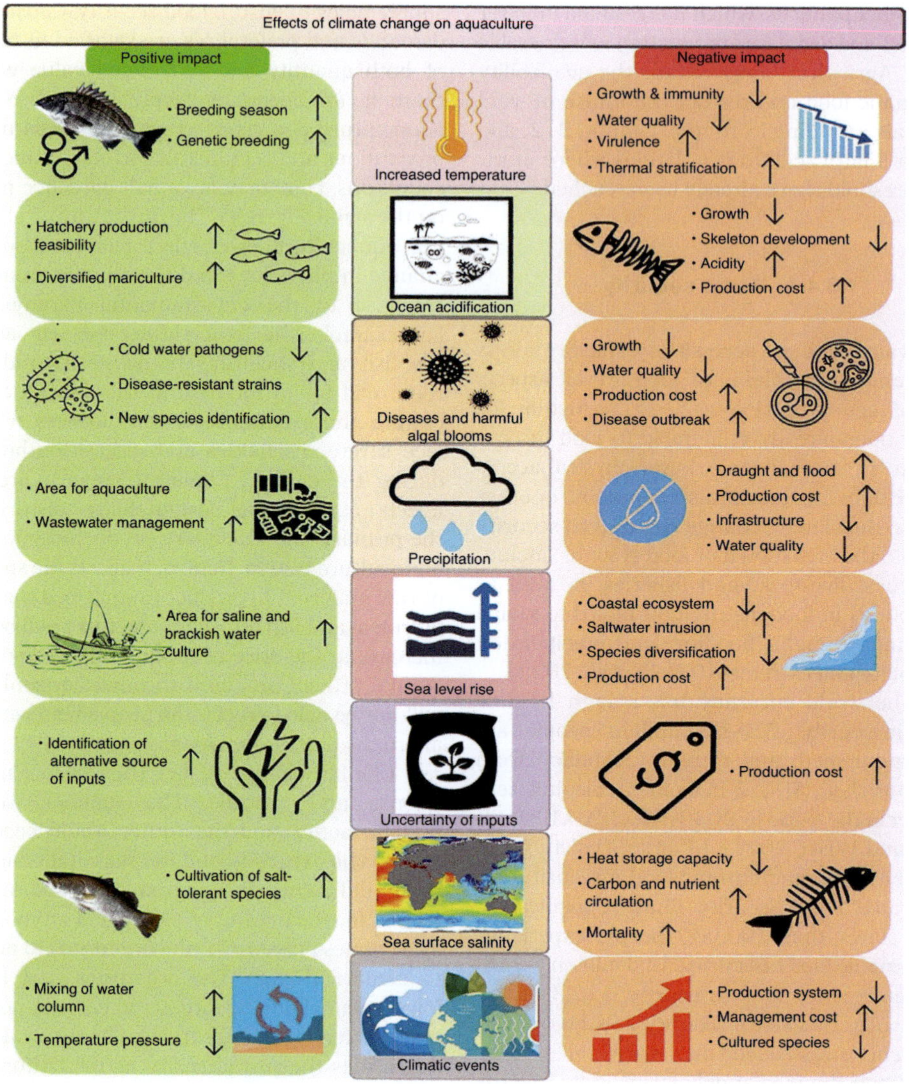

Fig. 8.1. Effects of climate change on aquaculture.

to spend a higher cost on the treatment of the disease which leads to a reduction in the profit and diminishes the socioeconomic condition of the farmers.

Climate change results in a change in the pattern of precipitation rate around the globe, severely affecting agriculture farmers who are fully dependent on rainwater for irrigation. It also affects the aquaculture sector, where the aquaculture farmers depend on rainfed waterbodies for their culture. The decline in precipitation also results in the exploitation of the groundwater for their usage, which further leads to the lowering of the water table.

The agriculture and capture fisheries are the pillars of the input supply for aquaculture, as the ingredients for the aqua feeds are mainly taken from agro-related products, and the ingredients for the protein and the seed from the wild broodstock are taken from the capture fisheries (e.g. fishmeal). Due to climate change, the sustainability of those food production sectors will be under a big question mark, which may result in aquaculture being unsustainable and collapsing the share of the sector to food security.

Food security is not only dependent upon food production; food supply also plays a significant role. Climate change causes extreme weather patterns and natural disasters. These disasters are a hindrance to the transportation of the food harvested and logistics. It also increases the demand in the market for the food product, and so there will be an elevation in the price. This, in turn, hinders people below the poverty line from consuming nutritious food and thus leads to malnutrition. Natural disasters like floods and cyclones would become frequent events as a result of climate change. These disasters make agriculture and aquaculture more challenging and result in crop failures, which results in economic loss and time loss to the farmers. The same thing affects rural communities and the working class in the same sectors more seriously. In comparison to large-scale producers, small-scale farmers will be more vulnerable to the effects of climate change because they will face higher production costs in farm management and a lack of support systems to help them recover (Schjolden, 2004; Sult *et al.*, 2005; Hamdan *et al.*, 2012 ; King and Harrington, 2018).

8.4.3 Impacts of aquaculture on climate change

According to Barange *et al.* (2018), the aquaculture industry's share of worldwide GHG emissions in 2010 was projected to be 385 million tons, or 7% of the emissions from the agricultural sector. According to estimates, the worldwide fisheries industry increased its GHG emissions by 28% between 1990 and 2011, mostly as a result of expanding fishing activities (Parker *et al.*, 2018). Animal production, especially livestock farming, is thought to account for 45% of agriculture's total net contribution (Preto *et al.*, 2015). It is the sector's largest contributor, contributing mostly to CH_4, CO_2, and N_2O (De Silva and Soto, 2009; Preto *et al.*, 2015; Zolnikov, 2018). However, CO_2 is the primary GHG released by aquaculture due to the regular respiration of aquatic plants and animals (De Silva and Soto, 2009). According to Handisyde *et al.* (2017), the consequences of climate change on aquaculture producers are anticipated to vary according to the freshwater, brackish, and marine culture habitats. However, because aquaculture's life cycle analysis is still in its early stages, there is still a gap in our understanding of the pathways and contribution of aquaculture output to global GHG emissions and this has to be filled by more thorough investigations.

8.4.4 Positive impacts of climate change

Longer growing seasons and the production of warmer water species, such as tilapia, oysters, mussels, and giant tiger prawns, may be encouraged by warmer periods, particularly in temperate regions (Pickering *et al.*, 2011; Troell *et al.*, 2017; Guyondet *et al.*, 2018; Collins *et al.*, 2020). Warmer regions are anticipated to encourage the growth of aquaculture output in colder places like the Arctic (Chan *et al.*, 2019). Additionally, warmer seasons can present chances to cultivate different species and support ongoing advancements in aquatic organisms' genetic modifications (Gubbins *et al.*, 2013; Bueno and Soto, 2017). Through higher production outputs and job possibilities, these prospects will support social sustainability; similarly, they will support economic sustainability through higher profits and lower management costs in these domains. Doing this will need the adoption of cutting-edge molecular biology techniques as well as useful genetic enhancement strategies in aquaculture. The fact that large-scale aquaculture producers may become entirely or partially dependent on hatcheries for the generation of sprats could have a positive effect on the hatchery owners. Furthermore, as these hatcheries will need a lot of labor to meet the increasing demand this might lead to additional job possibilities for the local community and support the social and economic sustainability of aquaculture production. On the other hand, the emergence of unfavorable conditions owing to vibriosis and winter ulcer, two diseases that harm Atlantic salmon, may cause them to progressively go extinct (Sae-Lim *et al.*, 2017). This could be beneficial to the development of this particular species of fish. Positively, sea level rise may expand the regions that may support the brackish water culture of valuable species like mud crabs and prawns (Kibria *et al.*, 2017). This could contribute to the sustainability of aquaculture output by creating new options, especially for coastal locations. Changes in species composition, the abundance and distribution of creatures, ecosystem productivity, and fish species' phenological shifting are all anticipated outcomes of sea level rise. Storms and other extreme weather events will probably be important in redistributing nutrients and water columns that were previously limited to specific columns because of thermal stratification (Seggel and De Young, 2016). This could help aquaculture operations become more environmentally sustainable. Furthermore, storms might have a critical role in lowering water temperatures and related hazards that could endanger both wild and cultivated organisms (Seggel and De Young, 2016).

8.4.5 Monitoring stress in aquatic animals

The organism's capacity to endure environmental disruption from homeostasis, which is commonly referred to as stress, is just as important to its success as niche adaptation (Clark and Peck, 2009). All living things perceive temperatures over their optimal range as heat stress (HS), which disrupts cellular equilibrium and can cause severe growth and development retardation or even death. When cells are subjected to high temperatures or other forms of stress, the production of a class of functionally related proteins known as heat shock proteins (HSP) increases. One important aspect of the heat shock (stress) response (HSR) is the sharp upregulation of the HSPs. The production of HSPs, such as HSP70, has been described as a species-specific stress management strategy; examples of these include *Sperata seenghala* (Mohanty *et al.*, 2018), freshwater catfish *Rita rita* (Mohanty *et al.*, 2010b), Indian catfish *Clarias batrachus*, Indian major carps *Labeo rohita*, *Catla catla*, and *Cirrhinus mrigala*; exotic carp *Cyprinus carpio* var. communis and the murrel *Channa striatus*; and the climbing perch *Anabas testudineus* (Mohanty and Mohanty, 2009). Standardizing methods that are appropriate for assessing stress brought on by the effects of climate change and global warming on aquatic life from capture fisheries and aquaculture systems is necessary. The expression of HSP70 has been documented in numerous fish species, it could potentially function as a suitable tool for overseeing the effects of thermal stress and global warming. Nevertheless, while HSP70 proteins are expressed in other contexts, it is imperative to pinpoint the heat shock (stress) transcription factors (HSFs) that are specifically linked to global warming, thermal stress, and climate change.

8.5 Mitigations and Adaptation

Systems for agriculture will need to change in order to become more productive, more efficient in their use of inputs, more stable and less variable in their outputs, and more resistant to shocks, risks, and long-term climate variability. The base of natural resources must not be exhausted in order to complete this change. Additionally, it will need to involve a rise in carbon sinks and a drop in GHG emissions, both of which will greatly aid in the mitigation of climate change. Fisheries and aquaculture adaptations to climate change are becoming more important, as a result of irreversible changes brought about by increased warming that call for immediate intervention from the ocean to enhance and expedite climate mitigation and adaptation strategies.

Climate-smart agriculture and aquaculture refers to a method for the sustainable growth of these two industries that increases productivity while simultaneously lowering GHG emissions from these two industries and ensuring the nutritional security of millions of people. In order to create resilience and deal with climate change as effectively and efficiently as possible, farming communities, ecosystems, and populations as a whole may find it helpful to adopt mitigation and adaptation strategies (Zolnikov, 2019). The goal of mitigation is to slow down or even stop the rate of climate change (Mohanty *et al.*, 2010a; IPCC, 2014). Aquaculture's primary source of GHG emissions is thought to be feed production. Building resilience to the outcomes and the ability to take advantage of new opportunities in a sustainable and ethically accountable manner are the main goals of adaptation (Bueno and Soto, 2017). According to Yazdi and Shakouri (2010), it entails anticipating changes and accounting for them in both short- and long-term planning and decision making.

8.6 Challenges to Adaptation Mechanism

Adaptation is thought to be the most practical short-term response to climate change, but there may be a number of obstacles in the way of its implementation. Failure to recognize that producers will be impacted differently by climate change depending on their particular geography, environment, level of awareness about climate change, population, dynamics of their communities, state of the economy, and their present situation (Marshall *et al.*, 2009). The producers and stakeholders in the aquaculture sector will not have much profit from adaptation techniques unless these issues are taken into consideration (Zolnikov, 2019). Accurate future projections of different aquaculture sectors will likely determine the producers' ability to respond to climate change. Policymakers and aquaculture producers may be misled by unrealistic and biased estimates of potential threats linked to climate change, prompting them to make poor decisions and precautions.

8.7 Conclusions

Over the past few decades, there has been a substantial advancement in our understanding of climate change, but many issues remain. The challenge of reducing the effects of climate change and adjusting to them will involve global cooperation among a variety of specialists in different sectors. The contribution of the average person will be crucial in mitigating the effects of climate change and safeguarding the planet from its associated risks. Climate change has a significant impact on fish producers, fishermen, and coastal communities. Because they are at the forefront, marshes, estuaries, and deltas are the most susceptible to climate change. Mitigation strategies are critically needed to resist and mitigate these escalating risks, as well as to adjust to their effects. As a healthy aquatic habitat is essential for livelihoods and global food security, there is an urgent need to spread awareness, find mitigation measures, and most importantly resilience; finding and developing climate-resilient plants, animals, and fish to ensure a brighter future for the planet Earth.

Acknowledgments

The authors are thankful to the Indian Council of Agricultural Research, New Delhi, for their support and encouragement.

References

Akegbejo-Samsons, Y. (2009) Climate change impacts on fisheries production in land-water interface. *IOP Conference Series: Earth and Environmental Science* 6, 292022. DOI: 10.1088/1755-1307/6/29/292022.

Baker, S.M., Baker, P., Heuberger, D. and Sturmer, L. (2005) Short-term effects of rapid salinity reduction in seed clams (*Mercenaria mercenaria*). *Journal of Shellfish Research* 24(1), 29–33. DOI: 10.2983/0730-8000(2005)24[29:SEORSR]2.0.CO;2.

Barange, M., Bahri, T., Beveridge, M.C.M., Cochrane, A.L., Funge-Smith, S. *et al.* (2018) *Impacts of Climate Change on Fisheries and Aquaculture, Synthesis of Current Knowledge, Adaptation and Mitigation Options.* FAO Fisheries and Aquaculture Technical Paper 627. Food and Agriculture Organization of the United Nations, Rome, Italy.

Blanchard, J.L., Watson, R.A., Fulton, E.A., Cottrell, R.S., Nash, K.L. *et al.* (2017) Linked sustainability challenges and trade-offs among fisheries, aquaculture and agriculture. *Nature Ecology & Evolution* 1, 1240–1249. DOI: 10.1038/s41559-017-0258-8.

Bradley, M., van Putten, I. and Sheaves, M. (2015) The pace and progress of adaptation: Marine climate change preparedness in Australia's coastal communities. *Marine Policy* 53, 13–20. DOI: 10.1016/j.marpol.2014.11.004.

Brander, K.M. (2007) Global fish production and climate change. *Proceedings of the National Academy of Sciences of the United States of America* 104(50), 19709–19714. DOI: 10.1073/pnas.0702059104.

Bueno, P.B. and Soto, D. (2017) *Adaptation Strategies of the Aquaculture Sector to the Impacts of Climate Change*. FAO Fisheries and Aquaculture Circular No. 1142 (FIAA/C1142 (En). Food and Agriculture Organization of the United Nations, Rome, Italy. DOI: 10.1111/raq.12140. (accessed 25 April 2024).

Chan, F.T., Stanislawczyk, K., Sneekes, A.C., Dvoretsky, A., Gollasch, S. *et al.* (2019) Climate change opens new frontiers for marine species in the Arctic: Current trends and future invasion risks. *Global Change Biology* 25, 25–38. DOI: 10.1111/gcb.14469.

Clark, M.S. and Peck, L.S. (2009) Triggers of the HSP70 stress response: Environmental responses and laboratory manipulation in an Antarctic marine invertebrate (*Nacella concinna*). *Cell Stress & Chaperones* 14, 649–660. DOI: 10.1007/s12192-009-0117-x.

Clements, J.C. and Chopin, T. (2017) Ocean acidification and marine aquaculture in North America: Potential impacts and mitigation strategies. *Reviews in Aquaculture* 9, 326–341. DOI: 10.1111/raq.12140.

Cochrane, K., De Young, C., Soto, D.T. and Bahri, D.T. (2009) *Climate Change Implications for Fisheries and Aquaculture: overview of Current Scientific Knowledge*. FAO Fisheries and Aquaculture Technical Paper no.530, Rome, Italy. Available at: https://openknowledge.fao.org/handle/20.500.14283/i0994e (accessed 25 April 2024).

Collins, C., Bresnan, E., Brown, L., Falconer, L., Guilder, J. *et al.* (2020) Impacts of climate change on aquaculture. *Marine Climate Change Impacts Partnership (MCCIP) Science Review* 2020, 482–520. DOI: 10.14465/2020.arc21.aqu.

Cooley, S.R., Lucey, N., Kite-Powell, H. and Doney, S.C. (2012) Nutrition and income from molluscs today imply vulnerability to ocean acidification tomorrow. *Fish and Fisheries* 13, 182–215. DOI: 10.1111/j.1467-2979.2011.00424.x.

Dabbadie, L., Anguilar-Manjarrez, J., Beveridge, M.C.M., Bueno, P.B., Ross, L.G. *et al.* (2018) Effects of climate change on aquaculture: Drivers, impacts and policies. In: Barange, M., Bahri, T., Beveridge, M.C.M., Cochrane, K.L., Funge-Smith, S. *et al.* (eds) *Impacts of Climate Change on Fisheries and Aquaculture, Synthesis of Current Knowledge, Adaptation and Mitigation Options*. FAO Fisheries and Aquaculture Technical Paper No.627. Food and Agriculture Organization of the United Nations, Rome, Italy, pp. 449–463. Available at: https://openknowledge.fao.org/handle/20.500.14283/i9705en (accessed 25 April 2024).

De Silva, S.S. and Soto, D. (2009) Climate change and aquaculture: Potential impacts, adaptation and mitigation. In: Cochrane, K., De Young, C., Soto, D.T. and Bahri, D.T. (eds) *Climate Change Implications for Fisheries and Aquaculture: Overview of Current Scientific Knowledge*. FAO Fisheries and Aquaculture Technical Paper No. 530, FAO, Rome, Italy, pp. 151–212. Available at: https://openknowledge.fao.org/handle/20.500.14283/i0994e (accessed 25 April 2024).

Doney, S.C., Ruckelshaus, M., Duffy, J.E., Barry, J.P., Chan, F. *et al.* (2012) Climate change impacts on marine ecosystems. *Annual Review of Marine Science* 4, 11–37. DOI: 10.1146/annurev-marine-041911-111611.

FAO, IFAD, UNICEF, WFP and WHO (2020) *The State of Food Security and Nutrition in the World 2020. Transforming Food Systems for Affordable Healthy Diets*. Food and Agriculture Organization of the United Nations, Rome, Italy. Available at: https://openknowledge.fao.org/server/api/core/bitstreams/9a0fca06-5c5b-4bd5-89eb-5dbec0f27274/content (accessed 25 April 2024).

FAO, IFAD, UNICEF, WFP and WHO (2021) *The State of Food Security and Nutrition in the World 2021. Transforming Food Systems for Food Security, Improved Nutrition and Affordable Healthy Diets for All*. Food and Agriculture Organization of the United Nations, Rome, Italy. DOI: 10.4060/cb4474en.

Ficke, A.D., Myrick, C.A. and Hansen, L.J. (2007) Potential impacts of global climate change on freshwater fisheries. *Reviews in Fish Biology and Fisheries* 17, 581–613. DOI: 10.1007/s11160-007-9059-5.

Fleming, A., Hobday, A.J., Farmery, A., van Putten, E.I., Pecl, G.T. *et al.* (2014) Climate change risks and adaptation options across Australian seafood supply chains – a preliminary assessment. *Climate Risk Management* 1, 39–50. DOI: 10.1016/j.crm.2013.12.003.

Franke, A. and Clemmesen, C. (2011) Effect of ocean acidification on early life stages if Atlantic herring (*Clupeaharengus* L. *Biogeosciences* 8(12), 3697–3707. DOI: 10.5194/bg-8-3697-2011.

Frommel, A.Y., Maneja, R., Lowe, D., Malzahn, A.M., Geffen, A.J. *et al.* (2012) Severe tissue damage in Atlantic cod larvae under increasing ocean acidification. *Nature Climate Change* 2, 42–46. DOI: 10.1038/nclimate1324.

Frommel, A.Y., Maneja, R., Lowe, D., Pascoe, C.K., Geffen, A.J. *et al.* (2014) Organ damage in Atlantic herring larvae as a result of ocean acidification. *Ecological Applications* 24, 1131–1143. DOI: 10.1890/13-0297.1.

Gazeau, F., Alliouane, S., Bock, C., Bramanti, L., Correa, M.L. *et al.* (2014) Impact of ocean acidification and warming on the Mediterranean mussel (*Mytilus galloprovincialis*). *Frontiers in Marine Science* 1, 62. DOI: 10.3389/fmars.2014.00062.

Gubbins, M., Bricknell, I. and Service, M. (2013) Impacts of climate change on aquaculture. *Marine climate change impacts partnership (MCCIP) Science Review* 2013, 318–327. DOI: 10.14465/2013.arc33.318-327.

Gupta, R., Singh, R.L. and Singh, P. (2015) Quantification of phytochemicals and evaluation of antioxidant potential of ethanolic leaf extract of Terminalia bellerica, Terminalia chebula and Emblica offinalis vis-a-vis Triphala. *International Journal of Pharmaceutical Sciences Review and Research* 32(2), 14–22. DOI: 10.18231/j.ijpca.2023.032.

Gupta, R., Shukla, S., Gupta, A., Singh, R.L. and Singh, P. (2023) Estimation of phytochemical constituents and analysis of antioxidant activity in different parts of Cassia fistula plant. *International Journal of Pharmaceutical Chemistry and Analysis* 10(3), 194–201. DOI: 10.18231/j.ijpca.2023.032.

Guyondet, T., Comeau, L.A., Bacher, C., Grant, J., Rosland, R. *et al.* (2018) Climate change influences carrying capacity in a coastal embayment dedicated to shellfish aquaculture. *Estuaries and Coasts* 38, 1593–1618. DOI: 10.18231/j.ijpca.2023.032.

Hamdan, R., Othman, A. and Kari, F. (2015) Climate change effects on aquaculture production performance in Malaysia: An environmental performance analysis. *International Journal of Business and Society* 16(3), 364–385. DOI: 10.18231/j.ijpca.2023.032.

Hamdan, R., Kari, F., Othman, A. and Samsi, S.M. (2012) Climate change, socio-economic and production linkages in east malaysia aquaculture sector. In: *International Conference on Future Environment and Energy IPCBEE 28*. IACSIT Press, Singapore.

Handisyde, N., Telfer, T.C. and Ross, L.G. (2017) Vulnerability of aquaculture-related livelihoods to changing climate at the global scale. *Fish and Fisheries* 18, 466–488. DOI: 10.18231/j.ijpca.2023.032.

HLPE (2020) *Food Security and Nutrition: Building A Global Narrative Towards 2030*. A Report by the High Level Panel of Experts on Food Security and Nutrition of the Committee on World Food Security, Rome, Italy. Available at: https://openknowledge.fao.org/server/api/core/bitstreams/8357b6eb-8010-4254-814a-1493faaf4a93/content (accessed 25 April 2024).

Hoegh-Guldberg, O., Mumby, P.J., Hooten, A.J., Steneck, R.S., Greenfield, P. *et al.* (2007) Coral reefs under rapid climate change and ocean acidification. *Science* 318(5857), 1737–1742. DOI: 10.1126/science.1152509.

IPCC Watson, R.T. and the Core Writing Team, (eds) (2001) *Climate change 2001: Synthesis report. A Contribution of Working Groups I, II, and III to the Third Assessment Report of the Intergovernmental Panel on Climate Change*. Cambridge University Press, Cambridge, UK, and New York. Available at: https://www.ipcc.ch/site/assets/uploads/2018/05/SYR_TAR_full_report.pdf (accessed 26 April 2024).

IPCC Field, C.B., Barros, V.R., Dokken, D.J., Mach, K.J. and Mastrandrea, M.D. (eds) (2014) *Climate change 2014: Impacts, adaptation, and vulnerability. part A: global and sectoral aspects.* Contribution of Working Group II to the Fifth Assessment Report of the Intergovernmental Panel on Climate Change. Cambridge University Press, Cambridge, UK, and New York. Available at: https://www.ipcc.ch/site/assets/uploads/2018/02/ipcc_wg3_ar5_full.pdf (accessed 26 April 2024).

IPCC Masson-Delmotte, V., Zhai, P., Pörtner, H.O., Roberts, D. and Skea, J. (eds) (2018) *Global warming of 1.5°C.* An IPCC special report on the impacts of global warming of 1.5°C above pre-industrial levels and related global greenhouse gas emission pathways, In the Context of Strengthening the Global Response to the Threat of Climate Change, Sustainable Development, and Efforts to Eradicate Poverty. Cambridge University Press, Cambridge, UK, and New York. Available at: https://www.ipcc.ch/site/assets/uploads/sites/2/2022/06/SR15_Full_Report_HR.pdf (accessed 26 April 2024).

Kandu, P. (2017) Papua new guinea. impacts of climate variations on local fisheries and aquaculture resources in PNG. In: Ramos, J.E. (ed.) *Ecological Risk Assessment of Impacts of Climate Change on Fisheries and Aquaculture Resources*. APEC Ocean and Fisheries Working Group October 2017. APEC publication number: APEC#217-OF-04.2, pp. 45–49. Available at: https://www.apec.org/docs/default-source/Publications/2017/12/Ecological-Risk-Assessment-of-Impacts-of-Climate-Change-on-Fisheries-and-Aquaculture-Resources/217OFWGEcological-Risk-Assessment-of-Impacts-of-Climate-Change-on-Fisheries-and-Aquaculture-Resource.pdf

Kibria, G., Haroon, Y.A.K. and Dayanthi, N. (2017) Climate change impacts on tropical and temperate fisheries, aquaculture, and seafood security and implications—a review. *Livestock Research Rural Development* 29(1), 1–16. Available at: http://www.lrrd.org/lrrd29/1/kibr29012.htm (accessed 26 April 2024).

King, A.D. and Harrington, L.J. (2018) The inequality of climate change from 1.5 to 2°C of global warming. *Geophysical Research Letters* 45, 5030–5033. DOI: 10.1029/2018GL078430.

Kroeker, K.J., Gaylord, B., Hill, T.M., Hosfelt, J.D., Miller, S.H. *et al.* (2014) The role of temperature in determining species' vulnerability to ocean acidification: A case study using *Mytilus galloprovincialis*. *PloS One* 9(7), e100353. DOI: 10.1371/journal.pone.0100353.

Lemasson, A.J., Hall-Spencer, J.M., Fletcher, S., Provstgaard-Morys, S. and Knights, A.M. (2018) Indications of future performance of native and non-native adult oysters under acidification and warming. *Marine Environmental Research* 142, 178–189. DOI: 10.1016/j.marenvres.2018.10.003.

Lindsey, R. (2024) Climate change: Atmospheric carbon dioxide. Available at: https://www.climate.gov/news-features/understanding-climate/climate-change-atmospheric-carbon-dioxide (accessed 26 April 2024).

Mangan, S., Urbina, M.A., Findlay, H.S., Wilson, R.W. and Lewis, C. (2017) Fluctuating seawater pH/pCO_2 regimes are more energetically expensive than static pH/pCO_2 levels in the mussel *Mytilus edulis*. *Proceedings of the Royal Society B* 284, 20171642. DOI: 10.1098/rspb.2017.1642.

Marcogliese, D.J. (2008) The impact of climate change on the parasites and infectious diseases of aquatic animals. *Revue Scientifique et Technique—Office International Des Epizooties* 27(2), 467–484. DOI: 10.20506/rst.27.2.1820.

Marshall, N.A., Marshall, P.A., Tamelander, J., Obura, D., Malleret-King, D. *et al.* (2009) *A Framework for Social Adaptation to Climate Change: Sustaining Tropical Coastal Communities and Industries*. IUCN, Gland, Switzerland. Available at: https://portals.iucn.org/library/efiles/documents/2010-022.pdf (accessed 26 April 2024).

Mohanty, S. and Mohanty, B.P. (2009) Global climate change: A cause of concern. *National Academy Science Letters* 32(5 & 6), 149–156. DOI: 10.20506/rst.27.2.1820.

Mohanty, B.P., Mohanty, S., Sahoo, J.K. and Sharma, A.P. (2010a) Climate change: Impacts on fisheries and aquaculture. In: Simard, S. (ed.) *Climate Change and Variability*. Intech Open, London. Available at: https://www.intechopen.com/chapters/11440 (accessed 26 April 2024).

Mohanty, B.P., Bhattacharjee, S., Mondal, K. and Das, M.K. (2010b) HSP70 expression profiles in white muscles of riverine catfish *Rita rita* show promise as biomarker for pollution monitoring in tropical rivers. *National Academy Science Letters* 33(5 & 6), 177–182.

Mohanty, B., Mahanty, A., Ganguly, S., Sankar, T.V., Chakraborty, K. *et al.* (2014) Amino acid compositions of 27 food fishes and their importance in clinical nutrition. *Journal of Amino Acids* 2014, 269797. DOI: 10.1155/2014/269797.

Mohanty, B.P., Sankar, T.V., Ganguly, S., Mahanty, A., Anandan, R. *et al.* (2016) Micronutrient composition of 35 food fishes from India and their significance in human nutrition. *Biological Trace Element Research* 174, 448–458. DOI: 10.1007/s12011-016-0714-3.

Mohanty, B.P., Mahanty, A., Mitra, T., Parija, S.C. and Mohanty, S. (2018) Heat shock proteins in stress in teleosts. In: Asea, A.A.A. and Kaur, P. (eds) *Regulation of Heat Shock Protein Responses. Heat Shock Proteins*, Vol. 13. Springer, Cham, Switzerland, pp. 71–94. DOI: 10.1007/978-3-319-74715-6_4.

Munday, P.L., Dixson, D.L., Donelson, J.M., Jones, G.P., Pratchett, M.S. *et al.* (2009) Ocean acidification impairs olfactory discrimination and homing ability of a marine fish. *Proceedings of the National Academy of Sciences of the United States of America* 106(6), 1848–1852. DOI: 10.1073/pnas.0809996106.

Myers, S.S., Smith, M.R., Guth, S., Golden, C.D., Vaitla, B. *et al.* (2017) Climate change and global food systems: Potential impacts on food security and undernutrition. *Annual Review of Public Health* 38, 259–277. DOI: 10.1146/annurev-publhealth-031816-044356.

Ngoan, L.D. (2018) Effects of climate change in aquaculture: Case study in Thua Thien Hue Province, Vietnam. *Biomedical Journal of Scientific & Technical Research* 10(1), 7551–7552. DOI: 10.26717/BJSTR.2018.10.001892.

Nguyen, L.A., Pham, T.B., Bosma, R., Verreth, J., Leemans, R. *et al.* (2018) impact of climate change on the technical efficiency of striped catfish, *Pangasianodon hypophthalmus* farming in the Mekong Delta, Vietnam . *Journal of the World Aquaculture Society* 49(3), 570–581. DOI: 10.1111/jwas.12488.

O'Neill, B., van Aalst, M., Zaiton Ibrahim, Z., Ford, L.B., Bhadwal, S. *et al.* (2022) Key risks across sectors and regions. In: Portner, H.O., Roberts, D.C., Tignor, M., Poloczanska, E.S., Mintenbeck, K. *et al.* (eds) *Climate Change 2022: Impacts, Adaptation, and Vulnerability. Contribution of Working Group II to the Sixth Assessment Report of the Intergovernmental Panel on Climate Change*. Cambridge University Press, Cambridge, UK, and New York, pp. 2411–2538. Available at: https://report.ipcc.ch/ar6/wg2/IPCC_AR6_WGII_FullReport.pdf (accessed 26 April 2024).

Parker, R.W.R., Blanchard, J.L., Gardner, C., Green, B.S., Hartmann, K. *et al.* (2018) Fuel use and greenhouse gas emissions of world fisheries. *Nature Climate Change* 8, 333–337. DOI: 10.1038/s41558-018-0117-x.

Paukert, C.P., Lynch, A.J. and Whitney, J.E. (2016) Effects of climate change on North American inland fishes: Introduction to the special issue. *Fisheries* 41(7), 329–330. DOI: 10.1038/s41558-018-0117-x.

Pickering, T.D., Ponia, B., Hair, C.A., Southgate, P.C., Poloczanska, E.S. *et al.* (2011) Vulnerability of aquaculture in the tropical pacific to climate change. In: Bell, J.D., Johnson, J.E. and Hobday, A.J. (eds) *Vulnerability of Tropical Pacific Fisheries and Aquaculture to Climate Change*. Secretariat of the Pacific Community, Noumea, New Caledonia, pp. 647–731. Available at: https://coastfish.spc.int/component/content/article/412-vulnerability-of-tropical-pacifc-fisheries-and-aquaculture-to-climate-change.html (accessed 26 April 2024).

Preto, B.L., Henares, M.N.P., Kimpara, J.M. and Valenti, W.C. (2015) Greenhouse gas emissions from aquaculture. Available at: https://www.globalseafood.org/advocate/greenhouse-emissions-from-aquaculture/#:~:text=Aquaculture%20might%20have%20potential%20for%20carbon%20fixation%20due%20to%20the,carbon%20sequestration%2C%20generating%20carbon%20credits (accessed 26 April 2024).

Rodrick, G.B. (2008) Effects of temperature, salinity and pesticides on oyster hemocyte activity. *Florida Water Resource Journal* 4–14. Available at: https://fwrj.com/techarticles/0708%20FWRJ_tech%201.pdf (accessed 26 April 2024).

Rutkayová, J., Vácha, F., Maršálek, M., Beneš, K., Civišová, H. *et al.* (2018) Fish stock losses due to extreme floods – findings from pond-based aquaculture in the Czech Republic. *Journal of Flood Risk Management* 11, 351–359. DOI: 10.1111/jfr3.12332.

Sae-Lim, P., Kause, A., Mulder, H.A. and Olesen, I. (2017) BREEDING and genetics symposium: Climate change and selective breeding in aquaculture. *Journal of Animal Science* 95, 1801–1812. DOI: 10.2527/jas2016.1066.

Schjolden, A. (2004) *Towards Assessing Socioeconomic Impacts of Climate Change in Norway-Sensitivity in the Primary Sectors: Fisheries, Agriculture, Agriculture and Forestry*. CICERO Report 2004:03. Center for International Climate and Environmental Research, Oslo, Norway. Available at: https://pub.cicero.oslo.no/cicero-xmlui/bitstream/handle/11250/192030/CICERO_Report_2004-03.pdf?sequence=1&isAllowed=y (accessed 26 April 2024).

Seggel, A. and De Young, C. (2016) *Climate Change Implications for Fisheries and Aquaculture: Summary of the Findings of the Intergovernmental Panel on Climate Change Fifth Assessment Report*. FAO Fisheries and Aquaculture Circular no.C1122. Food and Agriculture Organization of the United Nations, Rome, Italy. DOI: 10.1111/jfr3.12332. (accessed 26 April 2024).

Silbiger, N.J. and Sorte, C.J.B. (2018) Biophysical feedbacks mediate carbonate chemistry in coastal ecosystems across spatiotemporal gradients. *Scientific Reports* 8, 796. DOI: 10.1038/s41598-017-18736-6.

Singh, P., Singh, R.L. and Kakkar, P. (2016) Antioxidant, DNA damage protective and hepatoprotective activities of *Amorphophyllus campanulatus*. *International Journal of Pharmacy and Pharmaceutical Sciences* 8(3), 330–338.

Stévant, P., Rebours, C. and Chapman, A. (2017) Seaweed aquaculture in Norway: Recent industrial developments and future perspectives. *Aquaculture International* 25, 1373–1390. DOI: 10.1007/s10499-017-0120-7.

Stewart, H.A., Aboagye, D.L., Ramee, S.W. and Allen, P.J. (2019) Effects of acute thermal stress on acid–base regulation, haematology, ion-osmoregulation and aerobic metabolism in channel catfish (*Ictalurus punctatus*). *Aquaculture Research* 50, 2133–2141. DOI: 10.1111/are.14093.

Sult, V.T., Aldon, M.E.T., Tendencia, I.T., Ortiz, A.M.J., Alayon, S.B. *et al.* (2005) In: *Regional Technical Consultation on the Aquaculture of P. vannamei and Other Exotic Shrimps in Southeast Asia,*

Manila, Philippines, 1–2 March 2005, Aquaculture Department Southeast Asian Fisheries Development Center, Tigbauan, Iloilo, Philippines. Available at: https://repository.seafdec.org.ph/handle/10862/1717 (accessed 26 April 2024).

Thomsen, J., Stapp, L.S., Haynert, K., Schade, H., Danelli, M. *et al.* (2017) Naturally acidified habitat selects for ocean acidification-tolerant mussels. *Science Advances* 3, e1602411. DOI: 10.1126/sciadv.1602411.

Toussaint, M., Gyalog, G., Hough, C., Ytteborg, E., Toussaint, M. *et al.* (2018) The effects of climate change upon aquaculture. *Fish Farming Technology, International Aquafeed* 2018, 18–22. Available at: https://climefish.eu/wp-content/uploads/sites/18/2018/10/AquaFeed2.pdf (accessed 26 April 2024).

Troell, M., Eide, A., Isaksen, J., Hermansen, Ø. and Crépin, A.S. (2017) Seafood from a changing Arctic. *Ambio* 46, 368–386. DOI: 10.1007/s13280-017-0954-2.

Weatherdon, L.V., Magnan, A.K., Rogers, A.D., Sumaila, U.R. and Cheung, W.W.L. (2016) Observed and projected impacts of climate change on marine fisheries, aquaculture, coastal tourism, and human health: An update. *Frontiers in Marine Science* 3, 36. DOI: 10.3389/fmars.2016.00048.

Wittmann, A.C. and Pörtner, H.O. (2013) Sensitivities of extant animal taxa to ocean acidification. *Nature Climate Change* 3(11), 995–1001. DOI: 10.1038/nclimate1982.

Yazdi, S.K. and Shakouri, Dr.B. (2010) The effects of climate change on aquaculture. *International Journal of Environmental Science and Development* 1(4), 378–382. DOI: 10.7763/IJESD.2010.V1.73.

Zhang, P.F., Zhao, T., Zhou, L., Han, G.D., Shen, Y.W. *et al.* (2019) Thermal tolerance traits of the undulated surf clam *Paphia undulata* based on heart rate and physiological energetics. *Aquaculture* 498, 343–350. DOI: 10.1016/j.aquaculture.2018.08.037.

Zolnikov, T.R. (2018) *Autoethnographies on the Environment and Human Health*. Palgrave Macmillan, Cham, Switzerland. DOI: 10.1016/j.aquaculture.2018.08.037.

Zolnikov, T.R. (2019) *Global Adaptation and Resilience to Climate Change*. Palgrave Pivot Cham, Springer Nature Switzerland, Cham, Switzerland. DOI: 10.1016/j.aquaculture.2018.08.037.

9 Climate Change Impact on Human Health

Samar Singh, Himani Chhatwal and Ashutosh Pandey*
National Institute of Plant Genome Research (NIPGR), New Delhi, India

Abstract
Variations in environmental conditions caused by natural or anthropogenic activities have been exhibiting their impacts on climate change in the form of heat waves, drought, acid rain, floods, and storms throughout the decades. The resulting climate change has perturbated the interrelated threats to the ecosystem. Climate change affects human health, causing mental illness, undernutrition, and cardiovascular, respiratory, and infectious diseases. This outcome is observed in terms of worldwide disturbance in social dynamics such as health status, socioeconomic status, and public health infrastructure. This chapter explores the adverse climate conditions and highlights footprints on human health. Strategies to reduce the catastrophic impact of climate change on humans have been discussed in the latter sections.

Keywords: Acid rain, Drought, Heat waves, Malnutrition, UV irradiation

9.1 Introduction

Climate change has been observed as the biggest threat to human health and survival by the World Health Organization (WHO) for many decades. Since the 1950s, global climatic conditions have been altered due to anthropogenic activities. In particular, the burning of fossil fuels leads to the release of harmful gases, such as carbon dioxide (CO_2) and natural gases. Carbon dioxide (CO_2) is one of the greenhouse gases (GHGs) primarily released from the burning of oil and coal. Natural gases such as methane, nitrous oxide, and fluorinated gases like hydrofluorocarbons, sulfur hexafluoride, perfluorocarbons, and nitrogen trifluoride are synthetic and strong GHGs that are emitted from various industrial applications, household appliances, and processes that, in turn, cause greenhouse gas emissions, ozone depletion, and UV irradiation. These are the primary causes of unprecedented global warming, which is one of the contributors to climate change (van der Linden *et al.*, 2017).

Climate change severely affects several environmental conditions, such as reduced air quality, water quality, and heat waves, ultimately affecting human health. Nowadays, weather patterns have become less predictable due to climate change. Due to the increasing global temperatures, ice sheets and glaciers in polar regions are melting at a faster rate. Due to this, the sea levels have risen across the world. This resultant rise in sea level, along with the expanding ocean waters, causes flooding and soil erosion. These climatic catastrophes cause serious fatalities, including destruction of infrastructure, increase in malnutrition, respiratory illnesses, and vector-borne diseases

*Corresponding author: ashutosh@nipgr.ac.in

DOI: 10.1079/9781800625037.0009

(Bull *et al.*, 2020). The World Health Organization (WHO) and other international organizations, like Intergovernmental Panel on Climate Change (IPCC), have thoroughly documented the potential effects of climate change on health (Watts *et al.*, 2015).

Climate change has harmed people's physical and mental health, affecting the human population globally. The population is suffering from conditions like heat exhaustion and heat stroke, which show symptoms like cramped muscles, warm, dry skin, high fever, rapid heartbeat, loss of appetite, nausea, vomiting, headache, anxiety, and dizziness (Majra and Gur, 2009; Bittner *et al.*, 2014; Franchini and Mannucci, 2015). Many waterborne diseases, such as amoebiasis, cryptosporidiosis, cyclosporiasis, leptospirosis, cholera, dengue fever, malaria, typhoid fever, hepatitis, and diarrhea, are at a surge during weather changes, leading to serious health issues in humans (Cissé, 2019; Anas *et al.*, 2021). The prevalence of vector-borne diseases has increased due to disease vectors' expanding areas or increased reproductive potential rates. Climate change-induced air pollution also causes severe human diseases, including lung cancer, pneumonia, cataracts, chronic obstructive pulmonary disease, lung stroke, heart attacks, and ischemic heart diseases (Goshua *et al.*, 2022). The severity of some infectious diseases has already undergone significant and frequent changes over the past 10 years (Baker-Austin *et al.*, 2013; Garrett *et al.*, 2013; Burge *et al.*, 2014). It has already been reported that climate plays a crucial role in the life cycles and transmission of infectious disease-causing agents, affecting human health, the agricultural system, as well as the animal world (Liang and Gong, 2017). Recent reports suggest that climate change could bring an additional 250,000 deaths annually between 2030 and 2050, mostly from malaria, malnutrition, diarrhea, and heat stress (WHO, 2021). WHO projected that health issues due to climate change will cost around USD 2–4 billion/year by 2030 (i.e. excluding costs in health-determining industries like agriculture, water, and sanitation). It has been observed that increasing urban population in underdeveloped and low-income countries, leading to crowded areas and dense populations, invites severe infectious diseases. It is estimated that global temperature has increased by almost 1.1°C between 1850 and 1900 due to GHG emissions (Engelbrecht and Monteiro, 2021). The last decade has seen an increasing number of extreme weather events (Masson-Delmotte *et al.*, 2021) that have had significant health impacts. In the next few decades, the average global temperature will reach 1.5°C, affecting the entire planet.

Computing models and observations from the earth's surface, air, and space are utilized to track and study past, present, and future climatic variations. Important climate variation indicators are evidenced by climate data records, which include land quality, rising sea levels, increase in ocean temperature, and ice melting in the polar regions as well as in mountain glaciers. These variation indicators are recorded in consideration with the other parameters including changes in the frequency and intensity of extreme weather events like heat waves, cyclones, floods, wildfires, precipitation, droughts, and modifications in vegetation cover and cloud.

In this chapter, the scientific evidence regarding the cause of climate change and its impact on human health is summarized. The current knowledge and advancements regarding potential responses, adaptations, and mitigation strategies for the adverse effects of climate change on human health are majorly highlighted.

9.2 Consequences and Impact of Climate Change on Human Health

Climate change has an adverse impact on weather patterns, ecosystems, and environmental conditions, endangering human health and well-being. These consequences affect population movement, increase exposure to extreme events, and provide suitable ecological conditions for the spread of infectious diseases, thereby harming human health and livelihoods (Table 9.1). The most vulnerable individuals in society are disproportionately impacted by the resulting pressures on social and health status, and climate change amplifies existing disparities (Fig. 9.1). The following conditions demonstrate severe human health problems.

Table 9.1. Human diseases are listed here by the impact of climate change.

S. No.	Diseases	References
I	Waterborne diseases due to water pollution/temperature rise	
1.	Cholera	Rita, 2009
2.	Leptospirosis	Anas *et al.*, 2021
3.	Malaria	Anas *et al.*, 2021
4.	Dengue	Cissé, 2019
5.	Hemorrhagic Fever with Renal Syndrome diseases	Bittner *et al.*, 2014
6.	Arbovirus disease	Franchini and Mannucci, 2015
7.	Lymphatic filariasis	Majra and Gur, 2009
8.	Diarrhea	Franchini and Mannucci, 2015
9.	Ross River fever	Harley *et al.*, 2001
10.	Typhoid	Ren-Jian *et al.*, 2012
11.	Gastroenteritis	Romanello *et al.*, 2022
12.	Cryptosporidium infection	Ren-Jian *et al.*, 2012
13.	Chikungunya fever	Ren-Jian *et al.*, 2012
14.	St Louis Encephalitis virus	Romanello *et al.*, 2022
15.	West Nile virus	Patz and Hahn, 2013
16.	Hantavirus pulmonary syndrome (HPS)	Ferro *et al.*, 2020
17.	African trypanosomiasis	Harvell *et al.*, 2002
18.	Lyme disease	Moreno, 2006
19.	Yellow fever	Harvell *et al.*, 2002
20.	Plague	Moreno, 2006
21.	Tick-borne encephalitis	Chen and Hoek, 2020
II	Infectious respiratory diseases due to Air Pollution	
22.	Asthma	Osborne *et al.*, 2017
23.	Chronic obstructive pulmonary disease	Gauderman *et al.*, 2015
24.	Lung cancers	Gauderman *et al.*, 2015
25.	Respiratory allergies	Kekeh *et al.*, 2020
26.	Endocrine and metabolic diseases	Chen and Hoek, 2020
III	Wildfires and disease occurrence	
27.	Burns and injuries	Xu *et al.*, 2020
28.	Mental health effects, premature deaths	Xu *et al.*, 2020

9.2.1 Rising temperature and consequences on human health

Climate change is increasing the population's exposure to heat, and it has been noted that the frequency, duration, and magnitude of extreme temperature events are rising globally. The effect of climate change on health is growing adversely as the world focuses on the ongoing acute health crisis. Glaciers and ice sheets are melting due to the earth's warming atmosphere, decreasing heat reflection, and hastening global warming. Sea levels are rising because of seawater thermal expansion and ice sheets and glaciers melting. According to reports, the sea levels rose by 1.7 mm annually between 1971 and 2010, and the rate has doubled since 1993 (Thompson *et al.*, 2019). The unprecedented

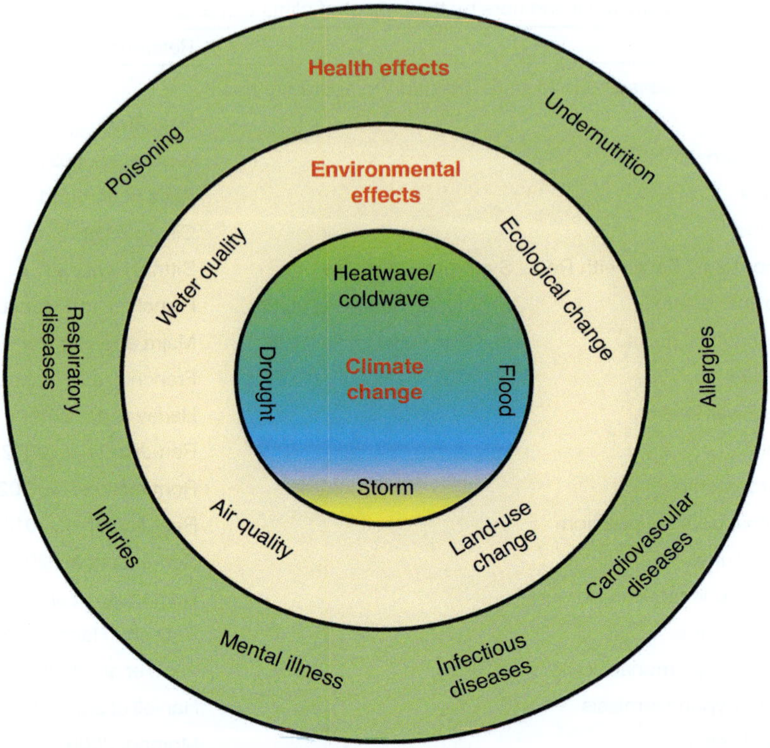

Fig. 9.1. Climate change and its factors involved in natural disturbances and impacts on human health. The inner circle represents the elements of climate change, such as heat waves, drought, floods, and storms; the middle circle (shows how climate change drastically affects the environmental factors of water and air quality, destroying the land and unbalancing the biological activities of humans), and the outer circle illustrates that climate change contributes to health crises, including respiratory and infectious diseases, mental health disorders, allergies, and foodborne illnesses.

heat wave that hit communities in the Pacific region on the northwest side of the United States and Canada in June 2021 resulted in more than 1000 deaths, and the increasing glacier melting is putting Huaraz (Peru) at high risk of flooding (Philip *et al.*, 2021; Stuart-Smith *et al.*, 2021; Van Oldenborgh *et al.*, 2021). The health concerns of high temperatures and heat waves mainly affect younger children and older people from the total population. The total number of days when people over 65 and infants under 1 year old are exposed to heat wave events constitute a serious risk to their lives (De Perez *et al.*, 2018; Romanello *et al.*, 2021). The mortality risk from cerebrovascular, cardiovascular, pulmonary, and gastric illnesses increases after exposure to high heat (Song *et al.*, 2017). Intense heat can impact people's working capacity and reduce working hours. A rise in temperature affects the distribution of disease vectors in a spatiotemporal manner. When the environment gets hotter in low-latitude regions, insects find other habitats in middle- or high-latitude areas, leading to the geographic spread or shift of many diseases. Various heat-induced vector-borne infectious diseases, such as African trypanosomiasis, malaria, Lyme disease, yellow fever, plague, tick-borne encephalitis, and dengue fever, have been found to have spread to a broader geographic area (Harvell *et al.*, 2002).

Extremely higher temperatures may cause behavioral and mental disorders, as evidenced by the rise in depression, anxiety, and suicidal incidents on hotter days (Kim *et al.*, 2019; Khan *et al.*, 2021). Evidence also indicated that extremely low temperatures increase the risk of hypothermia and traffic accidents strongly associated with cardiorespiratory outcomes (Zhang *et al.*, 2019; Lee *et al.*, 2020). According to some studies, people are more vulnerable at the beginning of hot or cold seasons because their bodies are not used to dealing with extreme weather conditions (Zhao *et al.*, 2019). Climate change and trends like population growth and aging, urbanization, and socioeconomic development can either intensify or reduce heat-related hazards (Ebi *et al.*, 2021).

However, the temperature changes might also limit the movement of disease vectors, as observed in the case of *Aedes aegypti* mosquito that hosts the viruses causing dengue and yellow fever (Epstein, 2001). According to early findings, *A. aegypti* adults begin to die when the air temperature exceeds 40°C, while larvae die when the water temperature exceeds 34°C (Christophers, 1960). In those areas where temperatures exceed their thresholds, disease hosts like *A. aegypti* may vanish. Contrastingly, following the expansion of the habitats of ticks, mosquitoes, and midge vectors, more of these diseases have spread into the higher latitude areas. *Oncomelania hupensis*, the intermediate host of *Schistosoma japonicum*, expanded geographic distribution into new regions, including northern China, where winter temperatures continued to rise (Zhou *et al.*, 2010). Thus, increasing temperatures might affect the spread of disease by limiting the vector movement in the regions with higher temperatures.

9.2.2 Air pollution causing severe human diseases

There is a direct link between air pollution and climate change. The earth's air quality is negatively affected by climate change, which is just one side of the same coin. Pollutants like aerosols, methane, black carbon, and tropospheric ozone affect the amount of incoming sunlight. Fine particulate matter (PM2.5) inhalation causes health risks because it can enter the body's sensitive areas, causing serious health issues and even leading to premature death (WHO, 2013). The direct link between air pollution and human health is greatly influenced by changes in climatic factors (Hong *et al.*, 2019; Fann *et al.*, 2021). Air-stagnation events prevent the spreading and dispersion of air pollutants, while rainfall hastens their deposition. High temperature encourages the chemical structure rearrangement of some air pollutants (such as the photoinduced synthesis of ozone). Contrarily, environmental factors can affect a person's behavior, determining how they are exposed to air pollution and depending on the type of pollutants and duration of exposure. Tropospheric O_3 has also been linked to asthma in the children's age group and has been shown to have negative health effects that may cause premature mortality (Johnston *et al.*, 2012; Zheng *et al.*, 2015). According to the WHO and the Global Burden of Disease Study (Gulis *et al.*, 2021), air pollution affects approximately 7 million premature deaths annually; among them, 0.4 million are caused by O_3 pollution, and the remainder comes from household PM2.5; other pollutants, such as NO_2 and SO_2, are also harmful to human health (Johns and Linn, 2011; Mills *et al.*, 2016). The health effects of air pollutants have been determined using various techniques. One of these methods is clinical data studies, in which participants are exposed to air pollutants under circumstances simulating ambient exposures (Li *et al.*, 1996; Sehlstedt *et al.*, 2010). For example, other *in vivo* and *in vitro* laboratory methods have been used to explore the role of oxidative damage on pulmonary inflammatory response associated with air pollution exposure.

The association between air pollution and a variety of health problems, including hospitalization rates, cardiopulmonary mortality, maternal health, cognitive impairment, neurodevelopment in children, and elevated risk of hospitalizations for neurological disorders and diabetes among older people, has been demonstrated by a significant number of epidemiological studies using different designs to reveal either short-term or long-term effects (Lanki *et al.*, 2006; Calderón-Garcidueñas *et al.*, 2014; Cesaroni *et al.*, 2014; Balakrishnan *et al.*, 2018). Long-term chronic effects including respiratory problems and immune compromise have been identified to be due to climate change (Manisalidis *et al.*, 2020). It is important to note that people with conditions like asthma, pneumonia, diabetes, and cardiovascular and respiratory illnesses are particularly vulnerable to the effects of PM2.5 and PM10 size pollutants, which are strongly linked to various respiratory diseases (Boschi, 1999; Kappos *et al.*, 2004). Following their chemical and physical characteristics, the particles have toxic effects on humans and other animals. Polycyclic aromatic hydrocarbons, dioxins, benzene, and 1-butadiene are examples of organic components found in PM10 and PM2.5, while carbon, chlorides, nitrates, sulfates, and metals are examples of inorganic ones (Cheung *et al.*, 2011). For future projections, epidemiological studies, artificial intelligence, and clinical data are primary methods of modeling the health effects of exposure to air pollution in applied studies.

9.2.3 Water pollution and waterborne diseases in humans

Water availability for drinking and agricultural practices is an essential factor in building the complex relationship between climate variation, food security, nutrition value, and human health at the global level. According to the analysis, climate change and population changes may cause 1.4 billion more people to experience drought annually by the end of the century. Importantly, the geographical distribution of exposure is exceptionally localized and variable (e.g. in Asian and European continental groups), severely affecting water supply and quality. The availability of groundwater resources, critical in many regions, impacts long-term water availability and supply for many areas. However, all such estimates ultimately focus on surface water availability (Taylor *et al.*, 2013; Schewe *et al.*, 2014).

An increase in the frequency of hurricanes, storm surges, and floods will significantly impact human health. In late 2013, over 6000 people died in the Philippines because of Typhoon Haiyan, which is an example of the immediate risks associated with extreme climatic conditions. Disease outbreaks, burdens on mental health, and displacement are additional long- and short-term effects of flooding on well-being (Ahern *et al.*, 2005). Huge populations are at risk due to flooding, water shortage, and contaminated drinking water (Dasgupta, 2007; Nicholls *et al.*, 2011). Large-scale human communities have been affected due to contaminated water, increasing the spread of vector-borne diseases.

Cholera-causing bacteria, *Vibrio cholerae*, are found naturally in moist areas, especially in coastal and estuarine ecosystems. The survival capacity of these bacteria and their multiplication frequency depend on plankton density and environmental factors like water bodies and seawater temperature. Cholera outbreaks are favored by flooding combined with hot surface sea temperatures (Rita, 2009). The risk posed by cholera in the United States (Chesapeake Bay) was evaluated using an early warning system (EWS) based on climatic changes (Allen *et al.*, 2005). Notably, the standard of public health sector services, living standards of the population, accessibility to clean water and sanitary facilities, and political stability are fundamental factors influencing cholera outbreaks. A recent epidemic of cholera outbreaks occurred in Yemen in 2016, with over 2000 reported deaths (Qadri *et al.*, 2017). Numerous waterborne diseases are susceptible to climate change, and they are significantly correlated with an increase in extreme rainfall events and hurricanes. For instance, a sizable cryptosporidium outbreak in the Mississippi River floods in Milwaukee in 1993 and Katrina's aftermath in Louisiana in 2005 saw the spread of numerous toxins and the norovirus (Caminade *et al.*, 2019).

Harmful algal blooms (HABs) in marine waters have been directly linked to human poisoning by eating contaminated raw seafood, skin contact with contaminated water, and ingesting aerosolized biotoxins. There are emerging phycotoxins and risks of poisoning from drinking desalinated water contaminated with biotoxins, along with the human health conditions linked to known toxins produced by microalgae (Ferrante *et al.*, 2013). Amnesic shellfish poisoning caused by *Pseudo-nitzschia* spp. and *Nitzschia* sp. and the release of domoic acid and isomers show diseased symptoms in humans, like vomiting, nausea, dizziness, headache, diarrhea, short-term memory deficits, confusion, disorientation, and motor weakness. Severe cases result in seizures, cardiac arrhythmia, respiratory distress, coma, and possibly death. Neurotoxic shellfish poisoning and diarrhetic shellfish poisoning caused by

Karenia brevis and *Dinophysis* spp., respectively, are responsible for many health issues (Young *et al.*, 2020).

9.2.4 Wildfire leads to serious human health problems

Wildfire risk and destruction are exacerbated by hotter and drier conditions caused by climate change. This indicator measures the risk of wildfire exposure by combining satellite observations of active fires with human populations' exposure to extremely high wildfire danger. They measure the likelihood of fire initiation, spreading rate, severity, and difficulty faced in putting it out using a combination of relative humidity, air temperature, drought impacts, and wind speed. High temperatures, less precipitation, and drought can make wildfires more likely to lengthen the fire season and burning duration in many areas. Other climatic factors, like the El Niño–Southern Oscillation (ENSO), are also linked to recurrent drought and wildfires in coastal regions (Cardil *et al.*, 2021). Premature deaths, burns, injuries, and mental health effects are the direct effects of wildfire events (Xu *et al.*, 2020). In addition to the 339,000 deaths directly related to wildfires, more than this number of individuals passed away each year from choking air pollution, specifically poisonous gases (Johnston *et al.*, 2012). In Australia, the destructive bushfire season in 2019–2020 was all caused by climate change. Smoke from wildfires also contains toxic substances such as nitrogen oxides (NO_x) and PMs. The level of fire-induced PM has been linked repeatedly in studies to the risks of hospitalization and death from any cause, including cardiovascular and respiratory causes (Johnston *et al.*, 2012; Chen *et al.*, 2021; Ye *et al.*, 2021). Some substances, like benzene and free radicals, are multi-organically toxic and can cause lesions and even cancer of the reproductive, hemopoietic, and digestive systems (Johnston *et al.*, 2012). Climate change will likely worsen due to increased greenhouse gas emissions from burning wood each year and reduced biodiversity caused by global forest cover loss (Bowman *et al.*, 2020).

9.2.5 Impact of climate change on vector-borne diseases

According to a recent systematic study of domestic animals and human pathogens in Europe, nearly two-thirds were climate sensitive, with many being susceptible to more than one climate driver (McIntyre *et al.*, 2017). Infectious diseases have had a major impact throughout history, causing many animal deaths and losses. Records show plague outbreaks in ancient Rome (second century BCE) and Athens, and the Black Death in the 14th century killed a third to half of Europe's population (McMichael, 2012). The most recent Ebola outbreak between 2014 and 2016 devastated vulnerable populations in Guinea, Liberia, and Sierra Leone (WHO, 2014). The harmful effects of infectious diseases on human health and well-being are inevitably related to various unhygienic conditions, including freshwater inaccessibility, poor living standards, illiteracy, political instability, and conflict (Woolhouse and Gowtage-Sequeria, 2005).

Arthropods serve as the primary carriers of infectious diseases and are, for various reasons, particularly vulnerable to climatic changes. The arthropod's internal temperature is controlled by the surrounding temperature. The presence of smelly and dirty water bodies and high humidity levels are necessary to provoke their larval development stage. Until a certain point, the tendency of vector spreading rates to rise with temperature tends to decline (Koenraadt, 2019). At high temperatures, pathogens spreading through vectors develop and replicate more quickly according to the environmental conditions (this is known as the extrinsic incubation period or EIP) (Reisen *et al.*, 2014).

Temperature conditions significantly impact the growth and survival of vectors. Female *Anopheles* mosquitoes transmit human malaria, which is brought on by five species of plasmodium parasites. *Plasmodium falciparum*, which is common in tropical regions and sub-Saharan Africa, is responsible for about 90% of all malaria cases worldwide, causing the most severe clinical symptoms (Alonso and Noor, 2017). Aedes mosquito species are known to transmit Zika, arboviruses dengue, chikungunya, and yellow fever, with the Asian tiger mosquito (*Aedes albopictus*) and yellow fever mosquito (*Aedes aegypti*) being the primary effective vectors. On average, a 30-fold increase in global incidence over the past 50 years, and an estimated 100–390 million dengue disease incidents, are reported annually (WHO, 2009). Dengue is a mosquito-borne disease that spreads faster in unhygienic areas (Bhatt *et al.*, 2013). Due to global marketing trade, such as the import of plants and used tires from Asia, these vectors stuck on them and spread into new areas (Hawley *et al.*, 1987; Sprenger and Reiter, 1987; Benedict *et al.*, 2007; Medlock *et al.*, 2012).

The most prevalent encephalitic flavivirus is West Nile virus, which affects humans, horses, other mammals, and birds. Except for Antarctica, all continents have Culex mosquitoes as the primary vector of West Nile virus transmission. The significance of weather and climate in causing West Nile virus epidemics has already been covered in several studies (Epstein, 2001; Paz, 2015).

9.2.6 Climate change leads to biodiversity loss and varied pathogen invasion

Invasive species are an important driver in climate change and are responsible for the loss of biodiversity, deterioration of ecosystems, and impairment of ecosystem services globally. Ecosystem services being disrupted due to biological species invasions are known to have negative financial, cultural, and health effects on people. Some European invaders insects, including the muskrat (*Ondatra zibethicus*), the raccoon dog (*Nyctereutes procyonoides*), the thrips (*Heliothrips hemorrhoidalis* and *Frankliniella occidentalis*), and the Chinese mitten crab (*Eriocheir sinensis*), show drastic impacts on animal and human health (Pyšek and Richardson, 2010). For instance, various human health issues, such as irritation on the skin and allergies, are brought on by invasive foreign species. The process of biological invasions is mainly caused by skin-to-human illness outbreaks brought on by novel viruses, including the human immunodeficiency virus (HIV), monkeypox, and severe acute respiratory syndrome (SARS) (Baker *et al.*, 2022). These microbes breach the barriers separating their natural reservoirs from the human population, sparking the outbreak of novel infectious diseases.

9.3 Mitigating Strategies to Protect Human Health From Varying Climatic Conditions

Projections indicate that the human population faces many critical problems because of changing climatic conditions. These problems can be mitigated by adopting strategies like adaptive responses and resiliency, enhancing the socioeconomic status of vulnerable people, and establishing an early event warning system for better sustainability.

9.3.1 Adaptation and resilience to the direct effects of climate change on health

Adaptation to climate change is "the process of adjustment in adverse climate conditions and its effects, in human population to mitigate or exploit beneficial opportunities or services and in natural systems human intervention may promote adjustment to expected climate conditions" (Birkmann *et al.*, 2022). Storms, floods, droughts, and heat waves are extreme weather events directly impacting health. Many responses emphasize how crucial it is to improve the health system, but other sectors also require action. Resilience is a system's ability to adapt to change and absorb disturbance while maintaining the same structure, function, and composition (Zaccarelli *et al.*, 2008). Another term for resilience is a community's capacity to withstand external perturbations to social networks. In terms of resilience, a system can tolerate and respond to a disturbance, innovate, and change.

9.3.1.1 Extreme event early warning systems

Identifying the most exposed and vulnerable communities and improving early warning systems (EWS) are vital components of managing extreme weather events and their ongoing health effects. It consists of planning, speculating on potential health outcomes, launching quick and efficient response plans, focusing on vulnerable populations, and communicating preventive measures. Public health authorities must upgrade existing emergency programs and hold drills to prepare better for anticipated health risks from new extreme events like rising sea levels, saltwater intrusion into drinking water courses, and significant flooding from storm surges. These efforts to increase disaster preparedness must coexist with initiatives to improve local communities' resilience.

9.3.1.2 Actions to deplete the effect of heat

With climate change, extreme hot days and heat waves will occur more frequently and intensely for extended periods, increasing the risk of heat stress and death. As a result of the construction of buildings and paved surfaces retaining more heat than reflective, transpiring, shady, and air-flow-promoting vegetation-covered surfaces, the ill-effects are being exacerbated. This phenomenon is known as the urban heat island effect (Mohan *et al.*, 2022). Effective adaptation strategies would lower the fatalities caused by these heat waves; healthcare workers can be trained. Integrated heat wave early warning systems (HEWS) are available as adaptation options, especially for the most vulnerable populations (Pascal *et al.*, 2006; Ebi and Burton, 2008). In addition, adaptation measures are considered to improve the design of social care facilities, schools, other public spaces, and public transport to be more climate responsive.

9.3.1.3 Actions to reduce flood and storm effects

Flood adaptation strategies are generally structural or non-structural (Kundzewicz, 1999). Flooding can be prevented by using various infrastructures like reservoirs, dams, dykes, and floodways. Some urban designs may include low-lying areas susceptible to flooding and could briefly submerge during a major event. Many flood managers prioritize structural programs because they are the primary funding source for controlling floods. A few examples of non-structural measures are flood insurance, development policies, zoning laws, flood-plain regulations, building codes, floodproofing, tax incentives, emergency preparedness, flood forecasting, and post-flood recovery (Michel-Kerjan, 2010; UN Economic, 2012). Reduced flood damage and improved ecological functions of flood plains are the goals of non-structural adaptation options. The rate of spread of infectious disease is affected by how we shape and adapt to the environment through our impact on land use (agricultural activities and deforestation/reforestation). The construction of artificial water bodies or dams, perturbing natural environment and the measures taken to control infectious diseases like the development of vaccines, insecticide spraying, and the development of rapid diagnostic tests are important factors determining the spread of diseases (Kibret *et al.*, 2017).

9.3.2 Action for resilience to indirect impacts

Due to the complex causal chains and lack of predictability, adaptation to indirect effects presents complex challenges to policymaking (Lafferty, 2009). These intricated interactions can produce "surprising" circumstances in which the behavior within or between systems deviates qualitatively from predictions. The indirect effects present significant barriers to climate adaptation, and health responses necessitate integrated and cross-sectoral interventions (Zinsstag *et al.*, 2005).

9.3.2.1 Dynamic infectious disease risks

Disease risks will significantly change nationally and internationally due to interactions and changes in demographic patterns, human connectivity, climatic conditions, land use, and biodiversity (Sacks *et al.*, 2021). Climate change-related temperature variations are expected to cause a difference in the geographical distribution of African trypanosomiasis (Moore *et al.*, 2012). Loss of biodiversity may increase the spread of infectious diseases like schistosomiasis, Lyme disease, West Nile virus, and Hantavirus (Keesing *et al.*, 2010; Knust *et al.*, 2012). The risks of contagious diseases are variable and influenced by several intricate factors. Focusing on co-benefits, having no regrets, and resiliency are a few uncertainties that must be considered when developing adaptation responses to dynamic disease risks.

9.3.2.2 Surveillance and monitoring

As evidenced by different pandemics occurring occasionally, it is necessary to improve nations' capacity to track and address disease outbreaks. Various data records are needed for climate change adaptation of human health, including information on human health climate risks and specific diseases related to climate variable impacts. Health-related information and data can be collected from public health surveillance and monitoring systems. It can predict disease burdens, identify at-risk individuals and communities, comprehend disease patterns, and develop response plans and public health interventions (Semenza *et al.*, 2012).

9.4 Conclusions

Climate change results in a range of unfavorable conditions, including polluted water and air, drought, flood, acid rain, rising temperature, wildfire, and habitat loss. These conditions impact human health directly and indirectly through various serious disease situations. Heat-related illnesses include cardiovascular diseases, skin conditions, mental illness, anxiety, and behavioral changes. Polluted air causes different respiratory and lung diseases. Globally, populations are affected by various vector-borne illnesses like malaria, dengue, and other diseases. Resilience and adaptive responses by humans against adverse climate conditions lead to better sustainability. Other policies, such as large-scale plant rotation by the governing body, could be effective against dangerous climatic conditions. However, early warning systems and technological advancements can effectively protect human lives.

Acknowledgments

This work was supported by the core grant of the National Institute of Plant Genome Research and Core Research Grant (CRG) from SERB—Department of Science and Technology (CRG/2022/001178) to AP. SS and HC acknowledge the University Grant Commission, the Government of India for Senior and Junior Research Fellowships, respectively. The authors are thankful to DBT-eLibrary Consortium (DeLCON) for providing access to e-resources.

Declarations

Conflict of interest: Authors declare no conflict of interest.

References

Ahern, M., Kovats, R.S., Wilkinson, P., Few, R. and Matthies, F. (2005) Global health impacts of floods: Epidemiologic evidence. *Epidemiologic Reviews* 27(1), 36–46. DOI: 10.1093/epirev/mxi004.

Allen, S.A., Bertilsson, S. and Goedkoop, W. (2005) *Factors Determining Distribution and Abundance of Pathogenic Bacteria, with a Focus on Vibrio spp*. Available at: https://api.semanticscholar.org/CorpusID:53464242

Alonso, P. and Noor, A.M. (2017) The global fight against malaria is at crossroads. *The Lancet* 390(10112), 2532–2534. DOI: 10.1016/S0140-6736(17)33080-5.

Anas, M., Sami, M.A., Siddiqui, Z., Khatoon, K., Zeyad, M.T. *et al.* (2021) Impact of climate change on the incidence and transfer of food-and waterborne diseases. In: Lone, S.A. and Malik, A. (eds) *Microbiomes and the Global Climate Change*. Springer, Singapore, pp. 123–144. DOI: 10.1007/978-981-33-4508-9_9.

Baker-Austin, C., Trinanes, J.A., Taylor, N.G., Hartnell, R., Siitonen, A. *et al.* (2013) Emerging Vibrio risk at high latitudes in response to ocean warming. *Nature Climate Change* 3(1), 73–77. DOI: 10.1038/nclimate1628.

Baker, R.E., Mahmud, A.S., Miller, I.F., Rajeev, M., Rasambainarivo, F. *et al.* (2022) Infectious disease in an era of global change. *Nature Reviews. Microbiology* 20(4), 193–205. DOI: 10.1038/s41579-021-00639-z.

Balakrishnan, K., Ghosh, S., Thangavel, G., Sambandam, S., Mukhopadhyay, K. *et al.* (2018) Exposures to fine particulate matter ($PM_{2.5}$) and birthweight in a rural-urban, mother-child cohort in Tamil Nadu, India. *Environmental Research* 161, 524–531. DOI: 10.1016/j.envres.2017.11.050.

Benedict, M.Q., Levine, R.S., Hawley, W.A. and Lounibos, L.P. (2007) Spread of the tiger: Global risk of invasion by the mosquito *Aedes albopictus*. *Vector Borne and Zoonotic Diseases* 7(1), 76–85. DOI: 10.1089/vbz.2006.0562.

Bhatt, S., Gething, P.W., Brady, O.J., Messina, J.P., Farlow, A.W. *et al.* (2013) The global distribution and burden of dengue. *Nature* 496(7446), 504–507. DOI: 10.1038/nature12060.

Birkmann, J., Jamshed, A., McMillan, J.M., Feldmeyer, D., Totin, E. *et al.* (2022) Understanding human vulnerability to climate change: A global perspective on index validation for adaptation planning. *The Science of the Total Environment* 803, 150065. DOI: 10.1016/j.scitotenv.2021.150065.

Bittner, M.I., Matthies, E.F., Dalbokova, D. and Menne, B. (2014) Are European countries prepared for the next big heatwave?. *The European Journal of Public Health* 24(4), 615–619.

Boschi, N. (1999) Defining an educational framework for indoor air sciences education. In: Boschi, N. (ed.) *Education and Training in Indoor Air Sciences*. Springer, Dordrecht, the Netherlands, pp. 3–6.

Bowman, D.M., Kolden, C.A., Abatzoglou, J.T., Johnston, F.H., van der Werf, G.R. *et al.* (2020) Vegetation fires in the Anthropocene. *Nature Reviews Earth & Environment* 1(10), 500–515. DOI: 10.1038/s43017-020-0085-3.

Bull, F.C., Al-Ansari, S.S., Biddle, S., Borodulin, K., Buman, M.P. *et al.* (2020) World health organization 2020 guidelines on physical activity and sedentary behaviour. *British Journal of Sports Medicine* 54(24), 1451–1462. DOI: 10.1136/bjsports-2020-102955.

Burge, C.A., Mark Eakin, C., Friedman, C.S., Froelich, B., Hershberger, P.K. *et al.* (2014) Climate change influences on marine infectious diseases: Implications for management and society. *Annual Review of Marine Science* 6, 249–277. DOI: 10.1146/annurev-marine-010213-135029.

Calderón-Garcidueñas, L., Torres-Jardón, R., Kulesza, R.J., Park, S.-B. and D'Angiulli, A. (2014) Air pollution and detrimental effects on children's brain. The need for a multidisciplinary approach to the issue complexity and challenges. *Frontiers in Human Neuroscience* 8, 613. DOI: 10.3389/fnhum.2014.00613.

Caminade, C., McIntyre, K.M. and Jones, A.E. (2019) Impact of recent and future climate change on vector-borne diseases. *Annals of the New York Academy of Sciences* 1436(1), 157–173. DOI: 10.1111/nyas.13950.

Cardil, A., Rodrigues, M., Ramirez, J., de-Miguel, S., Silva, C.A. *et al.* (2021) Coupled effects of climate teleconnections on drought, Santa Ana winds and wildfires in Southern California. *The Science of the Total Environment* 765, 142788. DOI: 10.1016/j.scitotenv.2020.142788.

Chen, G., Guo, Y., Yue, X., Tong, S., Gasparrini, A. *et al.* (2021) Mortality risk attributable to wildfire-related $PM_{2.5}$ pollution: A global time series study in 749 locations. *The Lancet. Planetary Health* 5(9), e579–e587. DOI: 10.1016/S2542-5196(21)00200-X.

Cesaroni, G., Forastiere, F., Stafoggia, M., Andersen, Z.J., Badaloni, C. *et al.* (2014) Long term exposure to ambient air pollution and incidence of acute coronary events: Prospective cohort study and meta-analysis in 11 European cohorts from the ESCAPE project. *BMJ* 348, f7412. DOI: 10.1136/bmj.f7412.

Chen, J. and Hoek, G. (2020) Long-term exposure to PM and all-cause and cause-specific mortality: A systematic review and meta-analysis. *Environment International* 143, 105974. DOI: 10.1016/j.envint.2020.105974.

Cheung, K., Daher, N., Kam, W., Shafer, M.M., Ning, Z. *et al.* (2011) Spatial and temporal variation of chemical composition and mass closure of ambient coarse particulate matter (PM10–2.5) in the Los Angeles area. *Atmospheric Environment* 45(16), 2651–2662. DOI: 10.1016/j.atmosenv.2011.02.066.

Christophers, S.R. (1960) Aedes aegypti: The yellow fever mosquito. CUP Archive.
Cissé, G. (2019) Food-borne and water-borne diseases under climate change in low- and middle-income countries: Further efforts needed for reducing environmental health exposure risks. *Acta Tropica* 194, 181–188. DOI: 10.1016/j.actatropica.2019.03.012.
Dasgupta, S. (2007) *The Impact of Sea Level Rise on Developing Countries: A Comparative Analysis*, Vol. 4136. World Bank Publications, Canada. DOI: 10.1596/1813-9450-4136.
De Perez, E.C., Van Aalst, M., Bischiniotis, K., Mason, S., Nissan, H. *et al.* (2018) Global predictability of temperature extremes. *Environmental Research Letters* 13(5), 054017. DOI: 10.1596/1813-9450-4136.
Ebi, K.L. and Burton, I. (2008) Identifying practical adaptation options: An approach to address climate change-related health risks. *Environmental Science and Policy* 11(4), 359–369. DOI: 10.1596/1813-9450-4136.
Ebi, K.L., Capon, A., Berry, P., Broderick, C., de Dear, R. *et al.* (2021) Hot weather and heat extremes: Health risks. *The Lancet* 398(10301), 698–708. DOI: 10.1596/1813-9450-4136.
Engelbrecht, F. and Monteiro, P. (2021) Climate change: The IPCC's latest assessment report. *Quest* 17(3), 34–35. DOI: 10.1596/1813-9450-4136.
Epstein, P.R. (2001) West Nile virus and the climate. *Journal of Urban Health* 78, 367–371. DOI: 10.1093/jurban/78.2.367.
Fann, N.L., Nolte, C.G., Sarofim, M.C., Martinich, J. and Nassikas, N.J. (2021) Associations between simulated future changes in climate, air quality, and human health. *JAMA Network Open* 4(1), e2032064. DOI: 10.1001/jamanetworkopen.2020.32064.
Ferrante, M., Conti, G.O., Fiore, M., Rapisarda, V. and Ledda, C. (2013) Harmful algal blooms in the Mediterranean sea: Effects on human health. *Euro Mediterranean Biomedical Journal* 8(6), 25–34.
Ferro, I., Bellomo, C.M., López, W., Coelho, R., Alonso, D. *et al.* (2020) Hantavirus pulmonary syndrome outbreaks associated with climate variability in Northwestern Argentina, 1997-2017. *PLoS Neglected Tropical Diseases* 14(11), e0008786. DOI: 10.1371/journal.pntd.0008786.
Franchini, M. and Mannucci, P.M. (2015) Impact on human health of climate changes. *European Journal of Internal Medicine* 26(1), 1–5. DOI: 10.1016/j.ejim.2014.12.008.
Garrett, K.A., Dobson, A.D.M., Kroschel, J., Natarajan, B., Orlandini, S. *et al.* (2013) The effects of climate variability and the color of weather time series on agricultural diseases and pests, and on decisions for their management. *Agricultural and Forest Meteorology* 170, 216–227.
Gauderman, W.J., Urman, R., Avol, E., Berhane, K., McConnell, R. *et al.* (2015) Association of improved air quality with lung development in children. *New England Journal of Medicine* 372(10), 905–913.
Goshua, A., Akdis, C.A. and Nadeau, K.C. (2022) World Health Organization global air quality guideline recommendations: Executive summary. *Allergy* 77(7), 1955–1960. DOI: 10.1111/all.15224.
Gulis, G., Aringazina, A., Sangilbayeva, Z., Zhan, K., de Leeuw, E. *et al.* (2021) Population health status of the republic of Kazakhstan: Trends and implications for public health policy. *International Journal of Environmental Research and Public Health* 18(22), 12235. DOI: 10.3390/ijerph182212235.
Harley, D., Sleigh, A. and Ritchie, S. (2001) Ross river virus transmission, infection, and disease: A cross-disciplinary review. *Clinical Microbiology Reviews* 14(4), 909–932.
Harvell, C.D., Mitchell, C.E., Ward, J.R., Altizer, S., Dobson, A.P. *et al.* (2002) Climate warming and disease risks for terrestrial and marine biota. *Science* 296(5576), 2158–2162.
Hawley, W.A., Reiter, P., Copeland, R.S., Pumpuni, C.B. and Craig, G.B. (1987) Aedes albopictus in North America: Probable introduction in used tires from Northern Asia. *Science* 236(4805), 1114–1116.
Hong, C., Zhang, Q., Zhang, Y., Davis, S.J., Tong, D. *et al.* (2019) Impacts of climate change on future air quality and human health in China. *Proceedings of the National Academy of Sciences* 116(35), 17193–17200. DOI: 10.1073/pnas.1812881116.
Johns, D.O. and Linn, W.S. (2011) A review of controlled human SO_2 exposure studies contributing to the US EPA integrated science assessment for sulfur oxides. *Inhalation Toxicology* 23(1), 33–43.
Johnston, F.H., Henderson, S.B., Chen, Y., Randerson, J.T., Marlier, M. *et al.* (2012) Estimated global mortality attributable to smoke from landscape fires. *Environmental Health Perspectives* 120(5), 695–701. DOI: 10.1289/ehp.1104422.
Johnston, F.H., Henderson, S.B., Chen, Y., Randerson, J.T., Marlier, M. *et al.* (2012) Estimated global mortality attributable to smoke from landscape fires. *Environmental Health Perspectives* 120(5), 695–701.
Kappos, A.D., Bruckmann, P., Eikmann, T., Englert, N., Heinrich, U. *et al.* (2004) Health effects of particles in ambient air. *International Journal of Hygiene and Environmental Health* 207(4), 399–407.

Keesing, F., Belden, L.K., Daszak, P., Dobson, A., Harvell, C.D. *et al.* (2010) Impacts of biodiversity on the emergence and transmission of infectious diseases. *Nature* 468(7324), 647–652.

Kekeh, M., Akpinar-Elci, M. and Allen, M.J. (2020) Sea level rise and coastal communities. In: Akhtar, R. (ed.) *Extreme Weather Events and Human Health: International Case Studies*. Springer, Cham, Switzerland, pp. 171–184. DOI: 10.1007/978-3-030-23773-8_12.

Khan, A.M., Finlay, J.M., Clarke, P., Sol, K., Melendez, R. *et al.* (2021) Association between temperature exposure and cognition: A cross-sectional analysis of 20,687 aging adults in the United States. *BMC Public Health* 21(1), 1–12.

Kibret, S., Wilson, G.G., Ryder, D., Tekie, H. and Petros, B. (2017) The influence of dams on malaria transmission in sub-Saharan Africa. *EcoHealth* 14, 408–419.

Kim, Y., Kim, H., Gasparrini, A., Armstrong, B., Honda, Y. *et al.* (2019) Suicide and ambient temperature: A multi-country multi-city study. *Environmental Health Perspectives* 127(11), 117007. DOI: 10.1289/EHP4898.

Knust, B., Macneil, A. and Rollin, P.E. (2012) Hantavirus pulmonary syndrome clinical findings: Evaluating a surveillance case definition. *Vector Borne and Zoonotic Diseases* 12(5), 393–399. DOI: 10.1089/vbz.2011.0764.

Koenraadt, C.J.M. (2019) Arthropods and infectious diseases. In: Goddard, J. (ed.) *Infectious Diseases and Arthropods*. Humana Press, New Jersey Totowa, p. 296. DOI: 10.1016/S1473-3099(19)30639-5.

Kundzewicz, Z.W. (1999) Flood protection—sustainability issues. *Hydrological Sciences Journal* 44(4), 559–571. DOI: 10.1080/02626669909492252.

Lafferty, K.D. (2009) Calling for an ecological approach to studying climate change and infectious diseases. *Ecology* 90(4), 932–933. DOI: 10.1890/08-1767.1.

Lanki, T., Pekkanen, J., Aalto, P., Elosua, R., Berglind, N. *et al.* (2006) Associations of traffic related air pollutants with hospitalisation for first acute myocardial infarction: The HEAPSS study. *Occupational and Environmental Medicine* 63(12), 844–851. DOI: 10.1136/oem.2005.023911.

Lee, H., Myung, W., Kim, H., Lee, E.M. and Kim, H. (2020) Association between ambient temperature and injury by intentions and mechanisms: A case-crossover design with a distributed lag nonlinear model. *Science of The Total Environment* 746, 141261. DOI: 10.1016/j.scitotenv.2020.141261.

Liang, L. and Gong, P. (2017) Climate change and human infectious diseases: A synthesis of research findings from global and spatio-temporal perspectives. *Environment International* 103, 99–108. DOI: 10.1016/j.envint.2017.03.011.

Li, X.Y., Gilmour, P.S., Donaldson, K. and MacNee, W. (1996) Free radical activity and pro-inflammatory effects of particulate air pollution (PM10) *in vivo* and *in vitro*. *Thorax* 51(12), 1216. DOI: 10.1136/thx.51.12.1216.

Majra, J.P. and Gur, A. (2009) Climate change and health: Why should India be concerned? *Indian Journal of Occupational and Environmental Medicine* 13(1), 11–16. DOI: 10.4103/0019-5278.50717.

Manisalidis, I., Stavropoulou, E., Stavropoulos, A. and Bezirtzoglou, E. (2020) Environmental and health impacts of air pollution: A review. *Frontiers in Public Health* 8, 14. DOI: 10.3389/fpubh.2020.00014.

Masson-Delmotte, V., Zhai, P., Pirani, A., Connors, S.L., Péan, C. *et al.* (eds) (2021) *Climate Change 2021: The Physical Science Basis. Contribution of Working Group I to the Sixth Assessment Report of the Intergovernmental Panel on Climate Change*. Cambridge University Press, Cambridge, UK, p. 2391.

McIntyre, K.M., Setzkorn, C., Hepworth, P.J., Morand, S., Morse, A.P. *et al.* (2017) Systematic assessment of the climate sensitivity of important human and domestic animals pathogens in Europe. *Scientific Reports* 7(1), 7134. DOI: 10.1038/s41598-017-06948-9.

McMichael, A.J. (2012) Insights from past millennia into climatic impacts on human health and survival. *Proceedings of the National Academy of Sciences* 109(13), 4730–4737. DOI: 10.1073/pnas.1120177109.

Medlock, J.M., Hansford, K.M., Schaffner, F., Versteirt, V., Hendrickx, G. *et al.* (2012) A review of the invasive mosquitoes in Europe: Ecology, public health risks, and control options. *Vector Borne and Zoonotic Diseases* 12(6), 435–447. DOI: 10.1089/vbz.2011.0814.

Michel-Kerjan, E.O. (2010) Catastrophe economics: The national flood insurance program. *Journal of Economic Perspectives* 24(4), 165–186. DOI: 10.1257/jep.24.4.165.

Mills, I.C., Atkinson, R.W., Anderson, H.R., Maynard, R.L. and Strachan, D.P. (2016) Distinguishing the associations between daily mortality and hospital admissions and nitrogen dioxide from those of particulate matter: A systematic review and meta-analysis. *BMJ Open* 6(7), e010751. DOI: 10.1136/bmjopen-2015-010751.

Mohan, M., Bhati, S. and Sati, A.P. (2022) Urban heat island effect in India: Assessment, impacts, and mitigation. In: *Global Urban Heat Island Mitigation*. Elsevier, pp. 199–250. DOI: 10.1016/C2020-0-02681-2.

Moore, S., Shrestha, S., Tomlinson, K.W. and Vuong, H. (2012) Predicting the effect of climate change on African trypanosomiasis: Integrating epidemiology with parasite and vector biology. *Journal of the Royal Society, Interface* 9(70), 817–830. DOI: 10.1098/rsif.2011.0654.

Moreno, A.R. (2006) Climate change and human health in Latin America: Drivers, effects, and policies. *Regional Environmental Change* 6(3), 157–164. DOI: 10.1007/s10113-006-0015-z.

Nicholls, R.J., Marinova, N., Lowe, J.A., Brown, S., Vellinga, P. *et al.* (2011) Sea-level rise and its possible impacts given a 'beyond 4°C world' in the twenty-first century. *Philosophical Transactions of the Royal Society A: Mathematical, Physical and Engineering Sciences* 369(1934), 161–181. DOI: 10.1098/rsta.2010.0291.

Osborne, N.J., Alcock, I., Wheeler, B.W., Hajat, S., Sarran, C. *et al.* (2017) Pollen exposure and hospitalization due to Asthma exacerbations: Daily time series in a European city. *International Journal of Biometeorology* 61(10), 1837–1848. DOI: 10.1007/s00484-017-1369-2.

Pascal, M., Laaidi, K., Ledrans, M., Baffert, E., Caserio-Schönemann, C. *et al.* (2006) France's heat health watch warning system. *International Journal of Biometeorology* 50, 144–153. DOI: 10.1007/s00484-005-0003-x.

Patz, J.A. and Hahn, M.B. (2013) Climate change and human health: A one health approach. *Current Topics in Microbiology and Immunology* 366, 141–171. DOI: 10.1007/82_2012_274.

Paz, S. (2015) Climate change impacts on west nile virus transmission in a global context. *Philosophical Transactions of the Royal Society B: Biological Sciences* 370(1665), 20130561. DOI: 10.1098/rstb.2013.0561.

Philip, S.Y., Kew, S.F., Van Oldenborgh, G.J., Anslow, F.S., Seneviratne, S.I. *et al.* (2021) Rapid attribution analysis of the extraordinary heatwave on the pacific coast of the US and Canada June 2021. *Earth System Dynamics Discussions* 2021, 1–34. DOI: 10.1146/annurev-environ-033009-095548.

Pyšek, P. and Richardson, D.M. (2010) Invasive species, environmental change and management, and health. *Annual Review of Environment and Resources* 35, 25–55. DOI: 10.1146/annurev-environ-033009-095548.

Qadri, F., Islam, T. and Clemens, J.D. (2017) Cholera in Yemen - an old foe rearing its ugly head. *The New England Journal of Medicine* 377(21), 2005–2007. DOI: 10.1056/NEJMp1712099.

Sehlstedt, M., Dove, R., Boman, C., Pagels, J., Swietlicki, E. *et al.* (2010) Antioxidant airway responses following experimental exposure to wood smoke in man. *Particle and Fibre Toxicology* 7(1), 1–11. DOI: 10.1186/1743-8977-7-21.

Semenza, J.C., Suk, J.E., Estevez, V., Ebi, K.L. and Lindgren, E. (2012) Mapping climate change vulnerabilities to infectious diseases in Europe. *Environmental Health Perspectives* 120(3), 385–392. DOI: 10.1289/ehp.1103805.

Reisen, W.K., Fang, Y. and Martinez, V.M. (2014) Effects of temperature on the transmission of West Nile virus by *Culex tarsalis* (Diptera: Culicidae). *Journal of Medical Entomology* 43(2), 309–317. DOI: 10.1146/annurev-environ-033009-095548.

Ren-Jian, Z., Kin-Fai, H.O. and Zhen-Xing, S. (2012) The role of aerosol in climate change, the environment, and human health. *Atmospheric and Oceanic Science Letters* 5(2), 156–161. DOI: 10.1080/16742834.2012.11446983.

Rita, R. (2009) Cholera and climate: A demonstrated relationship. *Transactions of the American Clinical and Climatological Association* 120, 119.

Romanello, M., McGushin, A., Di Napoli, C., Drummond, P., Hughes, N. *et al.* (2021) The 2021 report of the Lancet countdown on health and climate change: Code red for a healthy future. *The Lancet* 398(10311), 1619–1662. DOI: 10.1016/S0140-6736(21)01787-6.

Romanello, M., Di Napoli, C., Drummond, P., Green, C., Kennard, H. *et al.* (2022) The 2022 report of the lancet countdown on health and climate change: Health at the mercy of fossil fuels. *The Lancet* 400(10363), 1619–1654. DOI: 10.1016/S0140-6736(22)01540-9.

Sacks, E., Yangchen, S. and Marten, R. (2021) COVID-19, climate change, and communities. *The Lancet Planetary Health* 5(10), e663–e664. DOI: 10.1016/S2542-5196(21)00257-6.

Schewe, J., Heinke, J., Gerten, D., Haddeland, I., Arnell, N.W. *et al.* (2014) Multimodel assessment of water scarcity under climate change. *Proceedings of the National Academy of Sciences* 111(9), 3245–3250. DOI: 10.1073/pnas.1222460110.

Song, X., Wang, S., Hu, Y., Yue, M., Zhang, T. *et al.* (2017) Impact of ambient temperature on morbidity and mortality: An overview of reviews. *The Science of the Total Environment* 586, 241–254. DOI: 10.1016/j.scitotenv.2017.01.212.

Sprenger, P. and Reiter, D. (1987) The used tire trade: A mechanism for the worldwide dispersal of container breeding mosquitoes. *Journal of the American Mosquito Control Association* 3, 494.

Stuart-Smith, R.F., Roe, G.H., Li, S. and Allen, M.R. (2021) Increased outburst flood hazard from lake palcacocha due to human-induced glacier retreat. *Nature Geoscience* 14(2), 85–90. DOI: 10.1038/s41561-021-00686-4.

Taylor, R.G., Scanlon, B., Döll, P., Rodell, M., van Beek, R. *et al.* (2013) Ground water and climate change. *Nature Climate Change* 3(4), 322–329. DOI: 10.1038/nclimate1744.

Thompson, P.R., Widlansky, M.J., Leuliette, E., Sweet, W., Chambers, D.P. *et al.* (2019) Sea level variability and change. *Bulletin of the American Meteorological Society* 100(9), S84. DOI: 10.1038/nclimate1744.

UN Economic (2012) *Reducing Vulnerability and Exposure to Disasters*. The Asia-Pacific Disaster Report. Social Commission for Asia and the Pacific, UN Office for Disaster Risk Reduction. Available at: https://www.unescap.org/publications/asia-pacific-disaster-report-2012-reducing-vulnerability-and-exposure-disasters (accessed 4 August 2024).

van der Linden, S., Leiserowitz, A., Rosenthal, S. and Maibach, E. (2017) Inoculating the public against misinformation about climate change. *Global Challenges* 1(2), 1600008. DOI: 10.1002/gch2.201600008.

Van Oldenborgh, G.J., Krikken, F., Lewis, S., Leach, N.J., Lehner, F. *et al.* (2021) Attribution of the Australian bushfire risk to anthropogenic climate change. *Natural Hazards and Earth System Sciences* 21(3), 941–960. DOI: 10.1038/nclimate1744.

Watts, N., Adger, W.N., Agnolucci, P., Blackstock, J., Byass, P. *et al.* (2015) Health and climate change: Policy responses to protect public health. *The Lancet* 386(10006), 1861–1914. DOI: 10.1016/S0140-6736(15)60854-6.

WHO Ebola Response Team (2014) Ebola virus disease in West Africa--the first 9 months of the epidemic and forward projections. *New England Journal of Medicine* 371(16), 1481–1495. DOI: 10.1056/NEJMoa1411100.

Woolhouse, M.E.J. and Gowtage-Sequeria, S. (2005) Host range and emerging and reemerging pathogens. *Emerging Infectious Diseases* 11(12), 1842. DOI: 10.3201/eid1112.050997.

World Health Organization (2009) *Dengue: Guidelines for Diagnosis, Treatment, Prevention and Control*. Special Programme for Research, Training in Tropical Diseases, World Health Organization. Department of Control of Neglected Tropical Diseases, World Health Organization Epidemic and Pandemic Alert, World Health Organization, Geneva, Switzerland. DOI: 10.1016/S0140-6736(15)60854-6.

World Health Organization (2013) *World Health Report 2013: Research for Universal Health Coverage*. World Health Organization, Geneva, Switzerland. DOI: 10.1016/S0140-6736(15)60854-6.

World Health Organization (2021) *COP26 Special Report on Climate Change and Health: The Health Argument for Climate Action*. World Health Organization, Geneva, Switzerland.

Xu, R., Yu, P., Abramson, M.J., Johnston, F.H., Samet, J.M. *et al.* (2020) Wildfires, global climate change, and human health. *New England Journal of Medicine* 383(22), 2173–2181. DOI: 10.1056/NEJMsr2028985.

Ye, T., Guo, Y., Chen, G., Yue, X., Xu, R. *et al.* (2021) Risk and burden of hospital admissions associated with wildfire-related $PM_{2.5}$ in Brazil, 2000-15: A nationwide time-series study. *The Lancet Planetary Health* 5(9), e599–e607. DOI: 10.1016/S2542-5196(21)00173-X.

Young, N., Sharpe, R.A., Barciela, R., Nichols, G., Davidson, K. *et al.* (2020) Marine harmful algal blooms and human health: A systematic scoping review. *Harmful Algae* 98, 101901. DOI: 10.1016/j.hal.2020.101901.

Zaccarelli, N., Petrosillo, I. and Zurlini, G. (2008) Retrospective analysis. In: Jørgensen, S.E. and Fath, B.D. (eds) *Vol.[4] of Encyclopedia of Ecology*, Vol. 5 vols. Elsevier, Oxford, UK, pp. 3020–3029.

Zhang, P., Wiens, K., Wang, R., Luong, L., Ansara, D. *et al.* (2019) Cold weather conditions and risk of hypothermia among people experiencing homelessness: Implications for prevention strategies. *International Journal of Environmental Research and Public Health* 16(18), 3259. DOI: 10.3390/ijerph16183259.

Zhao, Q., Li, S., Coelho, M.S., Saldiva, P.H., Hu, K. *et al.* (2019) Assessment of intraseasonal variation in hospitalization associated with heat exposure in Brazil. *JAMA Network Open* 2(2), e187901. DOI: 10.1001/jamanetworkopen.2018.7901.

Zheng, X., Ding, H., Jiang, L., Chen, S., Zheng, J. *et al*. (2015) Association between air pollutants and Asthma emergency room visits and hospital admissions in time series studies: A systematic review and meta-analysis. *PloS One* 10(9), e0138146. DOI: 10.1371/journal.pone.0138146.
Zhou, Y., Zhuang, J., Yang, M., Zhang, Z., Wei, J. *et al*. (2010) Effects of low temperature on the schistosome-transmitting snail Oncomelania hupensis and the implications of global climate change. *Molluscan Research* 30(2), 102–108. DOI: 10.11646/mr.30.2.5.
Zinsstag, J., Schelling, E., Wyss, K. and Mahamat, M.B. (2005) Potential of cooperation between human and animal health to strengthen health systems. *The Lancet* 366(9503), 2142–2145. DOI: 10.1016/S0140-6736(05)67731-8.

10 Climate Change Impact on Food Safety

Sravanthi M.[1*], Muralidhar Y.[2], Ramya P.[3] and Sudheer K.[4]
[1]State Level Diagnostic Laboratory, Sri Venkateswara Veterinary University, Tirupati, India; [2]Department of Veterinary Pharmacology & Toxicology, College of Veterinary Science, Tirupati, India; [3]Department of Veterinary Public Health & Epidemiology, College of Veterinary Science, Tirupati, India; [4]Department of Livestock Products Technology, College of Veterinary Science, Tirupati, India

Abstract
Climate change is a major threat to human health due to its impacts on human health, including human health and the environment. Foodborne pathogens, poisons, and chemical residues in food are a growing concern for food safety. Pesticides, heavy metals, and mycotoxins can jeopardize food safety. The lack of safe water to irrigate crops, frequent use of pesticides, the breakage of the cold chain leading to temperature abuse, or the frequent occurrence of floods that result in the leaching of chemical contaminants into natural water sources are a few of the concerns. These issues raise the possibility of intoxications, foodborne infections, heavy metal and chemical bioaccumulation, antibiotic resistance, and the emergence of superbugs. This chapter focuses on the detection of microorganisms in food samples, the current state of the art of microbial contamination monitoring in the context of climate change, and its impact on food safety in a multidisciplinary manner.

Keywords: Human health, foodborne pathogen, contaminants, antibiotic resistance, food safety

10.1 Introduction

Any biological agent(s) that causes a foodborne illness event can be called a foodborne pathogen. Foodborne pathogens—mostly bacteria, fungi, viruses, and parasites—may contaminate food during production, processing, storage, or transport before reaching the consumer. Foodborne illnesses can be categorized into foodborne infections and foodborne intoxications. Foodborne infections are direct effects due to the presence of pathogens in food, whereas foodborne intoxications arise when certain of these pathogens secrete toxins into the food. According to the US Food and Drug Administration (FDA), foodborne pathogens are classified based on the specific foods consumed (FDA, 2010). In the US, *Campylobacter* sp., primarily linked to raw or undercooked poultry, causes over 2 million infections annually. Almost a million cases of food poisoning are caused by *Salmonella*, which is frequently present in meat, poultry, and eggs. Foodborne illnesses can also be caused by other pathogens, such as *Listeria, Escherichia coli, Shigella, Clostridium botulinum, Yersinia, Vibrio cholerae, V. vulnificus, V. parahaemolyticus, Staphylococcus aureus, Bacillus spp.* and *Shigella* (found in uncooked meats,

*Corresponding author: sravanthi.mannem@gmail.com

DOI: 10.1079/9781800625037.0010

vegetables, unpasteurized milk, and soft cheese). *Adenoviruses, rotaviruses, noroviruses, astroviruses,* and *hepatitis A* viruses are foodborne viruses that damage human health. *Toxoplasma gondii, Giardia lamblia, Giardia parvum,* and *Cyclospora cayetanensis* are examples of parasites associated with foodborne diseases. Of these, *Cryptosporidium* and *G. lamblia* are responsible for 71% of waterborne infections. The most common bacteria, viruses, fungi, and parasites responsible for foodborne infections are listed here (Table 10.1).

Biochemical and physical factors such as saliva, gastric acid, peristalsis, intestinal mucous, and the barrier formed by intestinal epithelial cells (IEC) are among the many host defenses that foodborne pathogens must contend with. Additionally, the gut-associated lymphoid tissue (GALT) helps to mount varying degrees of immune responses. Successful pathogens are able to infect hosts and cause tissue damage because they have ways to counter and avoid many of these defenses. These pathogens can impact human health beyond gastrointestinal symptoms, leading to substantial systemic effects. Bacterial infections can result in secondary complications such as meningoencephalitis, endophthalmitis, osteomyelitis, brain abscesses, Guillain-Barré syndrome, and peritonitis (Blintisis, 2017).

10.2 Sources of Contamination of Food

Food contamination is intricately linked to its origin and environmental conditions during handling and production. Compounds like lead, polychlorinated biphenyls (PCBs), dioxins, heavy metals, and chemical poisons in our food supply are heavily influenced by the environment, impacting food quality due to environmental pollution. Agricultural practices involving pesticides, insecticides, and pharmaceuticals contribute to the contamination of food and water sources, posing health risks to consumers (Thompson and Darwish, 2019; Lebelo *et al.*, 2021). Food packing practices, essential for transportation and preservation, introduce the risk of chemicals migrating into the food, jeopardizing safety and quality. Contamination can occur at any point in the supply chain, from farming or fishing to consumption. Poor handwashing, unclean environments, and improper storage contribute to contamination during food handling, even in homes. Multiple sources, including environmental pollutants, agricultural methods, and packaging materials, can lead to food contamination. Addressing these diverse sources is crucial for ensuring food safety, protecting public health and preserving the integrity of the food supply chain. Efforts are needed to mitigate these risks and safeguard consumers from potential health concerns.

10.3 Monitoring of Microbial Contamination in Food Chain

Monitoring microbial contamination in the food chain is crucial for ensuring food safety and preventing the spread of foodborne illnesses. Fecal coliform bacteria contamination signals potential pathogens, raising public health and economic concerns (Abdelzaher *et al.*, 2010). To manage health risks, Microbial Source Tracking (MST) methods, encompassing phenotypic and genotypic approaches, have been developed to identify sources of fecal contamination (Scott *et al.*, 2003; Stoeckel and Harwood, 2007). MST approaches are classified as library-independent methods (LIMs) or library-dependent methods (LDMs), using phenotypic and genotypic traits to create study libraries. Phenotypic approaches have limitations like limited specificity and low sensitivity, while genotypic approaches are simple, quick, and sensitive. Molecular methods for source tracking, include ribotyping, pulsed-field gel electrophoresis, AFLP, rep-PCR, multilocus sequence typing, and host-specific molecular marker PCR, which are commonly used in most cases.

10.3.1 Ribotyping

A technique for DNA fingerprinting called ribotyping makes use of rRNA genes that are highly conserved. Restriction endonucleases are used to treat genomic DNA, and the resultant fragments are sorted *via* gel electrophoresis. To distinguish between different bacterial strains,

Table 10.1. Different classes of foodborne pathogens.

No	Name of the organism	Nature of the organism	Mechanism of action/effects/source
Bacteria			
1.	*Bacillus cereus*	Gram-positive spore-forming bacteria	Diarrheal toxin, Emetic toxin
2.	*Campylobacter jejuni*	Gram-negative bacteria	Distending cytolethal toxin (CDT)
3.	*Clostridium botulinum*	Gram-positive anaerobic, spore-forming bacteria	Botulinum neurotoxins Seven types: A, B, C, D, E, F, and G
4.	*Clostridium perfringens*	Gram-positive anaerobic, spore-forming bacteria	Alpha toxin causing myonecrosis
5.	*Cronobacter sakazakii*	Gram-negative bacteria	Infant septicemia, meningitis, and necrotizing enterocolitis
6.	*Escherichia coli (STEC)*	Gram-negative, non-spore-forming rod	Shiga toxin
7.	*Listeria monocytogenes*	Gram-positive rods	Listeriolysin
8.	*Salmonella species (typhimurium, choleraesuis, and enteritidis)*	Gram-negative rods	Endotoxins and exotoxins
9.	*Shigella species*	Gram-negative rods	Shiga toxins (cytotoxins)
10.	*Staphylococcus aureus*	Gram-positive cocci	Hemolysin-α (Hla or α-toxin), Hemolysin-β, leukotoxins
11.	*Vibrio parahaemolyticus*	Gram-negative rods	thermostable direct hemolysin (TDH)
12.	*Yersinia enterocolitica*	Gram-negative rods	Pore-forming toxins, YaxA and YaxB
13.	*Aeromonas hydrophilia*	Gram-negative rods	Aerolysin-related cytotoxic enterotoxin (Act)
14.	*Coxiella burnetti*	Gram-negative coccobacillus	-
15.	*Francisella tularensis*	Gram-negative coccobacillus	-
Viruses			
16.	*Hepatitis A*	Hepatovirus	Non-enveloped, single positive-stranded RNA genome
17.	*Noroviruses*	Calciviridae	Non-enveloped, single-stranded RNA genome
18.	*Poliomyelitis*	Picornaviridae	Non-enveloped, single-stranded, RNA genome
19.	*Rotavirus*	Reoviridae	Segmented double-stranded RNA virus
Fungi			
20.	*Aspergillus flavus*	Aflatoxins	Grains, groundnuts, maize
21.	*Fusarium species*	Deoxynivalenol, Vomitoxin	Wheat, maize, oats, rice
22.	*Alternaria species*	Alternariol, tenuazonic acid	Maize, rice, wheat, hay
Parasites			
23.	*Cyclospora cayetanensis*	Coccidian parasite	Contaminated water and vegetables
24.	*Toxoplasma gondii*	Protozoan parasite (Intra-cellular)	Improperly cooked meat; accidental ingestion of contaminated food

Continued

Table 10.1. Continued

No	Name of the organism	Nature of the organism	Mechanism of action/effects/source
25.	*Trichinella spiralis*	Viviparous nematode parasite	Improperly cooked meat
26.	*Entamoeba histolytica*	Protozoan parasite	Contaminated water and food, spreads through feco-oral route
27.	*Anisakis spp.*	Nematode	Contaminated fish
28.	*Ascaris lumbricoides*	Nematode	Contaminated food
29.	*Clonorchis sinensis*	Trematode	Infected fish
30.	*Cryptosporidium parvum*	Protozoan parasite	Contaminated water, vegetables, fruits and unpasteurized milk
31.	*Diphyllobothrium latum*	Cestode	Raw or undercooked fish
32.	*Fasciola species*	Trematode	Contaminated water and vegetables
33.	*Giardia lamblia/duodenalis*	Flagellated protozoan Parasite	Contaminated water and food
34.	*Echinococcus granulosus*	Cestode	Contaminated water and food
35.	*Taenia solium*	Cestode	Undercooked pork

unique restriction fragment band patterns are identified using a probe specific to a conserved rRNA region (Carson *et al.*, 2001; Hartel *et al.*, 2002).

10.3.2 Pulse field gel electrophoresis

The gold standard for subtyping foodborne pathogens, PFGE, depends on purified genomic DNA that has undergone restriction digestion. Melted agarose is mixed with bacteria, and the resultant plugs undergo detergent-enzyme lysis and whole-genome digestion. Electrophoresis of large DNA fragments results in a banding pattern or DNA fingerprint. Because of its exceptional reproducibility and selective power, PFGE finds extensive use in the comparative typing of bacterial species (Foley *et al.*, 2006; Furukawa *et al.*, 2011).

10.3.3 AFLP analysis

The benefits of PCR amplification and restriction digest analysis are combined in AFLP analysis for genotyping. This technique involves digesting the target DNA with two distinct restriction enzymes and then ligating adaptor oligonucleotides to the sticky-ended DNA fragments that are produced. After amplifying a portion of these fragments using the PCR technique, an automated sequencer system is used to separate and identify the remaining fragments. With fluorescent dye-labeled primers (FAFLP) and an automated sequencer, AFLP can be modified for increased throughput and offers excellent sensitivity similar to PFGE profiling. High-resolution genotyping for outbreak investigation and source tracking of different foodborne diseases, such as *Salmonella*, pathogenic *E. coli*, and *Campylobacter*, has been done using this technology (Siemer *et al.*, 2005; Tokunaga *et al.*, 2007).

10.3.4 Rep-PCR

Rep-PCR creates extremely specific genomic fingerprints by using primers that correspond to repetitive DNA fragments that are randomly found throughout the bacterial genome. Targeting certain families of repeated elements, three techniques for repetitive sequence analysis have been used: REP-PCR, ERIC-PCR, and BOX-PCR. These repeating units are thought to be extremely conserved, and different-sized DNA segments are produced by amplifying different rep sites. By separating the fragments using agarose gel electrophoresis, it is possible to compare the resulting banding patterns that are particular to different bacterial strains.

10.3.5 MLST

Finding genetic diversity in several conserved genes is the essential basis of MLST, which is used to classify strains, identify clonal groups, and clarify the history of chromosomal background divergence. Using many genetic loci spread throughout the genome, the nucleotide sequences of areas of 500 base pairs or less are ascertained using this approach. Certain PCR primers are used for gene amplification, and an automated sequencer is often used for sequencing. The choice of gene targets is carefully considered while creating a new MLST system. For researching foodborne diseases, MLST on virulence genes and virulence-associated genes (MVLST) gives greater discriminatory power, even if housekeeping genes are a useful sequencing target for global phylogenetic investigations (Lomonaco *et al.*, 2008; Foley *et al.*, 2009).

10.3.6 Host-specific molecular marker PCR

As an alternative to library-dependent approaches, the direct detection of host-specific genetic markers has been proposed for the identification of fecal contamination origins in ambient waters. Suggested targets include a variety of distinct toxin genes or extra host-specific genes found by PCR. Based on their pathogenic properties and the hosts they target, bacteria can be identified.

10.4 Detection of Microorganisms in Food Samples

Detection of foodborne pathogens is essential for food safety and health. Several methods are available for the detection of microorganisms in food samples. The advantages and disadvantages of various methods employed for the detection of foodborne pathogens are listed in Table 10.2.

10.4.1 Conventional methods

Traditional methods for identifying foodborne bacterial pathogens in food involve culturing microorganisms on agar plates and using standard biochemical analyses. These methods are cost-effective but time-intensive, requiring culture media preparation, plate inoculation, and colony counting. They may have limited sensitivity, potentially leading to false negatives due to viable but non-culturable pathogens. Recently, there has been a surge in the development of rapid methods with heightened sensitivity and specificity to address these drawbacks. These methods play a crucial role in the food industry, identifying pathogens in both raw and processed foods quickly. They are more efficient, reduce labor, and minimize human errors. Rapid detection methods typically fall into three categories: nucleic acid-based, biosensor-based, and immunological-based methods (Zhao *et al.*, 2014)

10.4.2 Nucleic acid-based methods

The targeted pathogen's specific DNA or RNA sequences are identified by nucleic acid-based techniques. In this procedure, a synthetic oligonucleotide (often referred to as a primer or probe) that complements the target sequence is matched to the target nucleic acid sequence (Zhao *et al.*, 2014). Toxins produced by a variety of bacterial infections can lead to foodborne diseases (Singh *et al.*, 2001; Akbulut *et al.*, 2004). Nucleic acid-based techniques can be used to identify the genes linked to these toxins in these infections (Zhao *et al.*, 2014). Different types of nucleic acid-based methods are enlisted here.

10.4.2.1 Polymerase chain reaction

The primary approach for identifying foodborne pathogens is through the polymerase chain reaction (PCR). This method involves a three-step process: denaturation, annealing, and extension. Initially, DNA strands are separated at approximately 95°C, then specific primers bind at 55–65°C, and finally, an extension step duplicates the targeted DNA at 72°C using DNA polymerase. These stages repeat multiple times to generate numerous copies of the DNA. After amplification,

Table 10.2. Advantages and disadvantages of different techniques used for detection of microorganisms in food and water samples

No	Method used for detection	Advantages	Disadvantages
1.	Conventional method	Cost-effective	Time consuming and limited sensitivity
Nucleic acid-based methods			
2.	PCR	High sensitivity, specificity	Affected by PCR inhibitors. Difficult to differentiate viable and non-viable cells
3.	Multiplex PCR	Rapid and can detect multiple organisms in a single reaction	Optimization is difficult and expertise is required
4.	Real-time PCR	High sensitivity, specificity, does not require post-amplification products processing, real-time monitoring	High cost, difficult to distinguish between viable and non-viable cells, requires trained personnel
5.	NASBA	Low detection limits and rapid assay times, does not require thermal cycler	Requires viable microorganisms, difficulty in handling RNA
6.	LAMP	Low cost, easy to operate, does not require thermal cycling system	Primer design is complicated, insufficient to detect unknown or unsequenced targets
Biosensor-based methods			
7.	Optical biosensor	High sensitivity Label-free detection	High cost No quantitation
8.	Piezoelectric biosensor	High amplitude and frequency range	Depends on temperature Less energy efficiency
9.	Cell-based biosensors	Rapid, cost-effective, high sensitivity and specificity	High cost and high interference due to heterogeneity in cell populations
10.	Electrochemical sensors	System portability, rapid response, and high stability	Assay result depends on temperature and pH
11.	Microbial metabolism-based biosensors	Low cost, rapid response, and portability	Poor selectivity, low sensitivity
12.	Immunosensors	High sensitivity and low cost	Non-specific binding, increased complexity of the assay
13.	DNA-based biosensors	Simple to use, adaptable, customizable affinity, more affordable, longer-lasting lifespan	Easily affected by temperature
Immunological methods			

Continued

Table 10.2. Continued

No	Method used for detection	Advantages	Disadvantages
14.	ELISA	Quick, easy to perform and allows detection of bacterial toxins	Low sensitivity, false negative results and cross reactivity with closely related antigens
15.	Lateral flow immunoassay	Low cost, easy to operate, sensitive, specific and allow detection of bacterial toxins	Costly, requires labelling of antigens or antibodies

the DNA was visualized through agarose gel electrophoresis by staining with ethidium bromide. PCR is known for its sensitivity, specificity, and speed, and its sensitivity can be further enhanced by sequencing the amplified product.

10.4.2.2 *Multiplex PCR*

Unlike conventional PCR, multiplex PCR (mPCR) can detect multiple gene targets in a single reaction. The basic principle is similar to conventional PCR. However, multiple primers corresponding to different genes of various organs are used simultaneously in a single PCR reaction. Designing the primers for multiplex PCR is crucial, as all the primers should have the same annealing temperature and the concentration of primers should also be optimized accordingly to avoid the problem of primer dimer formation. In addition, the concentrations of magnesium, dNTPs, template, and Taq DNA polymerase should also be optimized for successful mPCR. Previously, mPCR was limited to detecting two or three pathogens. These days, mPCR is more sophisticated and capable of concurrently detecting four or more pathogens simultaneously. To detect *Salmonella enterica*, *Listeria monocytogenes*, *Escherichia coli*, and *Staphylococcus aureus* simultaneously, Boukharouba *et al.* (2022) devised a quick and easy mPCR that targets the *gad A*, *nuc*, *lis A* and *inv A* genes, respectively.

10.4.2.3 *Real-time or quantitative PCR*

Real-time PCR, or quantitative PCR, continuously monitors PCR product formation by detecting the fluorescence emitted from intercalating dyes or dual-labeled probes like TaqMan probes or molecular beacons. SYBR Green, a commonly used intercalating dye, binds non-specifically to double-stranded DNA, emitting fluorescence. TaqMan probes and molecular beacons, with reporter and quencher dyes at their ends, utilize fluorescence resonance energy transfer (FRET). The 5′–3′ exonuclease activity of Taq DNA polymerase cleaves TaqMan probes, separating the dyes and producing a fluorescence signal. TaqMan probes target specific nucleotide sequences in amplicons, while molecular beacons, with a stem-and-loop shape, emit fluorescence upon hybridizing with complementary sequences during conformational changes.

It is possible to detect some foodborne diseases using real-time PCR. The TaqMan real-time polymerase chain reaction (PCR) assay was created to detect and quantify 12 common pathogens in food and drinking water simultaneously in a single reaction. These pathogens include *Salmonella enterica*, *Enterococcus fecalis*, *Salmonella O157:H7*, *Listeria monocytogenes*/*ivanovii*, *Vibrio parahaemolyticus*, *β-streptococcus hemolyticus*, *Yersinia enterocolitica*, *Enterococcus fecalis*, *Shigella* spp., *Proteus mirabilis*, *Vibrio fluvialis*, *Staphylococcus aureus*, and *Campylobacter jejuni*. The limit of detection for the assay in meat samples was 10^4 CFU/g for the other strains and 10^3 CFU/g for *V. parahaemolyticus*. No cross-reaction was seen between the pathogens (Liu *et al.*, 2019).

10.4.2.4 *NASBA (Nucleic acid sequence-based amplification)*

NASBA involves amplifying nucleic acids at a constant temperature, particularly RNA, wherein single-stranded RNA undergoes conversion to cDNA through reverse transcription. This entire process occurs within an incubator without the need for a thermocycler. The isothermal reaction happens at 41°C and involves the use of two specific primers, namely complementary and antisense primers, along with three enzymes: avian myeloblastosis virus (AMV) reverse transcriptase, T7 RNA polymerase, and RNase H. The amplification of the target material can be observed through agarose gel electrophoresis. This technique finds widespread application in detecting RNA viruses and pathogenic bacteria, such as *Campylobacter* spp., *Listeria monocytogenes*, *Vibrio cholerae*, and *Escherichia coli*, in food and environmental samples (Zhai *et al.*, 2019). An evolution of this method, known as real-time NASBA, employs fluorescently labeled probes like molecular beacons, enabling high-throughput analysis.

10.4.2.5 LAMP (Loop-mediated isothermal amplification)

LAMP was developed by Notomi *et al.* (2000), a highly specific, sensitive, and rapid technique for nucleic acid amplification. The LAMP technique relies on continuous cycling DNA synthesis driven by *Bst* DNA polymerase large fragments at a consistent temperature between 59°C and 65°C for 60 min. This method employs four primers, consisting of two inner and two outer primers, designed to target six specific regions within the DNA of interest. As a result of LAMP, the final products manifest as intricate cauliflower-shaped DNA structures featuring multiple loops and diverse-sized stem-loop DNAs. This process yields a substantial quantity of amplicons within the 60 min timeframe, typically reaching a 103-fold or higher increase compared to standard PCR. Detection of LAMP-generated amplicons can be achieved through agarose gel electrophoresis or using SYBR Green dye (Law *et al.*, 2014).

10.4.3 Biosensor-based methods

A biosensor is a device that converts various effects into measurable electrical signals. It consists of a bioreceptor and a transducer, which convert biological interactions into detectable electrical signals. Biosensors are essential for detecting foodborne pathogens, involving a live capture molecule, a signal-to-signal conversion mechanism, and an output information system. Antibodies are commonly used as biorecognition elements due to their specificity and strong bonding. The immobilization of the biorecognition element is crucial for specific applications. Signal detection is executed by transducers, which convert the signals into measurable forms for analysis and interpretation. Various types of biosensors employed for the detection of foodborne pathogens are listed below.

10.4.3.1 Optical biosensors

Optical biosensors, particularly those based on surface plasmon resonance (SPR), are widely used for detecting foodborne infections. There are two main forms: label-free and label-based. Label-free uses a transducer's interaction with the material, while label-based uses a label and produces the optical signal through luminescence, fluorescence, or colorimetric techniques (Damborský *et al.*, 2016). Bioreceptors are affixed to a thin metal surface, causing a strong resonance effect when pathogens adhere, shifting the required wavelength for electron resonance (Zhang, 2013). Ohk and Bhunia (2013) developed a multiplex optical fiber technique to identify foodborne bacterial diseases like *Salmonella enterica*, *Listeria monocytogenes*, and *E. coli O157:H7*. Optical biosensors have been developed to quickly identify pollutants, poisons, and harmful bacteria in food.

10.4.3.2 Piezoelectric biosensor

Piezoelectric biosensors use affinity interaction detection to identify pathogenic microbes in food samples without labels. They consist of biological elements and transducers, relying on piezoelectric materials like quartz. Gold electrodes are commonly used in biosensing applications. For protein detection, antibody-based immunoassays are commonly conducted using quartz crystal microbalance (QCM) techniques (Chen *et al.*, 2010). The efficiency of QCM biosensors in detecting protein targets depends on the immobilization process and the chosen detection strategy (Saravanan *et al.*, 2021). This system is particularly useful for identifying pathogenic microbes in food samples.

10.4.3.3 *Cell-based biosensors*

Cell-based biosensors (CBBs) combine live cells with electrodes or transducers to gather data on biologically active substances. They have high sensitivity, capture a wide range of signals, and enable noninvasive, long-term recording. CBBs provide insights into pathogens and toxins as well as their functionality (Lu *et al.*, 2021). Biological recognition mechanisms are integrated with signal-transducing components, allowing for the detection and analysis of changes in physiological parameters (Xu *et al.*, 2005). CBBs also facilitate the detection of enzymes and co-factors within cellular metabolic systems. Advancements in recombinant DNA technology have expanded the possibilities for tailoring microorganisms, enhancing their activity (Arora *et al.*, 2011), and efficient consumption or breakdown of new substrates.

10.4.3.4 *Electrochemical sensors*

The core of an electrochemical biosensor is the biorecognition element, fixed onto the electrode surface through physical or chemical methods. This element selectively pinpoints and captures the target molecule, leveraging its specific recognition function. The electrode, functioning as the signal converter, translates the surface-generated identification signal into an electrical form—current, voltage, or resistance—allowing for measurement and analysis. This process facilitates both qualitative and quantitative analysis of the target substance.

With their great sensitivity, these sensors have proven essential in the detection of a wide range of bacteria, including *Vibrio cholera* (Sharma *et al.*, 2006), *Staphylococcus aureus* (Jia *et al.*, 2014), *Clostridium perfringens* (Jiang *et al.*, 2014) and *E. coli* (Giovanni *et al.*, 2015). Based on the data observed—current, impedance, potential, or conductance, respectively—electrochemical biosensors can be classified as amperometric, impedimetric, potentiometric, or conductometric. In the annals of biosensor development, electrochemical biosensors were the first known type to be commercialized.

10.4.3.5 *Microbial metabolism-based biosensors*

Biochemical reactions within bacterial cells drive metabolic processes, enabling the functionality of amperometric biosensors. A clerk-type oxygen electrode measures oxygen utilization during bacterial proliferation, particularly for foodborne bacteria like *S. typhimurium* (Ruan *et al.*, 2002). Electrochemical transducers detect marker enzymes like β-D-glucuronide and β-D-galactosidase (β-GAL) in water samples. Previously, identifying *E. coli* with GUS required a laborious procedure based on spectrometric identification. However, an effective electro-oxidative technique has bypassed this process by immobilizing *Moraxella* species on carbon-attached electrodes. D-GAL catalyzes the conversion of lactose to glucose and galactose, enumerating coliforms in water. This enzyme has shown promise in identifying live *E. coli* cells by hydrolyzing 4-aminophenyl β-D-galactopyranoside to 4-aminophenol.

10.4.3.6 *Immunosensors*

Immunosensors are affinity-based biosensors used to identify antigens by forming a stable compound with antibodies. They are categorized into electrochemical devices, optical devices, and piezoelectric devices (Janik-Karpinska *et al.*, 2022). They can be labeled or label-free depending on the transducer and signal processing techniques. Label-free immunosensors do not require extra labeling, reducing analysis costs, sample complexity, and preparation time. They enable real-time detection of target-probe binding but may have non-specific adsorption issues. Labeled immunosensors use signal-generating labels like metal ions, fluorescent dyes, enzymes, and nanomaterials to promote electron transport. These labels exhibit less non-specific signal adsorption, increased flexibility, and increased sensitivity compared to label-free immunosensors. Labeled immunosensors may have less non-specific signal adsorption, increased flexibility, and increased sensitivity compared to label-free ones.

10.4.3.7 DNA-based biosensors

DNA-based biosensors have gained popularity for pathogen detection due to their use of probes designed for specific bacteria. These probes are affixed to a transducer's surface, and the detection of pathogenic bacteria occurs through hybridization, where the degree of hybridization determines the presence or absence of complementary sequences in the sample. There are two types of DNA biosensors: DNA hybridization biosensors, which use short single-stranded DNA probes to form duplexes with complementary target nucleic acid fragments, and aptamer biosensors, which use *in vitro*-selected short segments of single-stranded DNAs/RNAs. These biosensors are popular due to their affordability, chemical stability, small size, and structural design flexibility. Recent developments have combined aptamers with new nanomaterials to improve their functionality.

10.4.4 Immunological methods

Foodborne pathogen detection using immunological techniques depends on the interaction of antibodies and antigens. The sensitivity and specificity of these methods depend on the strength of this binding. Both monoclonal and polyclonal antibodies are used by them. Among these techniques, enzyme-linked immunosorbent assay (ELISA) and lateral flow immunoassay are well known and have been used recently to identify foodborne pathogens.

10.4.4.1 Enzyme-linked immunosorbent assay (ELISA)

ELISA is a widely used immunological method for detecting foodborne bacteria. Sandwich ELISA, a successful variation, uses a dual-antibody technique. The main antibody is immobilized on microtiter plate wells, and the target antigen, such as bacteria or foodborne poisons, attaches to the immobilized primary antibody. Unbound antigens are removed, and the antigen is bound by an enzyme-conjugated secondary antibody. This results in a complex with the antigen sandwiched between the two antibodies. To identify this combination, a colorless substrate is added. ELISA has been used in various investigations to quickly identify food-associated infections, such as pathogenic *Vibrio parahaemolyticus* in seafood (Kumar *et al.*, 2011). Toxins found in food, such as *Clostridium perfringens* toxins, staphylococcal enterotoxins, botulinum toxins, and *Escherichia coli* enterotoxins, are also frequently detected (Aschfalk and Müller, 2002; Zhao *et al.*, 2014).

10.4.4.2 Lateral flow immunoassay

Innovative lateral flow immunoassays, like immunochromatographic strips and dipsticks, enable rapid on-site identification of foodborne pathogens. These devices, comprising a sample pad, conjugate pad, nitrocellulose membrane, and absorbent pad on a plastic backing, use capillary action to drive the sample fluid. The sample combines with a colored particle-labeled conjugate on the conjugate pad and flows through the nitrocellulose membrane, where antigens or antibodies immobilize it. Colored particles attach to antibodies or antigens at the test line, depending on the analytes. Within 2 to 10 min of adding the sample, the color becomes visible. Two main types of lateral flow immunoassays exist: the sandwich assay for analytes with multiple epitopes and the competitive assay for analytes with a single epitope (Ngom *et al.*, 2010).

10.5 Good Hygienic and Manufacturing Practices

In order to guarantee that food is safe and appropriate for human consumption, good hygiene practices work to put into practice the fundamental food hygiene principles that apply to every stage of the food chain, from primary production to ultimate consumption.

10.5.1 Building and equipment

- Selecting an appropriate location for buildings and equipment to minimize potential hazards. Avoid areas prone to pollution, nearby industrial activities, flooding, or pest infestation.
- Ensuring efficient equipment installation for efficient maintenance and cleaning.

- Maintaining walls and floors to prevent contamination of food. Walls and floors must not harbor dirt, germs, or fungus that might contaminate food. They should be free from cracks, peeling paint, or toxic materials.
- Designing floors for effective drainage and cleaning.
- Utilize working surfaces that are easy to clean, regularly disinfected, and made of inert (non-reactive) materials.
- Installing sufficient, easily maintained windows with removable, cleanable insect-proof screens.
- Thoroughly clean, disinfect and properly maintain equipment and containers to prevent food contamination. It is important that containers and equipment are moveable and can be disassembled for regular upkeep, cleaning, disinfection, and observation.
- Provide adequate facilities for heating, cooling, cooking, refrigeration, freezing, storing refrigerated or frozen food, and monitoring food temperatures, depending on the type of food operations.

10.5.2 Waste

- Identifying and constructing waste containers.
- Ensure containers are clearly marked, expertly built, and impermeable.
- Implementing security measures for hazardous substances.

10.5.3 Facilities

- Water Quality and Separation: Ensure potable water meets standards and is stored appropriately. Establish separate systems for non-potable water.
- Personnel Hygiene Facilities: Provide wash basins, hot/cold water supply, lavatories, and changing areas.
- Lighting, Ventilation, and Air Quality: Ensure ventilation to minimize food contamination, control temperatures, manage odors and regulate humidity.
- Storage Facilities: Design separate spaces for food, ingredients, and non-food chemicals to prevent pest access and shelter.

10.5.4 Personal hygiene

- Maintain personal cleanliness and health monitoring to prevent contamination and disease transmission.
- Wear protective gear when necessary.
- Regularly monitor staff health for conditions like jaundice, diarrhea, vomiting, fever, and discharges.
- Ensure consistent hand hygiene to ensure food safety.

10.5.5 Transportation

To ensure food remains uncontaminated during transit, effective control measures are crucial, even if prior hygiene protocols were followed in the food chain.

- **Choosing Suitable Transport**: Evaluate and select transport vehicles or containers based on the food type and the transportation conditions it requires.
- **Standards for Transport Vehicles**: The transport vehicle and bulk containers should adhere to certain standards:
    - Prevent contamination of foods or packaging.
    - Maintain cleanliness and, if necessary, undergo disinfection and regular maintenance.
    - Facilitate segregation of different food items or separate food from non-food items when necessary.
    - Provide adequate protection against contamination, including dust and fumes.
    - Maintain appropriate conditions (temperature, humidity, atmosphere) to prevent microbial growth or spoilage that could render the food unsuitable for consumption.
    - Allow monitoring of essential conditions such as temperature and humidity when needed.
    - Clean and disinfect the vehicle between transporting different foods or between food and non-food items to prevent cross-contamination.

10.6 Food Spoilage

Food spoilage is a metabolic process that renders food unsuitable for human consumption due to changes in sensory properties. It can be caused by microbial contamination, insect infestation, or endogenous enzymes. Physical and chemical changes, such as tissue tearing or oxidation, can also accelerate spoilage. Once harvested or slaughtered, food deteriorates rapidly, leading to off-flavors, texture changes, and nutrient loss. Food spoilage not only causes economic losses but also contributes to foodborne illnesses worldwide due to microbial deterioration. For example, lactic acid bacteria can spoil beer by causing turbidity and unpleasant odors.

Food spoilage can result from multiple factors:

1. Microorganisms like bacteria, yeast and mold
2. Insects that carry various microorganisms to food
3. Indigenous enzymes such as catalase and proteinase in animal and plant products
4. Physical damage through radiation, drying, temperature, and pressure
5. Presence of high moisture, protein, fat, and unhygienic handling fostering microbial growth

10.6.1 Food spoilage by microorganisms

Microorganisms are vital within living organisms, holding both positive and negative impacts. Among these microbes, yeast, molds, and bacteria play crucial roles in food spoilage.

10.6.1.1 Yeast

Yeast, a fermentative fungus, is essential for bread and alcohol production. It thrives in both aerobic and anaerobic conditions and is commonly found in sugary or salty foods. Key yeast species, such as *Candida, Brettanomyces, Zygosaccharomyces,* and *Debaryomyces hansenii,* cause spoilage in various foods. These species release flavors, odors, and carbon dioxide, causing container issues and spoilage.

10.6.1.2 Molds

Molds, microscopic fungi with multicellular threads called hyphae, are essential for breaking down dead plants and animals but can also ruin food. *Zygomycetes, Mucor,* and *Rhizopus* are common molds found on fruits and bread, while Penicillium and similar genera are sources of antibiotics. Some molds can release mycotoxins, such as *spatulin, ochratoxin, citreoviridin,* and *penitrem,* which can cause spoilage. Genera like *Byssochlamys* promote deterioration in pasteurized fluids due to their heat-resistant spores. *Aspergillus* primarily ruins food products like grains, beans, groundnuts, tree nuts, and spices, apart from paper and leather. *Fusarium* spp. also produce mycotoxins, potentially causing health issues in harvested grains.

10.6.1.3 Bacteria

Bacteria that produce spores can persist in heat-treated food due to their ability to survive in high-temperature environments. These Gram-positive, thermophilic anaerobes display diverse behaviors when exposed to elevated temperatures, such as generating hydrogen sulfide, emitting hydrogen and carbon dioxide, and releasing gas and unpleasant odors. Lactic acid bacteria (LAB) are essential for fermentation but can ruin food in low-oxygen, low-temperature, and acidic environments by giving off unwanted flavors. *Pseudomonas* species, such as *P. fluorescens, P. fragi, P. lundensis,* and *P. viridiflava,* break down food produced from animal sources, releasing enzymes such as lipases and proteases, resulting in biofilms and sulfides and trimethylamine. Food deterioration is caused by various factors, including insects, natural enzymes, and temperature variations. Insects can carry toxins or pathogens that cause sickness, while natural enzymes found in plants and animals can affect food quality if not neutralized before storage. Variations in temperature can also cause food to spoil, as foods stored at low temperatures can partially freeze, harming cells, and degrading quality.

10.7 Food Preservation

Preservation methods typically aim to inhibit the growth of bacteria, fungi (like yeasts), and other microorganisms, while also delaying the oxidation of fats that cause rancidity. This process involves treating and handling food to impede or reduce spoilage, loss of quality, degradation in edibility, or decline in nutritional value, enabling extended periods of food storage. Employing a variety of techniques, food preservation acts to prevent or decelerate spoilage, thereby extending the duration for which food can be stored after harvesting or slaughter.

10.7.1 Methods of preservation

Numerous methods exist for preserving food, aiming to prevent, delay, or minimize spoilage. Preservatives play a crucial role in extending the shelf life of food, allowing it to be grown, stored, marketed, and maintained in the customer's home for a reasonable duration. The preservation process involves maintaining nutritional value, texture, and flavor, with some methods altering the character of preserved foods, such as in the case of cheese, yogurt, and pickled onions.

Among the ancient preservation techniques are drying, refrigeration, and fermentation. Modern methods include canning, pasteurization, freezing, irradiation, and chemical addition, with advancements in packaging materials contributing significantly to modern food preservation. Drying, an ancient technique, reduces water activity to prevent bacterial growth. Refrigeration slows microorganism growth and enzyme action, preserving food. Freezing is widely used for various foods, even those that do not require freezing in their unprepared state.

Salting removes moisture from meat through osmosis, often using sugar, nitrates, and nitrites. Crystallization involves preserving fruits in syrup or crystallized form with sugar. Smoking, one of the oldest preservation methods, exposes food to smoke, extending shelf life.

Additives, like antimicrobials or antioxidants, inhibit bacterial, fungal, or mold growth. Preservatives, a chemical method, include calcium propionate, sodium nitrite, sodium nitrate, sulfites, disodium EDTA, BHA, and BHT. Pickling, using an edible antimicrobial liquid, comes in chemical and fermented forms. Canning involves cooking food, sealing it in sterile containers, and sterilizing it by boiling. High-acid fruits like strawberries need no preservatives and minimal boiling, while low-acid foods like vegetables and meats require canning pressure (Gragg and Brashears, 2014).

10.8 Food Poisoning

Food poisoning, a type of foodborne illness, occurs when individuals consume contaminated food or beverages containing harmful elements like germs or toxins. The causative agents include bacteria, viruses, parasites, and toxins to environmental factors. Symptoms, including upset stomach, diarrhea, and vomiting, typically appear within hours or days. Most cases are self-limiting and tend to resolve without specific treatment. Infants and children are at risk of rapid dehydration, leading to serious complications.

Certain bacteria causing foodborne illnesses can be present in water sources and spread through contact with animals. Higher-risk individuals include infants, pregnant individuals, older adults, and those with weakened immune systems. Complications, though rare in healthy adults, may involve dehydration, systemic infections, meningitis, and sepsis. Pregnant individuals affected by *Listeria* may face miscarriage, stillbirth, or complications in newborns. Rare complications include arthritis, irritable bowel syndrome, Guillain-Barré syndrome, and, in extreme cases, breathing difficulties due to botulism-induced nerve damage.

10.9 Impact of Climate Change on Pathogen Prevalence

Observed, measurable changes in the climate over an extended period of time are referred to as "climate change." Extreme weather events, such as heat waves, droughts, violent storms, hurricanes, typhoons, dust storms, wildfires, melting snow, the disappearance of river deltas and coastal cities, as well as environmental changes like habitat fragmentation, bushmeat consumption, urbanization, and deforestation, are characteristics of changing climates (Nava *et al.*, 2017). Pathogens may be directly impacted by climate change in ways that affect their survival, reproduction, and life cycle, or they may be indirectly impacted in ways that affect their habitat, surrounding environment, or rivals. As a result, changes may occur in the amount as well as the seasonal and geographic distributions of diseases.

10.9.1 Temperature

Prolonged hot weather can elevate the average temperature of food environments and water bodies, which could be conducive to algal blooms and the reproductive cycles of microorganisms. For instance, during the warm summers of 2006, *Vibrio* spp., which are indigenous to the Baltic and North Sea, demonstrated an accelerated growth rate (Frank *et al.*, 2006). Similarly, the multiplication of *Salmonella* is maximum when the temperature rises in the range of 7°C to 37°C.

10.9.2 Rainfall

Rainfall has a significant impact on the growth of bacteria that cause waterborne illnesses. The growth of fecal pathogens is associated with the rainy season because heavy precipitation can stir up sediments in water, which can result in the build-up of fecal bacteria (Jofre *et al.*, 2010). On the other hand, unexpected precipitation following an extended drought may lead to a rise in pathogens and an epidemic of disease.

10.9.3 Humidity

Changes in humidity have an impact on waterborne illness viruses. For instance, the drying effect of surface water limits the ability of waterborne viruses to survive close to the water's surface.

10.9.4 Sunshine

Another significant climatic factor that can have an impact on the microorganisms causing infectious diseases is sunshine. For instance, during cholera outbreaks, temperature and daylight hours synergistically act to provide an environment that is conducive to *Vibrio cholerae* growth in aquatic settings (Islam *et al.*, 2009).

10.9.5 Impact of climate change on food and waterborne diseases

Foodborne and waterborne illnesses are major global public health concerns. According to Greer *et al.* (2008), they are more prevalent in the summer and become more so as the temperature and humidity rise. Waterborne infections, such as giardiasis, cryptosporidiosis, infections with pathogenic *E. coli*, *Shigella*, *cholera*, *Salmonella*, and viral *hepatitis* A, will rise in response to an increase in floods and severe rainfalls (Haines and Patz, 2004). During severe floods, there is an increased risk of drinking polluted water from drainage systems and sewage.

The most significant and frequent illness spread by contaminated water is cholera, which is caused by *Vibrio cholerae*. Overall, enteric and diarrheal cases rise significantly as a result of global warming; however, occasionally, it also causes the appearance of deadly epidemics, such as meningoencephalitis brought on by the thermophilic free-living amoeba *Naegleria fowleri* (Huizinga and McLaughlin, 1990). A gradual rise in temperature brought on by global warming may make fungus acclimatized to the increased heat, allowing them to infect humans and other animals that were previously unable to infect due to the hosts' higher body temperatures.

10.10 Prevention of Foodborne Infections

Preventing foodborne infections involve implementing a combination of good hygiene practices, proper food handling, and regulatory measures. Here are key strategies for preventing foodborne infections:

- Hand washing: Make sure to wash your hands well with soap and water, especially before eating, after using the restroom, and after handling food.
- Clean Produce: Rinse fruits and vegetables under running water before consuming or preparing them.
- Utensil Hygiene: Clean knives, cutting boards, and cooking utensils in soapy water, especially if they have come into touch with raw meats or unwashed vegetables.
- Cook Meat and Fish Correctly: To ensure proper cooking temperatures, use a meat thermometer. Whole meats and fish should be cooked to at least 145°F (63°C), ground meat to at least 160°F (71°C), and poultry to at least 165°F (74°C).
- Refrigerate or Freeze Leftovers: Put leftovers in the refrigerator immediately or freeze them in sealed containers.
- Safe Storage of Leftovers: Thaw frozen food in a microwave, overnight in the refrigerator, or in cold water in a leak-proof container on the counter. Reheat leftovers to an internal temperature of 165°F (74°C).
- Toss It When in Doubt: Even if it seems and smells fine, throw it away if you are not sure whether the food is safe.
- Get Rid of Moldy Food: Throw out baked items, soft fruits, vegetables, nuts, and nut-based products that have mold on them.
- Refrigerator Cleaning: Use a solution of 1 quart (0.9 l) of water and 1 tablespoon (15 ml) of baking soda every few months to clean the refrigerator's interior.
- Educate food handlers in food handling and preparation are train them in food safety practices.
- Implement surveillance systems to monitor and identify outbreaks of foodborne illnesses. Investigate and address the root causes of outbreaks to prevent future occurrences.
- Regulatory measures: Strictly enforce food safety regulations and standards and regularly inspect food establishments to ensure compliance with hygiene and safety practices.

10.11 Impacts of Climate Change on Chemical Hazards Such as Mycotoxins, Pesticides, and Heavy Metals

Mycotoxins, produced by fungi like Aspergillus, Penicillium, Fusarium, and Alternaria, are secondary metabolites that cause acute and chronic illnesses, including cancer. These toxins, including aflatoxin, zearalenone, trichothecenes, fumonisin, and ochratoxin, pose the greatest risk to human health and animal welfare due to their immunosuppressive and carcinogenic properties (Barkai-Golan and Paster, 2008; Nleya *et al.*, 2018; Agriopoulou *et al.*, 2020). Mycotoxigenic fungi infect staple crops like rice, maize, corn, almonds, coffee grains, dried fruits, feed crops, vegetables, and spices, leading to a loss of yield and compromised food safety. The hot and humid climates favor mycotoxin production, and climate change is believed to worsen food contamination (Paterson and Lima, 2010; Milićević *et al.*, 2019). Post-harvest conditions can increase mycotoxin production, and secondary variables such as the emergence of new virulent fungi, increased pest infestations, decreased efficacy of fungicides and insecticides, and changes in insect distribution also influence mycotoxin contamination (Paris *et al.*, 2015).

Increased temperatures encourage plant pests, insects, and pathogens to proliferate quickly. The anticipated shifts in the population, life cycle, and geographic distribution of agricultural pests and pathogens would probably increase the use, dosage, and frequency of pesticide and fungicide applications, and the potential for residues in food are ongoing concerns that are anticipated as a result of climate change (Delcour *et al.*, 2015). On the other hand, because of heavy rainfall, pesticides reach the groundwater and also runoff into water bodies as well as get too diluted to have any impact (Otieno *et al.*, 2013). There will be a rise in the application of pesticides to offset leaching-related losses due to altered precipitation and humidity patterns (Delcour *et al.*, 2015). With such a dramatic change in geographical distribution, agricultural land will likely experience an intensification in pesticide use, compounding the food safety risks.

Another significant issue with food safety is the contamination of food with heavy metals. Even at modest levels of exposure, lead (Pb), cadmium (Cd), chromium (Cr), mercury (Hg), and arsenic (As) are heavy metals that are substantial toxicants. In the seafood chain, heavy metals bioaccumulate at every trophic level. The two major routes of food contamination with heavy metals are naturally through metal corrosion and atmospheric deposition in soil and water, and artificially through anthropogenic activities like mining, fertilizer application, and pesticide use. Heavy metals enter feeds and animal diets as a result of feeds cultivated on contaminated soil and/or treated with contaminated water. Heavy metals in animal feed not only damage the health and productivity of the animals but also contaminate animal products and pose health concerns to humans (Chandorkar, 2023). Variations in temperature and precipitation are two aspects of climate change that impact the fate of heavy metals. The rise in sea level due to melting glaciers and an increase in unseasonal precipitation from insufficient snowfall causes an increased risk of flooding. It is anticipated that increased rainfall patterns, melting glaciers, and permafrost or frozen water in subsurface strata, are causing substances that were formerly buried, such as pesticides, heavy metals, dormant strains of dangerous bacteria and viruses, and microplastics to be released into the environment, both in intensity and frequency (Pörtner *et al.*, 2019). The absorption of these leached heavy metals by crops, which eventually make their way into the food chain, is another problem (Duchenne-Moutien and Neetoo, 2021).

10.12 Risk of Contamination of Food by Chemical Hazards

Chemical contamination in food poses a significant threat to the industry, causing health hazards such as headaches, nausea, cancer, developmental issues, and birth defects. Foodborne illnesses and outbreaks are frequently caused by chemical pollutants discovered in food, particularly when the amounts exceed safety regulations (Carter and Blizard, 2016). This contamination is primarily caused by pesticide use, food processing techniques, and environmental factors. Globalization, climate change, and food production techniques contribute to the increasing prevalence of chemical pollutants. Lead, mercury, and arsenic are a few contaminants that have been linked to increased risks of stomach, liver, and skin cancers, respectively (Zhao *et al.*, 2014; Lu *et al.*, 2015). Gastrointestinal disorders may arise from consuming food contaminated with pesticides or heavy metals (Song *et al.*, 2017). In 2010, Nigeria experienced a lead poisoning outbreak resulting in 400–500 child deaths due to accidental exposure to lead-contaminated food (Tirima *et al.*, 2018).

10.12.1 Chemical contaminants in food

Chemical contaminants, unintentionally present in food or feed, can emerge throughout various stages of production, processing, or transportation, often stemming from environmental pollution. These substances hold the potential to be harmful to both humans and animals.

Key categories of food and feed contaminants encompass:

1. Naturally occurring chemical contaminants: These are organic compounds created by living things. Raw food surfaces are inherently vulnerable to contamination from a wide range of bacteria, viruses, parasites, and other diseases. Sources of contamination include soil, sewage, living creatures, external surfaces, internal organs of meat-eating animals, and food derived from sick animals (Rather *et al.*, 2017; Gallo *et al.*, 2020). Important natural chemical contaminants are enlisted here:

    - Mycotoxins: Secondary metabolites produced by various fungi like *Penicillium*, *Fusarium*, and *Aspergillus*. Prevalence of moist conditions

associated with lower temperatures predisposes the formation of toxins and fungal growth.
- Marine biotoxins: These are the natural toxins produced by certain types of phytoplankton in the sea. Seafood is the major source of these toxins.

2. Environmental contaminants: Substances released into the air, water, or soil, primarily due to industrial or agricultural activities, can infiltrate the food and feed chain. The most reported chemical contaminants are:

   - Methyl mercury: Bacteria convert accumulated mercury in water bodies to produce methyl mercury, which is commonly found in fish (Guimaraes *et al.*, 2000). Methyl mercury is harmful to the nervous system and has negative health effects.
   - Polychlorinated biphenyls (PCBs): They are widely employed in industrial settings and have been found to accumulate in animal tissues, such as fish and meat. Most of these poisons are neurotoxic.
   - Polybrominated diphenyl ethers (PBDEs): Found in many food items, including fish, meats, poultry, dairy products, and even human breast milk, PBDEs can build up in the fat tissue of these things due to their widespread distribution in the environment (Rose *et al.*, 2010; Schecter *et al.*, 2010). They are hormone disruptors and neurotoxins.
   - Bisphenol A: This chemical, which is an endocrine disruptor, migrates from food and drink containers, particularly when the containers are heated, and causes contamination primarily through the diet.
   - Phthalates: Phthalates are a class of endocrine disruptors that are frequently present in dairy products, oils, fish, and shellfish, among other fatty foods.

3. Process-derived contaminants: Chemicals that form naturally in food and feed during industrial processes or cooking:

   - Acrylamide: High-temperature cooking techniques produce large amounts of this compound. Boiling food does not produce substantial amounts. Carbohydrate-rich foods are most susceptible.
   - 3-MCPD: Carcinogen found in cheese, bread, grilled meats, fish, baked products, and malt extracts.
   - Furans: Colorless liquid used in chemical manufacturing industries, found in canned or bottled foods. Furans and methyl furans can cause liver damage over time.

10.13 Future Challenges to Food Safety and Security

In the future, the increasing prevalence of foodborne infections and intoxication is anticipated due to climate change, posing a significant threat to food security. Addressing these challenges requires urgent multidisciplinary research and collaboration among food scientists, public health officials, epidemiologists, veterinarians, meteorologists, and statisticians. Enhanced contamination monitoring necessitates the adoption of high-throughput, automated techniques for identifying pathogens, poisons, and chemical residues in food. Improved tracing of foodborne illnesses should involve widespread adoption of genotype-based techniques like whole-genome sequencing and real-time PCR. Rapid dissemination of research findings and data sharing across sectors is crucial, along with a critical evaluation and enforcement of current food regulations and standards. Developing strategies for food safety in the face of natural disasters should be a top national priority globally, accompanied by regular public education initiatives to raise awareness about climate change's impact on food safety and overall health.

Climate change-induced modifications in pathogen survival and alterations in animal diseases may necessitate more frequent use of veterinary pharmaceuticals. This could result in elevated levels of veterinary drug residue in foods, contributing to increased antimicrobial resistance (AMR) in human and animal diseases. The intensified use of pesticides due to climate change raises concerns about higher residues in food, posing a growing threat to food safety. Climate change compromises food security by increasing the frequency and intensity of extreme weather events, leading to insecure food supplies. This results in people having access to less nutritious diets and more "unsafe foods," presenting chemical, microbiological, and other health hazards that contribute to increased malnutrition.

References

Abdelzaher, A.M., Wright, M.E., Ortega, C., Solo-Gabriele, H.M., Miller, G. *et al.* (2010) Presence of pathogens and indicator microbes at a non-point source subtropical recreational marine beach. *Applied and Environmental Microbiology* 76(3), 724–732. DOI: 10.1128/AEM.02127-09.

Agriopoulou, S., Stamatelopoulou, E. and Varzakas, T. (2020) Advances in occurrence, importance, and mycotoxin control strategies: Prevention and detoxification in foods. *Foods (Basel, Switzerland)* 9(2), 137. DOI: 10.3390/foods9020137.

Akbulut, D., Grant, K.A. and McLauchlin, J. (2004) Development and application of real-time PCR assays to detect fragments of the *Clostridium botulinum* types A, B, and E neurotoxin genes for investigation of human foodborne and infant botulism. *Foodborne Pathogens and Disease* 1(4), 247–257. DOI: 10.1089/fpd.2004.1.247.

Arora, P., Sindhu, A., Dilbaghi, N. and Chaudhury, A. (2011) Biosensors as innovative tools for the detection of food borne pathogens. *Biosensors and Bioelectronics* 28(1), 1–12. DOI: 10.1016/j.bios.2011.06.002.

Aschfalk, A. and Müller, W. (2002) *Clostridium perfringens* toxin types from wild-caught Atlantic cod (*Gadus morhua* L.), determined by PCR and ELISA. *Canadian Journal of Microbiology* 48(4), 365–368. DOI: 10.1139/w02-015.

Barkai-Golan, R. and Paster, N. (2008) Mouldy fruits and vegetables as a source of mycotoxins: Part 1. *World Mycotoxin Journal* 1, 147–159. DOI: 10.3920/WMJ2008.x018.

Blintisis, T. (2017) Food-borne pathogens. *AIMS Microbiology* 3(3), 529–563. DOI: 10.3934/microbiol.2017.3.529.

Boukharouba, A., González, A., García-Ferrús, M., Ferrús, M.A. and Botella, S. (2022) Simultaneous detection of four main foodborne pathogens in ready-to-eat food by using a simple and rapid multiplex PCR (mpcr) assay. *International Journal of Environmental Research and Public Health* 19(3), 1031. DOI: 10.3390/ijerph19031031.

Carson, C.A., Shear, B.L., Ellersieck, M.R. and Asfaw, A. (2001) Identification of fecal *Escherichia coli* from humans and animals by ribotyping. *Applied and Environmental Microbiology* 67(4), 1503–1507. DOI: 10.1128/AEM.67.4.1503-1507.2001.

Carter, C.J. and Blizard, R.A. (2016) Autism genes are selectively targeted by environmental pollutants including pesticides, heavy metals, bisphenol A, phthalates and many others in food, cosmetics or household products. *Neurochemistry International* 101(2016), 83–109. DOI: 10.1016/j.neuint.2016.10.011.

Chandorkar, S. (2023) *Climate Change and Food Safety*. IntechOpen, London. DOI: 10.5772/intechopen.112575.

Chen, H., Hu, Q.-Y., Yue-Zheng, Jiang, J.-H., Shen, G.-L. *et al.* (2010) Construction of supported lipid membrane modified piezoelectric biosensor for sensitive assay of cholera toxin based on surface-agglutination of ganglioside-bearing liposomes. *Analytica Chimica Acta* 657(2), 204–209. DOI: 10.1016/j.aca.2009.10.036.

Damborský, P., Švitel, J. and Katrlík, J. (2016) Optical biosensors. *Essays in Biochemistry* 60(1), 91–100. DOI: 10.1042/EBC20150010.

Delcour, I., Spanoghe, P. and Uyttendaele, M. (2015) Literature review: Impact of climate change on pesticide use. *Food Research International* 68, 7–15. DOI: 10.1016/j.foodres.2014.09.030.

Duchenne-Moutien, R.A. and Neetoo, H. (2021) Climate change and emerging food safety issues: A review. *Journal of Food Protection* 84(11), 1884–1897. DOI: 10.4315/JFP-21-141.

FDA (Food and Drug Administration) (2010) Food Safety Modernization Act (FSMA). Available at: https://www.fda.gov/food/guidance-regulation-food-and-dietary-supplements/food-safety-modernization-act-fsma (accessed 3 December 2023).

Foley, S.L., Lynne, A.M. and Nayak, R. (2009) Molecular typing methodologies for microbial source tracking and epidemiological investigations of gram-negative bacterial foodborne pathogens. *Infection, Genetics and Evolution: Journal of Molecular Epidemiology and Evolutionary Genetics in Infectious Diseases* 9(4), 430–440. DOI: 10.1016/j.meegid.2009.03.004.

Foley, S.L, White, D.G., McDermott, P.F., Walker, R.D., Rhodes, B. *et al.* (2006) Comparison of subtyping methods for differentiating *Salmonella enterica* serovar Typhimurium isolates obtained from food animal sources. *Journal of Clinical Microbiology* 44(10), 3569–3577. DOI: 10.1128/JCM.00745-06.

Frank, C., Littman, M., Alpers, K. and Hallauer, J. (2006) *Vibrio vulnificus* wound infections after contact with the Baltic sea, Germany. *European Communicable Disease Bulletin* 11(8), E060817. DOI: 10.2807/esw.11.33.03024-en.

Furukawa, T., Yoshida, T. and Suzuki, Y. (2011) Application of PFGE to source tracking of faecal pollution in coastal recreation area: A case study in Aoshima beach, Japan. *Journal of Applied Microbiology* 110(3), 688–696. DOI: 10.1111/j.1365-2672.2010.04918.x.

Gallo, M., Ferrara, L., Calogero, A., Montesano, D. and Naviglio, D. (2020) Relationships between food and diseases: What to know to ensure food safety. *Food Research International* 137, 109414. DOI: 10.1016/j.foodres.2020.109414.

Giovanni, M., Setyawati, M.I., Tay, C.Y., Qian, H., Kuan, W.S. *et al.* (2015) Electrochemical quantification of *Escherichia coli* with DNA nanostructure. *Advanced Functional Materials* 25, 3840–3846. DOI: 10.1002/adfm.201500940.

Gragg, S.E. and Brashears, M.M. (2014) Microbiological safety of meat. In: Dikeman, M. and Devine, C. (eds) *Hurdle Technology in Encyclopedia of Meat Sciences*, 2nd edn. Academic Press, London, pp. 345–347.

Greer, A., Ng, V. and Fisman, D. (2008) Climate change and infectious diseases in North America: The road ahead. *CMAJ: Canadian Medical Association Journal* 178(6), 715–722. DOI: 10.1503/cmaj.081325.

Guimarães, J.R.D., Ikingura, J. and Akagi, H. (2000) Methyl mercury production and distribution in river water-sediment systems investigated through radiochemical techniques. *Water, Air, and Soil Pollution* 124(1/2), 113–124. DOI: 10.1023/A:1005206109083.

Haines, A. and Patz, J.A. (2004) Health effects of climate change. *Journal of the American Medical Association* 291(1), 99–103. DOI: 10.1001/jama.291.1.99.

Hartel, P.G., Summer, J.D., Hill, J.L., Collins, J.V., Entry, J.A. *et al.* (2002) Geographic variability of *Escherichia coli* ribotypes from animals in Idaho and Georgia. *Journal of Environmental Quality* 31(4), 1273–1278. DOI: 10.2134/jeq2002.1273.

Huizinga, H.W. and McLaughlin, G.L. (1990) Thermal ecology of *Naegleria fowleri* from a power plant cooling reservoir. *Applied and Environmental Microbiology* 56(7), 2200–2205. DOI: 10.1128/aem.56.7.2200-2205.1990.

Islam, M.S., Sharker, M.A.Y., Rheman, S., Hossain, S., Mahmud, Z.H. *et al.* (2009) Effects of local climate variability on transmission dynamics of cholera in Matlab, Bangladesh. *Transactions of the Royal Society of Tropical Medicine and Hygiene* 103(11), 1165–1170. DOI: 10.1016/j.trstmh.2009.04.016.

Janik-Karpinska, E., Ceremuga, M., Niemcewicz, M., Podogrocki, M., Stela, M. *et al.* (2022) Immunosensors-the future of pathogen real-time detection. *Sensors* 22(24), 9757. DOI: 10.3390/s22249757.

Jia, F., Duan, N., Wu, S., Ma, X., Xia, Y. *et al.* (2014) Impedimetric aptasensor for *Staphylococcus aureus* based on nanocomposite prepared from reduced graphene oxide and gold nanoparticles. *Microchimica Acta* 181, 967–974. DOI: 10.1007/s00604-014-1195-8.

Jiang, D., Liu, F., Liu, C., Liu, L., Li, Y. *et al.* (2014) Induction of an electrochemiluminescence sensor for DNA detection of *Clostridium perfringens* based on rolling circle amplification. *Analytical Methods* 6, 1558–1562. DOI: 10.1039/C3AY41961D.

Jofre, J., Blanch, A.R. and Lucena, F. (2010) Water-borne infectious disease outbreaks associated with water scarcity and rainfall events. In: Sabater, S. and Barceló, D. (eds) *Water Scarcity in the Mediterranean. The Handbook of Environmental Chemistry*, Vol. 8. Springer, Berlin, Heidelberg, Germany. DOI: 10.1007/698_2009_22.

Kumar, B.K., Raghunath, P., Devegowda, D., Deekshit, V.K., Venugopal, M.N. *et al.* (2011) Development of monoclonal antibody based sandwich ELISA for the rapid detection of pathogenic *Vibrio parahaemolyticus* in seafood. *International Journal of Food Microbiology* 145(1), 244–249. DOI: 10.1016/j.ijfoodmicro.2010.12.030.

Law, J.W.-F., Ab Mutalib, N.-S., Chan, K.-G. and Lee, L.-H. (2014) Rapid methods for the detection of foodborne bacterial pathogens: Principles, applications, advantages and limitations. *Frontiers in Microbiology* 5, 770. DOI: 10.3389/fmicb.2014.00770.

Lebelo, K., Malebo, N., Mochane, M.J. and Masinde, M. (2021) Chemical contamination pathways and the food safety implications along the various stages of food production: A review. *International Journal of Environmental Research and Public Health* 18(11), 5795. DOI: 10.3390/ijerph18115795.

Liu, Y., Cao, Y., Wang, T., Dong, Q., Li, J. *et al.* (2019) Detection of 12 common food-borne bacterial pathogens by taqman real-time PCR using a single set of reaction conditions. *Frontiers in Microbiology* 10, 222. DOI: 10.3389/fmicb.2019.00222.

Lomonaco, S., Chen, Y. and Knabel, S.J. (2008) Analysis of additional virulence genes and virulence gene regions in listeria monocytogenes confirms the epidemiologic relevance of multi-virulence-locus sequence typing. *Journal of Food Protection* 71(12), 2559–2566. DOI: 10.4315/0362-028x-71.12.2559.

Lu, X., Ye, Y., Zhang, Y. and Sun, X. (2021) Current research progress of mammalian cell-based biosensors on the detection of foodborne pathogens and toxins. *Critical Reviews in Food Science and Nutrition* 61(22), 3819–3835. DOI: 10.1080/10408398.2020.1809341.

Lu, Y., Song, S., Wang, R., Liu, Z., Meng, J. *et al.* (2015) Impacts of soil and water pollution on food safety and health risks in China. *Environment International* 77, 5–15. DOI: 10.1016/j.envint.2014.12.010.

Milićević, D., Petronijević, R., Petrović, Z., Đjinović-Stojanović, J., Jovanović, J. *et al.* (2019) Impact of climate change on aflatoxin M1 contamination of raw milk with special focus on climate conditions in Serbia. *Journal of the Science of Food and Agriculture* 99(11), 5202–5210. DOI: 10.1002/jsfa.9768.

Nava, A., Shimabukuro, J.S., Chmura, A.A. and Luz, S.L.B. (2017) The impact of global environmental changes on infectious disease emergence with a focus on risks for Brazil. *ILAR Journal* 58(3), 393–400. DOI: 10.1093/ilar/ilx034.

Ngom, B., Guo, Y., Wang, X. and Bi, D. (2010) Development and application of lateral flow test strip technology for detection of infectious agents and chemical contaminants: A review. *Analytical and Bioanalytical Chemistry* 397(3), 1113–1135. DOI: 10.1007/s00215-010-3661-4.

Nleya, N., Adetunji, M.C. and Mwanza, M. (2018) Current status of mycotoxin contamination of food commodities in Zimbabwe. *Toxins* 10(5), 89. DOI: 10.3390/toxins10050089.

Notomi, T., Okayama, H., Masubuchi, H., Yonekawa, T., Watanabe, K. *et al.* (2000) Loop-mediated isothermal amplification of DNA. *Nucleic Acids Research* 28(12), E63. DOI: 10.1093/nar/28.12.e63.

Ohk, S.H. and Bhunia, A.K. (2013) Multiplex fiber optic biosensor for detection of listeria monocytogenes, *Escherichia coli* O157:H7 and *Salmonella enterica* from ready-to-eat meat samples. *Food Microbiology* 33(2), 166–171. DOI: 10.1016/j.fm.2012.09.013.

Otieno, P.O., Owuor, P.O., Lalah, J.O., Pfister, G. and Schramm, K.W. (2013) Impacts of climate-induced changes on the distribution of pesticides residues in water and sediment of Lake Naivasha, Kenya. *Environmental Monitoring and Assessment* 185(3), 2723–2733. DOI: 10.1007/s10661-012-2743-5.

Paris, M.P.K., Liu, Y.J., Nahrer, K. and Binder, E.M. (2015) Climate change impacts on mycotoxin production. In: Botana, L.M. and Sainz, M.J. (eds) *Climate Change and Mycotoxins*. Walter de Gruyter GmbH, Berlin, Germany, pp. 133–149.

Paterson, R.R.M. and Lima, N. (2010) How will climate change affect mycotoxins in food? *Food Research International* 43, 1902–1914. DOI: 10.1016/j.foodres.2009.07.010.

Pörtner, H.O., Roberts, D.C., Masson-Delmotte, V., Zhai, P., Tignor, M. *et al.* (2019) IPCC, 2019: Summary for policymakers. In: *IPCC Special Report on the Ocean and Cryosphere in a Changing Climate*. Cambridge University Press, Cambridge, UK, and New York, pp. 3–35. DOI: 10.1017/9781009157964.001.

Rather, I.A., Koh, W.Y., Paek, W.K. and Lim, J. (2017) The sources of chemical contaminants in food and their health implications. *Frontiers in Pharmacology* 8, 830. DOI: 10.3389/fphar.2017.00830.

Rose, M., Bennett, D.H., Bergman, A., Fängström, B., Pessah, I.N. *et al.* (2010) PBDEs in 2-5 year-old children from California and associations with diet and indoor environment. *Environmental Science & Technology* 44(7), 2648–2653. DOI: 10.1021/es903240g.

Ruan, C., Yang, L. and Li, Y. (2002) Rapid detection of viable *Salmonella typhimurium* in a selective medium by monitoring oxygen consumption with electrochemical cyclic voltammetry. *Journal of Electroanalytical Chemistry* 519(1–2), 33–38. DOI: 10.1016/S0022-0728(01)00711-2.

Saravanan, A., Kumar, P.S. and Hemavathy, R.V. (2021) Methods of detection of food-borne pathogens: A review. *Environmental Chemistry Letters* 19, 189–207. DOI: 10.1007/s10311-020-01072-z.

Schecter, A., Colacino, J., Patel, K., Kannan, K., Yun, S.H. *et al.* (2010) Polybrominated diphenyl ether levels in foodstuffs collected from three locations from the United States. *Toxicology and Applied Pharmacology* 243(2), 217–224. DOI: 10.1016/j.taap.2009.10.004.

Scott, T.M., Parveen, S., Portier, K.M., Rose, J.B., Tamplin, M.L. *et al.* (2003) Geographical variation in ribotype profiles of escherichia coli isolates from humans, swine, poultry, beef, and dairy cattle in Florida. *Applied and Environmental Microbiology* 69(2), 1089–1092. DOI: 10.1128/AEM.69.2.1089-1092.2003.

Sharma, M.K., Goel, A.K., Singh, L. and Rao, V.K. (2006) Immunological biosensor for detection of *Vibrio cholerae* O1in environmental water samples. *World Journal of Microbiology and Biotechnology* 22, 1155–1159. DOI: 10.1007/s11274-006-9156-y.

Siemer, B.L., Nielsen, E.M. and On, S.L.W. (2005) Identification and molecular epidemiology of *Campylobacter coli* isolates from human gastroenteritis, food, and animal sources by amplified fragment length polymorphism analysis and penner serotyping. *Applied and Environmental Microbiology* 71(4), 1953–1958. DOI: 10.1128/AEM.71.4.1953-1958.2005.

Singh, D.V., Matte, M.H., Matte, G.R., Jiang, S., Sabeena, F. *et al.* (2001) Molecular analysis of *Vibrio cholerae* O1, O139, non-O1, and non-O139 strains: Clonal relationships between clinical and environmental isolates. *Applied and Environmental Microbiology* 67(2), 910–921. DOI: 10.1128/AEM.67.2.910-921.2001.

Song, Q., Zheng, Y.J., Xue, Y., Sheng, W.G. and Zhao, M.R. (2017) An evolutionary deep neural network for predicting morbidity of gastrointestinal infections by food contamination. *Neurocomputing* 226, 16–22. DOI: 10.1016/j.neucom.2016.11.018.

Stoeckel, D.M. and Harwood, V.J. (2007) Performance, design, and analysis in microbial source tracking studies. *Applied and Environmental Microbiology* 73(8), 2405–2415. DOI: 10.1128/AEM.02473-06.

Thompson, L.A. and Darwish, W.S. (2019) Environmental chemical contaminants in food: Review of a global problem. *Journal of Toxicology* 2019, 2345283. DOI: 10.1155/2019/2345283.

Tirima, S., Bartrem, C., von Lindern, I., von Braun, M., Lind, D. *et al.* (2018) Food contamination as a pathway for lead exposure in children during the 2010–2013 lead poisoning epidemic in Zamfara, Nigeria. *Journal of Environmental Sciences* 67, 260–272. DOI: 10.1016/j.jes.2017.09.007.

Tokunaga, A., Kawano, M., Okura, M., Iyoda, S., Watanabe, H. *et al.* (2007) Identification of enterohemorrhagic *Escherichia coli* O157-specific DNA sequence obtained from amplified fragment length polymorphism analysis. *Microbiology and Immunology* 51(9), 883–888. DOI: 10.1111/j.1348-0421.2007.tb03970.x.

Xu, G., Ye, X., Qin, L., Xu, Y., Li, Y. *et al.* (2005) Cell-based biosensors based on light-addressable potentiometric sensors for single cell monitoring. *Biosensors & Bioelectronics* 20(9), 1757–1763. DOI: 10.1016/j.bios.2004.06.037.

Zhai, L., Liu, H., Chen, Q., Lu, Z., Zhang, C. *et al.* (2019) Development of a real-time nucleic acid sequence-based amplification assay for the rapid detection of *Salmonella* spp. from food. *Brazilian Journal of Microbiology* 50(1), 255–261. DOI: 10.1007/s42770-018-0002-9.

Zhang, G. (2013) Foodborne pathogenic bacteria detection: An evaluation of current and developing methods. *The Meducator* 1(24). DOI: 10.15173/m.v1i24.835.

Zhao, X., Lin, C.W., Wang, J. and Oh, D.H. (2014) Advances in rapid detection methods for foodborne pathogens. *Journal of Microbiology and Biotechnology* 24(3), 297–312. DOI: 10.4014/jmb.1310.10013.

11 Policies and Social Issues of Food Safety

Sudip Kumar Das[1]*, Olipriya Biswas[2] and Bikash Kanti Biswas[3]
[1]W.B. University of Animal & Fishery Sciences, Kolkata, India; [2]Department of Fishery Engineering, W.B.U.A.F.S., Kolkata, India; [3]DREF, W.B. University of Animal & Fishery Sciences, Kolkata, India

Abstract
Food safety policies are the sets of guidelines crucial in safeguarding public health and regulating the global food supply chain. These policies are set by different international and national organizations to address different food safety issues like foodborne illnesses, food fraud, food hygiene, labeling, and so on. Food safety policies also ensure adherence to safe practices, educate consumers, promote transparency, bring socioeconomic stability, and contribute to international trade and sustainability. The major social issues surrounding food safety are nutritional insecurity due to poverty and access disparities, socioeconomic losses, lack of consumer awareness, cultural and religious beliefs, consumer behaviors, variations in local food standards, public health concerns, and hampering sustainability. Resolving social issues of food safety requires a multifaceted approach, which in turn will result in providing safe and nutritious food to all.

Keywords: Policies, social issues, food safety, public health, foodborne illnesses, consumer

11.1 Introduction

Safe food has become a fundamental necessity of the present-day active and health-conscious "food citizens." From a social perspective, safe food supports international and national economy, promotes tourism and trade practices, aids in sustainable development, and contributes to food security. There has been huge progress in bringing more nuance to consumer understanding of food safety and food security in the recent past. Food safety is described as the concept that food will not cause harm to the consumer when it is prepared and/or eaten according to its intended use (Manning and Soon, 2016). Food safety has also been an enduring concern from ancient civilizations to the modern era, evolving in nature and complexity over the years.

The contemporary landscape of food safety issues presents challenges on an unprecedented scale. The notorious episodes of outbreaks of foodborne illnesses are a grim reminder of the importance of food safety (Das *et al.*, 2024). As per World Health Organization (WHO), about 600 million (1 in every 10 persons) cases of foodborne illness and 420,000 death cases are reported each year due to the consumption of unsafe food (WHO, 2021a, WHO, 2021b, 2022a, WHO, 2022b), leading to economic loss amounting to US$ 110 billion in low- and middle-income countries (World Bank, 2018; WHO, 2022a, WHO, 2022b). Among many, *E.*

*Corresponding author: sudipvet@gmail.com

DOI: 10.1079/9781800625037.0011

coli O157:H7 outbreak in the USA in the fall associated with the consumption of mesclun lettuce contaminated with cattle feces (Waltenburg *et al.*, 2022), BSE in British beef in the 1980s–1990s (Adinolfi *et al.*, 2016) and melamine in dairy products in China (Yamaguchi and Chiang, 2019) are some of the classical examples of public health concerns of food safety. In addition to foodborne illnesses, diseases like Tuberculosis (TB), Anthrax, Brucellosis, Listeriosis, Salmonellosis, Campylobacteriosis, Bovine Spongiform Encephalopathy (BSE) and other emerging and re-emerging zoonotic diseases, chemical contaminants and residues, environmental pollutants in food (e.g. microplastics, lead, mercury, and dioxins), antimicrobial resistance, and introduction of novel foods like genetically modified (GM) foods are evolving as the principal threats to food safety (Maldonado-Simán *et al.*, 2018; Miley, 2020; Patil *et al.*, 2023). Food fraud like the British horse meat scandal in 2013 (Smith and McElwee, 2021) or food adulteration or tampering (Li *et al.*, 2020; Montgomery *et al.*, 2020) and bioterrorism events (Gbashi *et al.*, 2021; Hu *et al.*, 2022) such as the 1984 Rajneeshee bioterror attack in the USA infecting 751 people in ten restaurants through salad bars contaminated with *Salmonella* Typhimurium (Katona and Carus, 2010) also have resulted in massive food safety issues. Globalization and easier travel access have also resulted in the spread of transboundary diseases, threatening food safety and public health (Abunna *et al.*, 2022; Mukherjee *et al.*, 2023). Additionally, climate change and disasters pose enormous threats to food safety (Duchenne-Moutien and Neetoo, 2021).

Furthermore, there has been a drastic rise in global consumption of perishable foods of animal origin like milk, meat, and eggs in the past decades (FAO, 2017). Food safety concerns associated with animal-origin foods are gaining prominence, primarily due to shifts in product processing methods, with many consumers showing a preference for minimally processed and ready-to-eat options. Heightened consumer awareness and increased scrutiny by news outlets, media platforms, and activists place significant responsibility on policymakers and food standard agencies to enact robust policy measures aimed at safeguarding public health.

Policies related to food safety constitute the foundation of public health protection (Garcia *et al.*, 2020). Early societies developed rudimentary food laws, standards, and policies for ensuring the quality and safety of animal-origin foods. Over the years, these food safety-related policies were refined in response to changing societal needs, industrialization, and trade liberalization (in the year 1991) of food products. The current domain of food safety policies is a dynamic arena where governments, international organizations like the World Health Organization (WHO), Food and Agriculture Organization (FAO), and local entities converge to establish standards, regulations, and monitoring mechanisms to safeguard the quality of food we eat. These policies aid in mitigating the risks associated with contamination, adulteration, improper handling, and other threats to food safety, thus ensuring that consumers can trust the food they consume.

The threats like social issues surrounding food safety mostly encompass a broad spectrum of challenges linked with poverty and food security in low- and middle-income countries (Carvalho, 2006; Tacoli *et al.*, 2013; Grace, 2015). One of the prominent issues revolves around unequal access to safe and nutritious food (SNF), creating a disparity that affects socioeconomically backward communities (Hallum *et al.*, 2020; Arjvand *et al.*, 2023). Individuals from the weaker sections of society with limited resources often grapple with compromised food safety due to inadequate storage facilities, insufficient knowledge about proper handling, and limited access to quality food products. Likewise, the challenges are compounded by stark disparities between developed and developing nations, both in terms of regulatory frameworks and prevalent food safety concerns (Faour-Klingbeil and Todd, 2019). The borderless nature of sociotechnical food systems exacerbates the difficulty for individual countries to effectively manage food safety issues, emphasizing the need for robust and uniform policies at the global scale (Henson and Caswell, 1999; Yamaguchi and Chiang, 2019).

Cultural practices and beliefs also influence the social dimensions of food safety (Lenart *et al.*, 2022). Traditional methods of food processing and preservation often clash with contemporary food safety standards, requiring a delicate balance between preserving cultural heritage and meeting modern food safety requirements

(Liu and Kwon, 2013; Sharman *et al.*, 2020). Moreover, a lack of awareness, information, and education related to food safety and misconceptions can lead to unsafe practices, putting consumers at risk (Bolek, 2020). This book chapter aims to discuss the various policy measures related to food safety and the growing concern over social issues of food safety.

11.2 Food Policies

A policy is a written set of guidelines, action plans, laws, or regulations and is implemented by an organization or government body as procedures, protocols, or standards. These policies aid in subjective and objective decision making (WHO, 2021a, WHO, 2021b). Basically, governments develop policies or regulations to implement laws or rules approved by the legislative bodies, development agencies or statutory bodies create and approve accredited standards. The food policy is the part of public policy that deals with the production, processing, distribution, purchasing, and consumption of food from farm to fork (Gauci, 2019). Food policies are based on many integrated principles. Risk analysis and risk assessment (Das *et al.*, 2019), monitoring through food safety management systems (Nguyen and Li, 2022), and implementation of standards through the statutory bodies are integral components of food safety policies. Regular inspections, testing procedures, and stringent penalties for non-compliance are required for implementing the food safety policies. The evolving nature of the food industry and emerging issues related to food safety necessitate continuous updates and revisions to these policies.

11.3 Action Framework of a Food Safety Policy

The action framework aids in designing a policy by proposing the key policy steps or sections. As per WHO, 2021aWHO, 2021b, the key food policy steps are divided into four sections of a "policy cycle" as given below:

1. Section 1: ***Policy Preparation***: This section outlines the key steps to be taken while preparing to develop a new policy or revise and improve an existing food policy.
2. Section 2: ***Policy Development***: This section describes the key steps in the policy development process. It also defines the purpose, scope, and other criteria for development of the policy.
3. Section 3: ***Policy Implementation***: This section reviews key steps to support and ensure policy enactment or implementation, highlighting that the implementation strategies are critical to the success of the policy.
4. Section 4: ***Policy Monitoring and Evaluation***: This section guides the process of policy monitoring and evaluation to determine the effective implementation of the policy.

11.4 Considerations for Framing or Revising a Food Safety Policy

As per (WHO, 2021a, WHO, 2021b), the basic considerations for preparing to frame or revise a food safety policy are:

1. Identification of public health priorities
2. Determination of the lead government authority and establishment of a working group of policymakers
3. Establishment of clear rules of engagement of stakeholders
4. Assessment of public food procurement, consumption, prevalent issues, and consumer behavior
5. Identification of opportunities and policy entry points
6. Advocating political support

11.5 Critical Need for Effective Food Safety Policies

Effective food safety policies are imperative for public health and well-being. Comprehensive food safety regulations are essential to establish and maintain high standards of hygiene, sanitation, and quality throughout the entire food production and distribution process. These policies aid in preventing foodborne illnesses and build consumer trust and confidence by addressing food fraud issues. Additionally, food safety policies contribute to socioeconomic stability by reducing economic losses and access disparities. Furthermore, these policies contribute to international trade and sustainability. Overall, food safety policies play a crucial role in addressing all food safety issues and fostering secure and safe food, which is fundamental to the well-being of consumers worldwide.

11.6 Statutory Bodies for Forming and Implementing Food Safety Policies

Different nations have instituted diverse food safety policies tailored to their livestock production practices, cultures, consumer needs, and public health considerations. The World Health Organization (WHO), Food and Agriculture Organization (FAO), World Organization for Animal Health (WOAH, formerly OIE), World Trade Organization (WTO) and other analogous bodies play pivotal roles in crafting and implementing policies, setting standards, framing regulations and laws, promoting best practices and ensuring harmonization at the global level (Qiu *et al.*, 2023; Zinsstag *et al.*, 2023). Sanitary and Phytosanitary (SPS) measures and Technical Barriers to Trade (TBT) are the important measures or agreements of WTO member countries that address food safety standards and regulations in the food trade (Wu *et al.*, 2023). Moreover, the Codex Alimentarius Commission (CAC), a joint WHO–FAO body plays the principal role in formulating food safety policies and setting food standards, regulations, and laws at the international level (Lee *et al.*, 2021; Nagel-Alne *et al.*, 2022). The International Organization for Standardization (ISO) also sets standards like quality management (ISO 9000) standard (Rogala and Wawak, 2021) and food safety management (ISO 22000) standard (Chen *et al.*, 2020, 2021), which are designed to control the quality and safety of foods in a food chain.

In developed countries like the United States of America, the United States Department of Agriculture (USDA) sets policies leading to standards and regulations related to food products. Likewise, the European Food Safety Authority (EFSA) frames policies, rules, regulations, and standards related to food products, including animal-origin foods in the European Union (EU) (Smith DeWaal *et al.*, 2021; Lee *et al.*, 2023). Furthermore, the Food Safety and Standards Authority of India (FSSAI), under the Ministry of Health and Family Welfare, Government of India, was established in 2011 as per the Food Safety and Standards Act (FSSA), 2006. The FSSAI is the statutory body in India that replaced the eight old existing laws related to food safety and now regulates the manufacture, storage, distribution, sale, and import of food products and establishes policies and mandatory standards to ensure food safety (Shukla *et al.*, 2014; Jena, 2018; Dhara *et al.*, 2021).

11.7 Tools for Addressing Food Safety Hazards

The physical, chemical, and biological hazards affecting food safety in a food supply chain are identified, minimized, controlled, and eliminated through the application of process control tools like risk analysis and food safety management systems. The important tools for ensuring safe practices in the food value chain are given below:

11.7.1 Risk analysis

A process with three components, risk assessment, risk management, and risk communication, is called risk analysis (Henson and Caswell, 1999). The three aspects of the risk analysis process are given below:

1. ***Risk assessment:*** A scientifically based process consisting of the steps—hazard identification, hazard characterization, exposure assessment, and risk characterization—is called risk assessment.
2. ***Risk management:*** Risk management is the process of weighing policy alternatives in consultation with all interested parties,

considering risk assessment and other factors relevant to the health protection of consumers and for the promotion of fair-trade practices and, if needed, selecting appropriate prevention and control options.

3. ***Risk communication:*** The interactive exchange of information and opinions throughout the risk analysis process is called risk communication.

Risk analysis for food safety is carried out by national, regional, and international food safety authorities. The three Joint FAO/WHO Expert Bodies—the Joint Expert Committee on Food Additives (JECFA), the Joint Meeting on Pesticide Residues (JMPR), and the Joint Expert Meeting on Microbiological Risk Assessment (JEMRA)—provide risk assessment based on the Codex standards. National food safety authorities are responsible for carrying out risk analysis at the country level (Das *et al.*, 2019).

11.7.2 Food safety management systems

The food safety management systems (FSMS) encompass organized approaches to ensure the safety and quality of food throughout the supply chain (Nguyen and Li, 2022). The FSMS are not only the legal requirements but also the tools for ensuring food safety. These systems involve comprehensive planning, monitoring, and documentation of the entire process of food production or processing (Das *et al.*, 2019). The policies on food safety are implemented through various FSMS *viz.*, Hazard Analysis Critical Control Point (HACCP), Hazard Analysis and Risk-based Preventive Control (HARPC), Good Manufacturing Practices (GMPs), Good Hygienic Practices (GHPs), Sanitation Standard Operating Procedure (SSOP), Longitudinally Integrated Safety Assessment (LISA), and so on (Kumar *et al.*, 2022; Pandita *et al.*, 2023).

11.8 Hazard Analysis Critical Control Point (HACCP)

The HACCP is a systemic approach that leads to the development of safe food supply for consumers through analysis of hazards that may come from the use of inappropriate raw materials, adoption of inappropriate processing steps and even that may occur from consumer abuse (Awuchi, 2023; Owusu-Apenten and Vieira, 2023). The HACCP is a factory and product-specific alarm system. The primary purposes of HACCP are:

1. To identify and control potential hazards in food.
2. To ensure that the finished product will be safe for consumers.

The concept was first developed in the 1960s by NASA (led by Dr Paul Lachance) in collaboration with Pillsbury Company of Minneapolis (led by Dr H. E. Bauman) and the U.S. Army Natick Research, Development and Engineering Center to ensure "zero defect" in the quality of food taken on space missions (Patel *et al.*, 2015).

11.8.1 Procedure for application of HACCP

There are 12 steps, including 5 prerequisites and 7 principles (Overbosch and Blanchard, 2023) of the HACCP program, which are given below (Fig. 11.1).

11.8.2 Seven principles of HACCP

The seven principles of HACCP (Awuchi, 2023; Motarjemi and Warren, 2023; Owusu-Apenten and Vieira, 2023) are described as follows:

11.8.2.1 Hazard identification

This involves determining the likely hazards to consumers that could be presented by a meat product. Hazard refers to unacceptable contamination, survival and growth of harmful microorganisms, undesirable chemicals, and physical agents in meat. Types of hazards can be physical, chemical, and biological. Hazards can also be classified as "material-related" or "process-related" hazards.

Constitute a HACCP team
↓
Describe the product
↓
Identify intended use
↓
Construct flow diagram
↓
On-site verification of flow diagram
↓
Identification of hazards at all steps of product processing
↓
Determine the critical control points (CCPs) for each identified hazard
↓
Establish the critical limit for each identified CCP
↓
Establish the monitoring procedure for each identified CCP
↓
Establish the corrective actions in case of deviations while monitoring the CCPs
↓
Establish verification procedure
↓
Effective record keeping and documentation

Fig. 11.1. Flow diagram showing procedure for application of HACCP.

11.8.2.2 Determine CCPs (CCPs)

The CCP means any point or procedure in a food system where control can be exercised, and a hazard can be minimized or prevented. There are two types of critical control points: CCP1 and CCP2. CCP1 is designated as effective CCP because if proper measures are taken, the hazard can be corrected. CCP2 is not absolute CCP because, at this point, hazards identified cannot be corrected, only minimized. Failure to control the real CCPs within the production process will lead to the presence of a hazard in the product (Chen *et al.*, 2019).

11.8.2.3 Establish critical limits for CCP

Critical limits are one or more prescribed tolerances which should be met to ensure that CCP effectively controls a health hazard. Critical limits, such as temperature, hygiene level, cooking time, or residue concentration, draw the line between acceptable and unacceptable operation of a meat plant. An example is cooling a meal to below 5°C, ensuring the absence of *Salmonella* in five samples of 25g per batch.

11.8.2.4 Monitoring

Monitoring refers to checking the effectiveness of the control at each CCP. It involves systemic observation, measurement, recording, and evaluation. For example, if the chilling step as indicated above is used as a CCP, then there is a need to monitor and record temp every half hour.

11.8.2.5 Corrective action

Corrective actions are taken when there is a deviation during the monitoring of the CCPs. Corrective actions ensure that the product is not injurious to health.

11.8.2.6 Verification

Review procedures performed to check the HACCP system's overall effectiveness in the plant is called verification.

11.8.2.7 Record keeping

It is the documentation procedure, which ensures that information resulting from the HACCP plan is available for validation, review, auditing, or other purposes.

11.9 Existing Food Safety Policies

There are many policies for addressing different issues of food safety. Continuous efforts are being made by the government and regulatory bodies to update and improve the existing policies related to food safety. Some of the existing policies addressing the issues of food safety from Indian and global perspectives are discussed below:

11.9.1 Indian perspective

The Food Safety and Standards Authority of India (FSSAI) provides standards and regulations to ensure that meat and meat products in India meet specific safety, hygiene, and quality requirements. The FSSAI specifications thus safeguard consumer health and promote transparency and accountability throughout the meat production and distribution chain in India. The FSSAI was established on August 5, 2011, under the Ministry of Health and Family Welfare, Government of India, as per the Food Safety and Standards Act (FSSA), 2006. The headquarters of FSSAI is situated in New Delhi, and it has six regional offices located in Delhi, Guwahati, Kolkata, Lucknow, Chandigarh, and Chennai. The FSSAI is the only statutory body in India that has overridden the eight old laws related to food safety and now regulates the manufacture, storage, distribution, sale and import of food products and establishes standards to ensure food safety (Jena, 2018; Dhara *et al.*, 2021). The food safety policies implemented by FSSAI in India collectively address different food safety issues such as suiting the diverse Indian population.

Food Safety and Standards Authority of India (FSSAI) was set up and empowered to regulate and ensure uniform food safety standards across the country. It covers all aspects of food safety from setting up of food business, licensing, consumer training, standard enforcement, and quality control (Jena, 2018; Dhara *et al.*, 2021). SNF at school was launched in September 2016, with the objective to create awareness among school children and through them community at large about "eating healthy," "eating safe," and "eating sustainably" (Bharti *et al.*, 2021). Other policies by FSSAI related to food safety addressing social issues are the Food Safety and Standards (Packaging and Labeling) regulations, Eat Right Initiative, Food Safety Training and Certification (FoSTaC), hygiene rating scheme for food retailers, National Food Security Act (NFSA) addressing access disparities by providing subsidized food grains to the eligible population through Public Distribution System (PDS), appointment of Food Safety Mitras (FSM), Food Safety on Wheels program, Detect Adulteration with Rapid Test (DART) booklet, and so on (FSSAI, 2016, 2023).

11.9.2 Global perspective

The Codex Alimentarius Commission (CAC) and the International Organization for Standardization (ISO) frame and enforce the policies and design standards on food safety and quality assurance. These international standards frequently advocate for food safety and quality management systems like HACCP, GMP, SSOP, GHP, and so on.

The ISO frames policies and sets standards like quality management (ISO 9000) standard (Rogala and Wawak, 2021) and food safety management (ISO 22000) standard (Chen *et al.*, 2020, 2021).

The Codex standards are given by the Codex Alimentarius Commission (CAC), established by a joint collaboration between the Food and Agriculture Organization (FAO) and the World Health Organization (WHO) (Wieck and Grant, 2021). The CAC was formed in the year 1963, and its headquarters are situated in Rome, Italy (LeJeune *et al.*, 2021). The CAC and its standards serve as globally recognized benchmarks ensuring safety, quality, and fair-trade practices. These standards prioritize consumer protection by setting stringent criteria for microbial contamination control, targeting harmful pathogens like *Salmonella spp.*, *E. coli*, and so on (Mahajan *et al.*, 2014). Additionally, they mandate strict guidelines for residues of veterinary drugs, pesticides, and other potential contaminants, ensuring that the meat and meat products meet health and safety parameters. Emphasizing quality assurance, the Codex standards define specific criteria related to freshness, appearance, texture, and flavor, maintaining the nutritional value and sensory attributes of meat products. Furthermore, the Codex standards enforce transparent labeling and packaging practices, requiring accurate information on product identity, ingredients, net quantity, and storage instructions. Addressing production practices, the Codex standards promote adherence to GMP and stringent hygiene controls throughout the slaughter, processing, storage, and distribution phases (Konings *et al.*, 2021). Overall, the Codex standards address food safety issues by facilitating international trade, increasing consumer confidence, safeguarding public health, harmonizing regulations across global markets, and ensuring consistent food quality and food security.

As per WHO, 2021aWHO, 2021b, the core messages of the "five keys to safer food" are (i) keep clean, (ii) separate raw and cooked, (iii) cook thoroughly, (iv) keep food at safe temperatures, and (v) use safe water and raw materials.

The objective of the "WHO Global Strategy for Food Safety 2022–2030" is for everyone everywhere to eat secure, wholesome food, thereby easing the burden of foodborne diseases. This strategy aims toward stronger food safety systems and global cooperation (WHO, 2022a, WHO, 2022b).

The United Nation's (UN) Sustainable Development Goals (SDGs) viz., SDG 2 targeting "zero hunger" aim at the provision of SNF, ensuring food security and ending hunger (Bizikova *et al.*, 2020). The World Food Programs (WFP) by the UN is the food assistance program operating in vulnerable regions, focusing on ensuring the provision of SNF to the needy and ensuring zero hunger (Patil *et al.*, 2023).

The Global Food Safety Initiative (GFSI) is a private-sector-led initiative that benchmarks and harmonizes food safety standards across the globe. The GFSI certification ensures that the food safety management system meets internationally recognized criteria (Gerardi, 2022).

Different countries have different policies related to food safety of meat and animal-origin foods, implemented by their statutory bodies. Some of the country-specific food safety standards related to animal-origin foods, especially meat and meat products, are listed in Table 11.1.

11.10 Social Issues of Food Safety

As per Eagle *et al.* (2022), a social issue refers to "a condition that harms one or more social entities or individuals, has at least one social cause, social effect, and consequently social remedies." Poverty and access disparities lead to nutritional insecurity (Tacoli *et al.*, 2013), huge socioeconomic losses due to food spoilage, wastage, and foodborne illnesses (Areche *et al.*, 2022; Misiou and Koutsoumanis, 2022; Karanth *et al.*, 2023), lack of consumer awareness and education (Redmond and Griffith, 2005; Jevšnik *et al.*, 2008; Bolek, 2020), cultural considerations (Liu and Kwon, 2013; Lenart *et al.*, 2022), consumer behaviors (Losasso *et al.*, 2012), variations in local food systems (Béné, 2020; Enthoven and Van den Broeck, 2021), public health concerns (Rudra *et al.*, 2022), hampering sustainability, and so on, are the common social issues associated with food safety and security. The different social issues related to food safety are discussed as follows:

11.10.1 Poverty and access disparities leading to nutritional insecurity to safe food

The widening inequalities and persistent poverty and access disparities force the weaker section of the community to consume inferior and unsafe food to satisfy their

Table 11.1. Food safety standards of different countries.

Sl. No.	Statutory body	Name of the policy regulation related to food safety	Role	References
1.	Food Safety and Standard Authority of India (FSSAI)	Food Safety and Standards (Food Products Standards and Food Additives) Regulations	Sets specific standards for meat products in India regarding quality, safety, and labeling	(Dhara *et al.*, 2021; Gadekar *et al.*, 2021)
2.	United States Department of Agriculture (USDA)	Food Safety and Inspection Service (FSIS) Regulations	Mandates sanitary production practices, inspection protocols, labeling requirements. And recalls for meat products in the USA	(Keener, 2019; Charlebois *et al.*, 2021)
3.	European Union (EU)	Regulation (EC) No. 853/2004	Establishes hygienic rules for the production and placing on the market of products of animal origin, including meat products, within the EU	(Micheli *et al.*, 2019; Cenci-Goga *et al.*, 2020)
4.	Australian Standard	Australia New Zealand Food Standards Code	Provides guidelines and standards for meat products in Australia, covering aspects like labeling, composition and safety	(Johnson and Parker, 2022; Nagel-Alne *et al.*, 2022)
5.	British Standards Institution (BSI)	British Standards for Meat Products	Offers guidelines and standards concerning the production, processing, and packaging of meat products in the UK	(Gerges *et al.*, 2019; Sgarro *et al.*, 2024)
6.	Canadian Food Inspection Agency (CFIA)	Safe Food for Canadians (SFC) Regulations	Ensures that meat and meat products produced or imported into Canada meet specific safety, quality, and labeling requirements	(Cadieux *et al.*, 2019; Dyer *et al.*, 2020)

hunger. The indigenous people and the low-income communities like tribal communities are the most vulnerable populations, which face challenges to access food that meets safety standards. Nutritional insecurity poses a major social issue, which in turn gives rise to food safety concerns (Tacoli *et al.*, 2013). Nobel laureate Amartya Sen has aptly mentioned in his pathbreaking work *Poverty and Famines* that even though food was widely available in the market, famines arose as a result of inaccessibility to food by a larger section of the society because of poverty. He highlighted that a person's entitlement to safe food was influenced by their socioeconomic status, endowments of land, employment status, trade assets, and rights within the society (Sen, 1982). In this context, Sen's "Capability approach" is one of the primary substitutes for the conventional monetary system. Items are evaluated depending on their contribution to an individual's capacities. Nourishment supplies essential capability, leading to "being adequately nourished," contingent on factors such as metabolic rate, existence of parasites, and similar conditions. Similarly, as per (FAO, 2023), during the COVID-19 pandemic in the year 2020, around 2.37 billion people—nearly one in three people—faced food security problems and had difficulty in accessing SNF.

11.10.2 Socioeconomic losses linked to food safety

Food safety issues, particularly foodborne illnesses, cause an enormous socioeconomic burden on consumers (Bolek, 2020). The economic loss due to food safety concerns includes medical expenses, productivity losses due to missed man-hours, and value of premature deaths but exclude travel costs in obtaining medical care and time lost in caring the sick persons (Vogt, 2005). Economic losses linked to food safety mostly occur due to food spoilage, wastage, and foodborne illnesses (Areche *et al.*, 2022; Misiou and Koutsoumanis, 2022; Karanth *et al.*, 2023).

A huge economic loss amounting to US$ 110 billion occurs toward medical expenses related to foodborne illness in low- and middle-income countries (World Bank, 2018; WHO, 2022a, WHO, 2022b). According to the estimates of the World Health Organization (WHO), around 33 million disability-adjusted life years (DALY) are lost each year due to the consumption of contaminated, spoiled, and unsafe food (WHO, 2021a, WHO, 2021b, 2022a, WHO, 2022b). Similarly, the BSE outbreak in the UK led to the "beef war." Banning the worldwide export of beef from the UK by the EU for a period of 10 years (1996–2006) incurred a heavy economic loss to the beef industry in the UK and affected the livelihoods of people engaged in the beef sector (Alarcon *et al.*, 2023). The above-mentioned information is indicative of the fact that the socioeconomic losses from foodborne illnesses can no longer be ignored or neglected.

11.10.3 Climate changes affecting the food system

Changes in climate could impact the food supply by altering temperatures, increased concentrations of greenhouse gases in the atmosphere, shifts in precipitation patterns, and heightened instability in weather conditions. Temperature fluctuations influence the amount and distribution of rainfall, evaporation rates, and temperature ranges. Precipitation levels and evaporation rates, in turn, influence soil moisture, water resource management, as well as the health of plants and animals, and susceptibility to pest infestations. These factors, combined with changes in maximum temperature thresholds, might exacerbate or mitigate existing environmental pressures and introduce novel challenges. Modifications in such environmental pressures have implications for crop yields, the status of natural resources, and the overall sustainability of the food system.

Due to the intricate interplay between food systems and ecosystems, the anticipated impact of climate change on the future food system is highly challenging to gauge and current estimates remain speculative. Nonetheless, there is a broad consensus suggesting that crop yields will decline in tropical regions while increasing by a similar margin in temperate zones. If these projections materialize, climate variations could exacerbate the existing disparity in food availability between affluent and impoverished nations. Historical evidence suggests that large-scale redistribution of food from surplus to deficient regions is improbable. A more plausible scenario entails a heightened prevalence of hunger in regions with food deficits and increased rates of obesity in surplus areas, especially if governments implement price controls or adopt anti-trade policies, whether on a domestic or international scale.

11.10.4 Lack of consumer awareness and education leading to unsafe food practices

Without adequate knowledge, individuals may unknowingly engage in unsafe food practices, leading to an increased risk of foodborne illnesses (Redmond and Griffith, 2005; Jevšnik *et al.*, 2008; Bolek, 2020). Lack of consumer awareness and poor educational backgrounds of the consumers may result in misinterpretation of labels, improper storage, and insufficient food hygiene, thereby compromising food safety. The common food handling mistakes by unaware consumers are serving contaminated foods, improper or inadequate cooking, overheating and reheating of food, food procurement from unsafe sources, inappropriate food preservation methods, and allowing too much time-lapse between production, processing, and consumption (Ergönül, 2013).

11.10.5 Cultural practices and beliefs influencing social dimensions of food safety

Cultural practices and beliefs significantly shape the social dimensions of food safety, influencing the consumption patterns, food handling, and processing and preservation practices within the communities (Zhou *et al.*, 2020). While cultural diversity enriches global food products, it can also contribute to food safety issues when traditional practices fail to meet modern food safety standards. The Halal, Jewish, and Jhatka slaughter practices are governed by the Muslim, Jewish, and Sikh religious norms, respectively (Aghwan and Regenstein, 2019), and may lead to food safety concerns if modern food quality and safety standards are not followed. Similarly, the Freibank System in Germany entailed the sale of <5% of total meat as "cheap or inferior meat," requiring treatment before sale, classed as "conditionally fit" for consumption ("*bedingt tauglich*") to the consumers from the poorest class (Anon, 1910; Peck, 1914). From 2001 to 2021, a total of 1604 cases of Trichinellosis were reported in Southeast Asia, mostly from Vietnam, Cambodia, Thailand, Laos, and Malaysia, due to consumption of raw and undercooked pork. Most outbreaks occurred in rural areas during celebrations or festivals (Yera *et al.*, 2022). Following certain orthodox food taboos, some religions declare certain foods as fit, while others find them unfit for human consumption.

In some cultures, long-stay culinary traditions involve practices like raw or undercooked meat consumption, consumption of game meat, eating fermented foods, and the intake of unpasteurized food products that compromise food safety. Moreover, community dining practices, sharing utensils, and contamination with bacteria and/or toxins during cultural and religious events can contribute to the spread of foodborne diseases due to non-adherence to food safety standards. Cultural reluctance to adopt hygienic measures or skepticism toward food safety regulations may further exacerbate food safety issues.

11.10.6 Consumer behavior affecting food safety

Consumer behavior and attitude play vital roles in combating food safety issues. Unhealthy eating behavior like eating raw or undercooked food, consumption of unconventional or poisonous game meat and insects, and cheaper and spoiled food result in adverse health effects (Losasso *et al.*, 2012). Five of the seven public health emergencies of international concern (PHEIC) as described by WHO are associated with eating factors. Hence, it is clear that eating behavior plays a substantial role in the outbreak of foodborne diseases (Zhou *et al.*, 2020). The locals of Yunnan province in China and Cape Province of South Africa have the habit of eating raw or undercooked pork liver. The pork meat and livers are often contaminated with the mature cyst containing larvae of the tapeworm *Taenia solium*. The intake of raw or undercooked pork results in cysticercosis in humans (Sithole *et al.*, 2020).

11.10.7 Variations in local food systems compromising food safety

The variations in local, regional, and national food safety regulations in different parts of the world give rise to some food safety concerns. Several local and national governments often make policies or laws in favor of the vast majority of the population that elected them and the private companies that invest in their parties (Nayak and Waterson, 2019; Béné, 2020). In India, the laws governing beef ban primarily stem from state government legislation and religious considerations. Several states including Maharashtra, Gujarat, Karnataka, Uttar Pradesh, and Odisha have implemented stringent regulations prohibiting the slaughter, sale, possession, and consumption of beef due to religious beliefs, particularly among the Hindu community. However, in states like West Bengal, Kerala, and some of the Northeastern states of

India, beef is legally not banned (Sunder, 2019; Dasgupta *et al.*, 2023).

Sometimes, the food safety regulations are modified to suit the requirements of multinational companies, without thinking about food safety concerns. The differences in farming methods, processing techniques, and storage facilities sometimes require modifications to the existing food safety policies for the locally processed foods. Limited access to modern technologies and lower purchasing power of the people involved in the food business may sometimes hinder standardized food safety measures (Nayak and Waterson, 2019).

11.10.8 Public health concerns of food safety

As per the World Health Organization (WHO), an estimated 7.69% (600 million; almost 1 in every 10 people) cases of foodborne illness in the world population (7.8 billion) and 7.5% (420,000) of all death cases (56 million) resulting in 33 million healthy life years lost are reported each year due to the consumption of contaminated, spoiled and unsafe food. The death figures due to foodborne diseases are equivalent to 31.1% of annual deaths due to road accidents (1.35 million) across the globe. With 125,000 child deaths every year, children under 5 years of age carry 40% of the foodborne illness burden globally (Lee and Yoon, 2021; WHO, 2021a, WHO, 2021b, 2022a, WHO, 2022b,; Yu *et al.*, 2021). The FDA and CDC also estimated that about 48 million cases of foodborne illness (1 in 6 Americans), 128,000 hospitalizations, and 3000 deaths occur in the USA due to unsafe food consumption (CDC, 2022; FDA, 2022). In Europe, 23 million people are affected, and 5000 deaths occur due to foodborne illnesses every year (Lee and Yoon, 2021).

Some latest instances of foodborne diseases affecting public health include 70 cases of Shigella toxin exposure including the death of a girl in Kerala, India due to infected *shawarma* (a Levantine meat preparation) in May 2022; *Staphylococcus aureus* poisoning among the bus drivers at Coachella Festival (>100 cases) due to contaminated dinner served at the Coachella Festival in California in May 2022 and Norovirus poisoning of oysters in Canada (>100 cases) in April 2022 (Rudra *et al.*, 2022). An investigation by the United States Department of Agriculture-Food Safety Inspection Service (USDA-FSIS) revealed that between 2007 and 2012, 163 outbreaks were identified associated with meat and poultry, which resulted in 4132 illnesses, 772 hospitalizations, and 19 deaths. Of those outbreaks, 150 were due to raw beef products in the market that did not follow the regulations, standards, and laws (Warmate and Onarinde, 2023). Another report stated that about 1060 illnesses and 216 deaths between 2017 and 2018, in the largest reported outbreak of *Listeria monocytogenes* ST6, were caused by processed meat product consumption from Tiger Brands, one of the big players in the food market of South Africa. This was caused by a lack of microbiological standards and a breach in standards related to meat products and resulted in a significant reduction in consumer trust in the brand (Smith *et al.*, 2019).

11.10.9 Food safety issues hampering sustainability

Food safety issues pose a threat to sustainability by contributing to food loss and public health problems, thereby affecting humans, animals, and the environment (Kumar *et al.*, 2020). Food spoilage, resulting in food losses and widespread foodborne illnesses, affects natural food resources and impedes social well-being. Discarding a lot of unsafe food to the environment affects other biotic and abiotic components of the whole ecosystem. The current systems of landfills and incineration to discard unsafe food are alarming threats to food security and environmental sustainability in the long run (Wang *et al.*, 2021).

11.10.10 Corruption or malfeasance

Corruption or malfeasance is the aspect of governance that garners the most scrutiny and is often conflated with governance itself. The exploitation of public authority for personal gain and malfeasance represents a classic instance of governmental breakdown with dire ramifications for the global food network. Northeastern

Argentina has suffered profound devastation due to corrupt governance, leading to thousands of children confronting severe malnourishment and a heightened risk of succumbing to starvation, despite abundant foreign food assistance in a country with a middle-income status. Nations characterized by higher levels of corruption typically exhibit lower per capita GDP, diminished human capital, prolonged bureaucratic processes for business operation authorization, restricted press freedom, and generally, less receptiveness to foreign imports. Corruption tends to be more pervasive in countries with stringent market regulations, particularly when bureaucratic officials wield discretionary power. Mitigating this phenomenon calls for measures such as diminishing discretionary barriers, bolstering the rule of law, and fostering heightened competition among civil servants.

11.11 Approaches for Resolving Social Issues of Food Safety

Resolving social issues related to food safety on a global scale demands a multifaceted approach incorporating policies, education, and international collaboration, which in turn will result in providing SNF to all. Ultimately, a cooperative and inclusive global effort is essential, recognizing that the social issues surrounding food safety are interconnected and require shared responsibility. By promoting transparency, sharing best food safety practices, and fostering a collective commitment to safety, the global agencies can mitigate the social issues related to food safety and build an equitable global food supply chain for all strata of the society. The major approaches for resolving the social issues related to food safety are given below:

11.11.1 Ensuring nutritional security

The world human population is expected to reach 9.7 billion by 2050 (Garcia *et al.*, 2020). With the increasing population, there exist challenges to ensure that all have access to safe, nutritious, and healthy food. Addressing social issues of food safety requires an unwavering commitment to ensuring nutritional security and provision of safe food for all. By implementing inclusive policies and establishing equitable food distribution systems, societies can alleviate disparities and mitigate health risks associated with unsafe food consumption by weaker sections of society. Ensuring nutritional security not only promotes physical well-being but also addresses the broader social issues of food safety, thus creating a foundation for healthier and more sustainable communities.

11.11.2 Uniform food safety policies and standards

Primarily, establishing uniform and universally accepted food safety policies and standards through organizations like Codex Alimentarius Commission can create a common ground, fostering consistency in food safety regulations worldwide. Emphasizing inclusivity and cultural sensitivity within food policies across the globe is crucial. The FSSAI standards are the uniform and mandatory standards in India that replaced the eight old existing laws related to food safety and now regulate the manufacture, storage, distribution, sale, and import of food products and establish standards to ensure food safety all over India (Shukla *et al.*, 2014; Jena, 2018; Dhara *et al.*, 2021). Similarly, in the United States of America, the United States Department of Agriculture (USDA) sets uniform standards and laws related to food products. The European Food Safety Authority (EFSA) frames uniform rules, laws, and policies related to food standards for all countries under the European Union (EU) (Smith DeWaal *et al.*, 2021; Lee *et al.*, 2023).

11.11.3 Education and awareness

Consumer education and awareness play pivotal roles in the resolution of social issues related to food safety. Informed consumers are more likely to adhere to safe practices (Ergönül, 2013). Implementing global consumer awareness campaigns, leveraging technology and local community networks, can bridge information gaps and dispel misconceptions surrounding food safety among consumers (Bolek, 2020). The different aspects of food safety like risks of foodborne illnesses, food preparation practices, and food hygiene measures should be taken into consideration while establishing policies related to educational or awareness programs (Badrie *et al.*, 2006). It is important to educate consumers by understanding consumer behavior and diverse food safety issues relevant to them (Ergönül, 2013).

11.11.4 International collaboration

International collaboration is imperative to address conflicts arising from economic disparities and differing regulatory frameworks. International organizations such as the World Health Organization (WHO) and Food and Agriculture Organization (FAO) and local entities like the Food Safety and Standards Authority of India (FSSAI), United States Department of Agriculture (USDA), and European Food Safety Authority (EFSA) should converge to facilitate knowledge exchange, capacity building programs and collaborative research to strengthen global food safety networks.

11.11.5 Traceability

Traceability acts as a crucial tool in resolving social issues of food safety by promoting accountability, enabling rapid responses to contamination, empowering consumers, reducing fraud, supporting FSMS, and fostering collaborations across the entire food supply chain, that is "from farm to fork" (Girish and Barbuddhe, 2020; Riccaboni *et al.*, 2021).

11.11.6 One Health approach

The "One Health" concept integrates human, animal, and environmental health, offering a holistic solution to resolve social issues of food safety (Dafale *et al.*, 2020). One Health is a transformative approach to increasing sustainable food production practices and improving the overall health and well-being of humans, animals, and the environment (Togami *et al.*, 2018; Garcia *et al.*, 2020). Through its interconnectedness, this approach addresses the shared risks posed by unsafe food. Implementing One Health approach enhances global resilience, mitigating the impact of foodborne illness and fostering safer and more secure food for consumers of all strata of the society.

11.12 Way Forward

The current scenario of occurrence of emerging and re-emerging foodborne illnesses has generated the need for continuous refinement and setting new generations of food safety policy. The future policies should focus on public health, have a preventive approach, and cope with food hazards from farm to fork by strengthening integrated risk management. The agencies within the food policies should promote greater participation in food systems at both individual and community levels. Safer and sustainable food systems are within reach in the near future with the support of government policies and more research. The policymakers should engage in pre-emptive consultation with the population directly affected by different food safety issues and consider their socioeconomic status, feeding habits, and behavior to bring unique solutions to bridge the conundrum of food safety and nutrition. Consumer awareness campaigns, hands-on training, leveraging technologies for early and real-time detection of food safety hazards, and quick and efficient communication of food safety issues between authorities and consumers in the coming days may act as crucial requirements for food safety. Collective efforts and shared responsibilities are essential for recognizing and resolving the social issues surrounding food safety. In order to unravel the societal and policy challenges inherent to food safety, insights from sociology, agriculture, food science, and social studies are imperative. The challenges and social issues must be resolved to ensure nutritional security and achieve real objectives of food safety.

11.13 Conclusions

Consumer priorities are gradually transitioning from food security to food safety, thereby elevating food safety into a multidimensional concept. Past incidents of foodborne illnesses underscore the imperative for robust food safety and public health regulatory frameworks to protect consumer well-being. It is imperative for governments to take decisive action in formulating and implementing food safety policies. These policies play a pivotal role in addressing all facets of food safety concerns and promoting a secure and safe food supply, which is essential for the overall well-being of consumers globally.

Natural resources serve as the cornerstone of food systems. Ensuring the sustainable management of each resource is paramount for the long-term sustainability of the entire system. While past achievements in food production have been partly achieved through the unsustainable exploitation of natural resources, efforts have also focused on mitigating pressures on land unsuitable for agriculture and on preserving biodiversity and wildlife by enhancing agricultural productivity. Achieving sustainability in future food systems necessitates internalizing environmental costs into private expenses. If such measures are implemented, with the resulting revenues directed toward repairing natural resource degradation and developing human-made resources to compensate for losses, significant structural and behavioral transformations within the food system would occur, bringing it closer to sustainability.

Numerous advanced technologies offer farms greater autonomy from local environmental conditions, alleviating key resource constraints and contributing to environmental improvement and risk reduction. Notably, significant opportunities exist for enhancing water-use efficiency. Public interventions aimed at supporting such initiatives should be accompanied by changes in production and market behavior. Enhanced irrigation technologies typically lead to increased yields and decreased water consumption, while also minimizing the proportion of applied water that remains unused by plants. Initiatives such as integrated pest management, soil and water conservation programs, and the use of green manure have been shown to reduce water runoff by up to 45% during periods of heavy rainfall and by 30% during drier periods. Planting nitrogen-fixing trees and crops, particularly during fallow periods, aids in restoring soil fertility and reducing the necessity for chemical fertilizers. Additionally, the adoption of productivity-enhancing technologies, such as improved crop varieties, not only boosts farm incomes but also provides an alternative revenue stream while diminishing reliance on pesticides.

The social issues linked with food safety are access disparities, cultural nuances, economic losses, lack of consumer awareness, variations in policy measures, sustainability issues, and public health concerns. These social issues of food safety necessitate a concentrated effort to bridge the knowledge gaps of consumers, promote inclusive policy approaches, and foster a culture or habit of safe food practices among consumers. Through international collaborations, technology adoption, research and development, and implementation of policies for food safety, societies can address the social issues surrounding food safety and build the foundation for a safer and more equitable food system.

References

Abunna, F., Mamo, G. and Megersa, B. (2022) One health – a holistic solution for sustainable management of globalization-driven public health challenges. *Ethiopian Veterinary Journal* 26(2), 107–131. DOI: 10.4314/evj.v26i2.7.

Adinolfi, F., Di Pasquale, J. and Capitanio, F. (2016) Economic issues on food safety. *Italian Journal of Food Safety* 5(1). DOI: 10.4081/ijfs.2016.5580.

Aghwan, Z.A. and Regenstein, J.M. (2019) Slaughter practices of different faiths in different countries. *Journal of Animal Science and Technology* 61(3), 111–121. DOI: 10.5187/jast.2019.61.3.111.

Alarcon, P., Wall, B., Barnes, K., Arnold, M., Rajanayagam, B. *et al.* (2023) Classical BSE in great Britain: Review of its epidemic, risk factors, policy and impact. *Food Control* 146, 109490. DOI: 10.1016/j.foodcont.2022.109490.

Anon (1910) FREIBANK meat. *The Lancet* 176(4553), 1637. DOI: 10.1016/S0140-6736(01)58051-4.

Areche, F.O., Gondal, A.H., Landeo, O.T., Flores, D.D.C., Rodríguez, A.R. *et al.* (2022) Innovative trends in reducing food waste and ensuring a more sustainable food system and environment. *CABI Reviews: Perspectives in Agriculture, Veterinary Science, Nutrition and Natural Resources*. DOI: 10.1079/cabireviews202217053.

Arjvand, F., Moeeni, M., Najafi, B. and Nosratnejad, S. (2023) Factors affecting inequality in the quality diets: A scoping review. *Value in Health Regional Issues* 37, 105–112. DOI: 10.1016/j.vhri.2023.05.003.

Awuchi, C.G. (2023) HACCP, quality, and food safety management in food and agricultural systems. *Cogent Food & Agriculture* 9(1), 2176280. DOI: 10.1080/23311932.2023.2176280.

Badrie, N., Gobin, A., Dookeran, S. and Duncan, R. (2006) Consumer awareness and perception to food safety hazards in trinidad, West Indies. *Food Control* 17(5), 370–377. DOI: 10.1016/j.foodcont.2005.01.003.

Béné, C. (2020) Resilience of local food systems and links to food security - a review of some important concepts in the context of COVID-19 and other shocks. *Food Security* 12(4), 805–822. DOI: 10.1007/s12571-020-01076-1.

Bharti, R., Marwaha, A., Badshah, T., Sengupta, R., Barmi, B. *et al.* (2021) Effectiveness of a nutritional education intervention focussed on iron among school children in National Capital Region and Mumbai. *Journal of Clinical and Diagnostic Research* 15(4), OC31–OC36. DOI: 10.7860/JCDR/2021/46024.14806.

Bizikova, L., Nkonya, E., Minah, M., Hanisch, M., Turaga, R.M.R. *et al.* (2020) A scoping review of the contributions of farmers' organizations to smallholder agriculture. *Nature Food* 1(10), 620–630. DOI: 10.1038/s43016-020-00164-x.

Bolek, S. (2020) Consumer knowledge, attitudes, and judgments about food safety: A consumer analysis. *Trends in Food Science & Technology* 102, 242–248. DOI: 10.1016/j.tifs.2020.03.009.

Cadieux, B., Goodridge, L.D. and Spink, J. (2019) Gap analysis of the canadian food fraud regulatory oversight and recommendations for improvement. *Food Control* 102, 46–55. DOI: 10.1016/j.foodcont.2019.03.012.

Carvalho, F.P. (2006) Agriculture, pesticides, food security and food safety. *Environmental Science & Policy* 9(7–8), 685–692. DOI: 10.1016/j.envsci.2006.08.002.

CDC (2022) *Food-Borne Germs and Illnesses*. Centers for Disease Control and Prevention, Atlanta, Georgia.

Cenci-Goga, B.T., Karama, M., Hadjichralambous, C., Sechi, P. and Grispoldi, L. (2020) Is EU regulation on the use of antioxidants in meat preparation and in meat products still cutting edge? *European Food Research and Technology* 246(4), 661–668. DOI: 10.1007/s00217-020-03433-y.

Charlebois, S., Juhasz, M., Music, J. and Vézeau, J. (2021) A review of Canadian and international food safety systems: Issues and recommendations for the future. *Comprehensive Reviews in Food Science and Food Safety* 20(5), 5043–5066. DOI: 10.1111/1541-4337.12816.

Chen, H., Chen, Y., Liu, S., Yang, H., Chen, C. *et al.* (2019) Establishment the critical control point methodologies of seven major food processes in the catering industry to meet the core concepts of ISO 22000:2018 based on the taiwanese experience. *Journal of Food Safety* 39(6), e12691. DOI: 10.1111/jfs.12691.

Chen, H., Liu, S., Chen, Y., Chen, C., Yang, H. *et al.* (2020) Food safety management systems based on ISO 22000:2018 methodology of hazard analysis compared to ISO 22000:2005. *Accreditation and Quality Assurance* 25(1), 23–37. DOI: 10.1007/s00769-019-01409-4.

Chen, H., Liou, B.-K., Hsu, K.-C., Chen, C.-S. and Chuang, P.-T. (2021) Implementation of food safety management systems that meets ISO 22000:2018 and HACCP: A case study of capsule biotechnology products of chaga mushroom. *Journal of Food Science* 86(1), 40–54. DOI: 10.1111/1750-3841.15553.

Dafale, N.A., Srivastava, S. and Purohit, H.J. (2020) Zoonosis: An emerging link to antibiotic resistance under "one health approach." *Indian Journal of Microbiology* 60(2), 139–152. DOI: 10.1007/s12088-020-00860-z.

Das, A., Bhattacharya, D., Nanda, P.K., Nath, S. and Das, A.K. (2024) Biopreservation in meat and meat products. In: Garcia-Gutierrez, E., Gomez-Torres, N. and Arbulu, S. (eds) *Novel Approaches in Biopreservation for Food and Clinical Purposes*. CRC Press, London, pp. 66–97. DOI: 10.1201/9781003267997-4.

Das, A.K., Nanda, P.K., Das, A. and Biswas, S. (2019) Hazards and safety issues of meat and meat products. In: Singh, R.L. and Mondal, S. (eds) *Food Safety and Human Health*. Academic Press, Cambridge, Massachusetts, pp. 145–168. DOI: 10.1016/B978-0-12-816333-7.00006-0.

Dasgupta, A., Majid, F. and Orman, W.H. (2023) The nutritional cost of beef bans in India. *Journal of Development Economics* 163, 103104. DOI: 10.1016/j.jdeveco.2023.103104.

Dhara, D., Biswas, S., Das, S.K. and Biswas, O. (2021) Status of food safety and food security in India in the perspective of FSSAI. *Indian Journal of Animal Health* 60(2-Spl), 167–173. DOI: 10.36062/ijah.2021.spl.01821.

Duchenne-Moutien, R.A. and Neetoo, H. (2021) Climate change and emerging food safety issues: A review. *Journal of Food Protection* 84(11), 1884–1897. DOI: 10.4315/JFP-21-141.

Dyer, J.A., Worth, D.E., Vergé, X.P.C. and Desjardins, R.L. (2020) Impact of recommended red meat consumption in Canada on the carbon footprint of Canadian livestock production. *Journal of Cleaner Production* 266, 121785. DOI: 10.1016/j.jclepro.2020.121785.

Eagle, L., Dahl, S., Pelsmacker, P. and Taylor, C. (eds) (2022) *The SAGE Handbook of Marketing Ethics*. Sage, London. DOI: 10.4135/9781529739725.

Enthoven, L. and Van den Broeck, G. (2021) Local food systems: Reviewing two decades of research. *Agricultural Systems* 193, 103226. DOI: 10.1016/j.agsy.2021.103226.

Ergönül, B. (2013) Consumer awareness and perception to food safety: A consumer analysis. *Food Control* 32(2), 461–471. DOI: 10.1016/j.foodcont.2013.01.018.

FAO (2017) The future of food and agriculture: trends and challenges. *The Future of Food and Agriculture: Trends and Challenges* 4(4), 1951–1960.

FAO (2023) *The State of Food Security and Nutrition in the World 2023*. FAO, Rome, Italy.

Faour-Klingbeil, D. and Todd, E.C.D. (2019) Prevention and control of foodborne diseases in middle-East North African countries: Review of national control systems. *International Journal of Environmental Research and Public Health* 17(1), 70. DOI: 10.3390/ijerph17010070.

FDA (2022) *What You Need to Know about Food-Borne Illnesses*. U.S. Food and Drug Administration, Silver Spring, Maryland.

FSSAI (2016) *Meat and Fish Products*. Available at: https://fssai.gov.in/upload/uploadfiles/files/Manual_Meat_Fish_09_01_2017(1).pdf (accessed 29 July 2024).

FSSAI (2023) *Chapter 2, Food Product Standards, Meat and Meat Products*. Available at: https://www.fssai.gov.in/upload/uploadfiles/files/6_%20Chapter%202_5%20(Meat%20and%20Meat%20products).pdf (accessed 29 July 2024).

Gadekar, Y.P., Das, A.K., Thomas, R., Banerjee, R., Gireesh-Babu, P. *et al.* (2021) Importance of food safety management in meat value chain. *Indian Journal of Animal Health* 60(2-Spl), 119–130. DOI: 10.36062/ijah.2021.spl.01621.

Garcia, S.N., Osburn, B.I. and Jay-Russell, M.T. (2020) One health for food safety, food security and sustainable food production. *Frontiers in Sustainable Food Systems* 4, 1.

Gauci, C. (2019) Food and nutrition policy in Malta. *Reference Module in Food Science*. DOI: 10.1016/B978-0-08-100596-5.22584-5.

Gbashi, S., Adebo, O., Adebiyi, J.A., Targuma, S., Tebele, S. *et al.* (2021) Food safety, food security and genetically modified organisms in Africa: A current perspective. *Biotechnology & Genetic Engineering Reviews* 37(1), 30–63. DOI: 10.1080/02648725.2021.1940735.

Gerardi, A. (2022) Global food safety initiative (GFSI): Underpinning the safety of the global food chain, facilitating regulatory compliance, trade and consumer trust. In: Knowles, M.E., Anelich, L., Boobis, A.R. and Popping, B. (eds) *Present Knowledge in Food Safety: A Risk-Based Approach through the Food Chain*. Elsevier, Amsterdam, Netherlands, pp. 1089–1098. DOI: 10.1016/B978-0-12-819470-6.00058-5.

Gerges, T., Asuoty, E. and Selim, O. (2019) Impact of packaging method on microbial flora and biogenic amines in chicken meat. *Animal Health Research Journal* 7(1), 1–12. DOI: 10.1016/B978-0-12-819470-6.00058-5.

Girish, P.S. and Barbuddhe, S.B. (2020) Meat traceability and certification in meat supply chain. In: *Meat Quality Analysis*. Elsevier, London, pp. 153–170. DOI: 10.1016/B978-0-12-819233-7.00010-0.

Grace, D. (2015) Food safety in low and middle income countries. *International Journal of Environmental Research and Public Health* 12(9), 10490–10507. DOI: 10.3390/ijerph120910490.

Hallum, S.H., Hughey, S.M., Wende, M.E., Stowe, E.W. and Kaczynski, A.T. (2020) Healthy and unhealthy food environments are linked with neighbourhood socio-economic disadvantage: An innovative geospatial approach to understanding food access inequities. *Public Health Nutrition* 23(17), 3190–3196. DOI: 10.1017/S1368980020002104.

Henson, S. and Caswell, J. (1999) Food safety regulation: An overview of contemporary issues. *Food Policy* 24(6), 589–603. DOI: 10.1016/S0306-9192(99)00072-X.

Hu, X., Xu, B., Xiao, Y., Liang, S., Zhang, C. *et al.* (2022) Overview and prospects of food biosafety. *Journal of Biosafety and Biosecurity* 4(2), 146–150. DOI: 10.1016/j.jobb.2022.11.001.

Jena, P.R. (2018) Socio-economic implications of food safety standards in India. *World Food Policy* 4(2), 133–152. DOI: 10.18278/wfp.4.2.8.

Jevšnik, M., Hlebec, V. and Raspor, P. (2008) Consumers' awareness of food safety from shopping to eating. *Food Control* 19(8), 737–745. DOI: 10.1016/j.foodcont.2007.07.017.

Johnson, H. and Parker, C. (2022) An impossible task? Australian food law and the challenge of novel meat analogues. *Federal Law Review* 50(3), 331–370. DOI: 10.1177/0067205X221107411.

Karanth, S., Feng, S., Patra, D. and Pradhan, A.K. (2023) Linking microbial contamination to food spoilage and food waste: The role of smart packaging, spoilage risk assessments, and date labeling. *Frontiers in Microbiology* 14, 1198124. DOI: 10.3389/fmicb.2023.1198124.

Katona, P. and Carus, S. (2010) The history of bioterrorism, biowarfare and biocrimes. In: Katona, P., Sullivan, J.P. and Intriligator, M.D. (eds) *Global Biosecurity: Threats and Responses*. Routledge, New York, pp. 41–60. DOI: 10.4324/9780203860243.

Keener, K.M. (2019) Food regulations. In: Kutz, M. (ed.) *Handbook of Farm, Dairy and Food Machinery Engineering*. Elsevier, Amsterdam, the Netherlands, pp. 15–44. DOI: 10.1016/B978-0-12-814803-7.00002-6.

Konings, E.J.M., Roux, A., Reungoat, A., Nicod, N., Campos-Giménez, E. *et al.* (2021) Challenge to evaluate regulatory compliance for nutrients in infant formulas with current state-of-the-art analytical reference methods. *Food Control* 119, 107423. DOI: 10.1016/j.foodcont.2020.107423.

Kumar, A., Mangla, S.K., Kumar, P. and Karamperidis, S. (2020) Challenges in perishable food supply chains for sustainability management: A developing economy perspective. *Business Strategy and the Environment* 29(5), 1809–1831. DOI: 10.1002/bse.2470.

Kumar, V., Kaur, J., Sharma, K., Kumar, S. and Sharma, R. (2022) Role of analytical techniques in food quality control and safety. In: Chandra, P. and Panesar, P.S. (eds) *Nanosensing and Bioanalytical Technologies in Food Quality Control*. Springer, Singapore, pp. 319–356. DOI: 10.1007/978-981-16-7029-9_12.

Lee, H. and Yoon, Y. (2021) Etiological agents implicated in foodborne illness world wide. *Food Science of Animal Resources* 41(1), 1–7. DOI: 10.5851/kosfa.2020.e75.

Lee, J.G., Lee, Y., Kim, C.S. and Han, S.B. (2021) Codex alimentarius commission on ensuring food safety and promoting fair trade: harmonization of standards between Korea and Codex. *Food Science and Biotechnology* 30(9), 1151–1170. DOI: 10.1007/s10068-021-00943-7.

Lee, S.Y., Lee, D.Y., Jeong, J.W., Kim, J.H., Yun, S.H. *et al.* (2023) Current technologies, regulation, and future perspective of animal product analogs - a review. *Animal Bioscience* 36(10), 1465–1487. DOI: 10.5713/ab.23.0029.

LeJeune, J.T., Zhou, K., Kopko, C. and Igarashi, H. (2021) FAO/WHO joint expert meeting on microbiological risk assessment (JEMRA): Twenty years of international microbiological risk assessment. *Foods* 10(8), 1873. DOI: 10.3390/foods10081873.

Lenart, I., Lakner, Z., Kovacs, L. and Kasza, G. (2022) Consumers across cultures: A comparative study about consumers' role in food safety-related academic literature. *British Food Journal* 124(12), 4610–4625. DOI: 10.1108/BFJ-09-2021-1026.

Li, D., Zang, M., Li, X., Zhang, K., Zhang, Z. *et al.* (2020) A study on the food fraud of national food safety and sample inspection of China. *Food Control* 116, 107306. DOI: 10.1016/j.foodcont.2020.107306.

Liu, P. and Kwon, J. (2013) The exploration of effects of chinese cultural values on the attitudes and behaviors of Chinese restaurateurs toward food safety training. *Journal of Environmental Health* 75(10), 38–46.

Losasso, C., Cibin, V., Cappa, V., Roccato, A., Vanzo, A. *et al.* (2012) Food safety and nutrition: Improving consumer behaviour. *Food Control* 26(2), 252–258. DOI: 10.1016/j.foodcont.2012.01.038.

Mahajan, R., Garg, S. and Sharma, P.B. (2014) Global food safety: Determinants are codex standards and WTO's SPS food safety regulations. *Journal of Advances in Management Research* 11(2), 176–191. DOI: 10.1108/JAMR-01-2013-0007.

Maldonado-Simán, E., González-Ariceaga, C.C., Rodríguez-de Lara, R. and Fallas-López, M. (2018) Potential hazards and biosecurity aspects associated with food safety. In: Holban, A.M. and Grumezescu, A.M. (eds) *Food Control and Biosecurity*. Elsevier, London, pp. 25–61. DOI: 10.1016/B978-0-12-811445-2.00002-7.

Manning, L. and Soon, J.M. (2016) Food safety, food fraud and food defense: A fast evolving literature. *Journal of Food Science* 81(4), R823–R834. DOI: 10.1111/1750-3841.13256.

Micheli, M.R., Rossi, A., Rossi, G., Rosamilia, A. and Guidi, E. (2019) Farm products' direct sale in accordance with national and EC regulations. *Italian Journal of Food Safety* 8(1), 7119. DOI: 10.4081/ijfs.2019.7119.

Miley, K. (2020) Global health biosecurity in a vulnerable world—an evaluation of emerging threats and current disaster preparedness strategies for the future. In: Masys, A.J., Izurieta, R. and Reina Ortiz, M. (eds) *Global Health Security. Advanced Sciences and Technologies for Security Applications*. Springer, Cham, Switzerland, pp. 79–102. DOI: 10.1007/978-3-030-23491-1_5.

Misiou, O. and Koutsoumanis, K. (2022) Climate change and its implications for food safety and spoilage. *Trends in Food Science & Technology* 126, 142–152. DOI: 10.1016/j.tifs.2021.03.031.

Montgomery, H., Haughey, S.A. and Elliott, C.T. (2020) Recent food safety and fraud issues within the dairy supply chain (2015-2019). *Global Food Security* 26, 100447. DOI: 10.1016/j.gfs.2020.100447.

Motarjemi, Y. and Warren, B.R. (2023) Hazard analysis and critical control point system (HACCP). In: Andersen, V., Lelieveld, H.L.M. and Motarjemi, Y. (eds) *Food Safety Management*. Elsevier, London, pp. 799–818. DOI: 10.1016/B978-0-12-820013-1.00017-6.

Mukherjee, R., Vidic, J., Auger, S., Wen, H.-C., Pandey, R.P. *et al.* (2023) Exploring disease management and control through pathogen diagnostics and one health initiative: A concise review. *Antibiotics* 13(1), 17. DOI: 10.3390/antibiotics13010017.

Nagel-Alne, G.E., Murphy, E., McCauslin, B., Hauge, S.J., Schrøder-Petersen, D.L. *et al.* (2022) Meat safety legislation and its opportunities and hurdles for innovative approaches: A review. *Food Control* 141, 109160. DOI: 10.1016/j.foodcont.2022.109160.

Nayak, R. and Waterson, P. (2019) Global food safety as a complex adaptive system: Key concepts and future prospects. *Trends in Food Science & Technology* 91, 409–425. DOI: 10.1016/j.tifs.2019.07.040.

Nguyen, T.T.B. and Li, D. (2022) A systematic literature review of food safety management system implementation in global supply chains. *British Food Journal* 124(10), 3014–3031. DOI: 10.1108/BFJ-05-2021-0476.

Overbosch, P. and Blanchard, S. (2023) Principles and systems for quality and food safety management. In: Andersen, V., Lelieveld, H.L.M. and Motarjemi, Y. (eds) *Food Safety Management*. Elsevier, London, pp. 497–512. DOI: 10.1016/B978-0-12-820013-1.00018-8.

Owusu-Apenten, R., Vieira, E., Owusu-Apenten, R., Vieira, E., Owusu-Apenten, R. *et al.* (2023) Food safety management, GMP & HACCP. In: *Elementary Food Science*. Food Science Text Series, Springer, Cham, Switzerland, pp. 217–236. DOI: 10.1007/978-3-030-65433-7_10.

Pandita, G., Bhosale, Y.K. and Choudhary, P. (2023) Sous vide: A proposition to nutritious and superior quality cooked food. *ACS Food Science & Technology* 3(4), 592–599. DOI: 10.1021/acsfoodscitech.2c00393.

Patel, D., Oliver, S.P., Almeida, R.A. and Vedamuthu, E.R. (2015) Management systems for safety and quality. In: Chandan, R.C., Kilara, A. and Shah, N.P. (eds) *Dairy Processing and Quality Assurance*. Wiley, Chichester, UK, pp. 553–599. DOI: 10.1002/9781118810279.

Patil, S.M., Ramu, R., Shirahatti, P.S., Prasad, M.N.N., Kumari, V.B.C, *et al.* (2023) Global food safety and quality: A worldwide scenario. In: Islam, S., Thangadurai, D., Sangeetha, J. and Chowdhury, Z.Z. (eds) *Food Microbial and Molecular Biology: From Fundamentals to Applications*. Apple Academic Press, New York, pp. 3–34.

Peck, H. (1914) Private slaughter-houses, public abattoirs and freibanks. *Journal of the Royal Sanitary Institute* 35(1), 62–68.

Qiu, Y., Guitian, J., Webster, J.P., Musallam, I., Haider, N. *et al.* (2023) Global prioritization of endemic zoonotic diseases for conducting surveillance in domestic animals to protect public health. *Philosophical Transactions of the Royal Society of London. Series B, Biological Sciences* 378(1887), 20220407. DOI: 10.1098/rstb.2022.0407.

Redmond, E.C. and Griffith, C.J. (2005) Consumer perceptions of food safety education sources: Implications for effective strategy development. *British Food Journal* 107(7), 467–483. DOI: 10.1002/9781118810279.

Riccaboni, A., Neri, E., Trovarelli, F. and Pulselli, R.M. (2021) Sustainability-oriented research and innovation in 'farm to fork' value chains. *Current Opinion in Food Science* 42, 102–112. DOI: 10.1016/j.cofs.2021.04.006.

Rogala, P. and Wawak, S. (2021) Quality of the ISO 9000 series of standards-perceptions of quality management experts. *International Journal of Quality and Service Sciences* 13(4), 509–525.

Rudra, S.G., Basu, S. and Chanda, A. (2022) Editorial: Food storage, spoilage and shelf life: Recent developments and insights. *Frontiers in Sustainable Food Systems* 6, 953983.

Sen, A. (1982) Poverty and famines: An essay on entitlement and deprivation. *Population and Development Review* 8(2), 418.

Sgarro, M.F., Maggiolino, A., Forte, L., Matera, R., Bifulco, G. *et al.* (2024) Effect of dietary administration of red orange and lemon extract on volatile compounds: Profile and sensory parameters of lamb meat. *Italian Journal of Animal Science* 23(1), 101–113.

Sharman, N., Wallace, C.A. and Jespersen, L. (2020) Terminology and the understanding of culture, climate, and behavioural change – impact of organisational and human factors on food safety management. *Trends in Food Science & Technology* 96, 13–20. DOI: 10.1016/j.tifs.2019.12.005.

Shukla, S., Shankar, R. and Singh, S.P. (2014) Food safety regulatory model in India. *Food Control* 37, 401–413. DOI: 10.1016/j.foodcont.2013.08.015.

Sithole, M.I., Bekker, J.L. and Mukaratirwa, S. (2020) Consumer knowledge and practices to pork safety in two taenia solium cysticercosis endemic districts in eastern cape province of South Africa. *BMC Infectious Diseases* 20(1), 107. DOI: 10.1186/s12879-020-4839-9.

Smith, A.M., Tau, N.P., Smouse, S.L., Allam, M., Ismail, A. *et al.* (2019) Outbreak of *listeria monocytogenes* in South Africa, 2017-2018: laboratory activities and experiences associated with whole-genome sequencing analysis of isolates. *Foodborne Pathogens and Disease* 16(7), 524–530. DOI: 10.1089/fpd.2018.2586.

Smith DeWaal, C., Roberts, C. and Plunkett, D. (2021) The legal basis for food safety regulation in the USA and EU. In: Morris, J.G. and Vugia, D.J. (eds) *Food-Borne Infections and Intoxications*. Elsevier, London, pp. 563–581. DOI: 10.1016/B978-0-12-819519-2.00005-0.

Smith, R. and McElwee, G. (2021) The "horse-meat" scandal: Illegal activity in the food supply chain. *Supply Chain Management* 26(5), 565–578. DOI: 10.1108/SCM-08-2019-0292.

Sunder, J. (2019) Religious beef: Dalit literature, bare life, and cow protection in India. *Interventions* 21(3), 337–353. DOI: 10.1080/1369801X.2018.1558097.

Tacoli, C., Fisher, S. and Budoor, B. (2013) *Urban Poverty, Food Security and Climate Change (no.37)*. International Institute for Environment and Development, London.

Togami, E., Gardy, J.L., Hansen, G.R., Poste, G.H., Rizzo, D.M. *et al.* (2018) Core competencies in one health education: What are we missing? *NAM Perspectives* 8(6). DOI: 10.31478/201806a.

Vogt, D.U. (2005) *Food Safety Issues in the 109th Congress*. CRS Report for Congress, Washington, DC.

Waltenburg, M.A., Schwensohn, C., Madad, A., Seelman, S.L., Peralta, V. *et al.* (2022) Two multistate outbreaks of a reoccurring Shiga toxin-producing *Escherichia coli* strain associated with romaine lettuce: USA, 2018–2019 . *Epidemiology and Infection* 150, e16. DOI: 10.1017/S0950268821002703.

Wang, Y., Yuan, Z. and Tang, Y. (2021) Enhancing food security and environmental sustainability: A critical review of food loss and waste management. *Resources, Environment and Sustainability* 4, 100023. DOI: 10.1016/j.resenv.2021.100023.

Warmate, D. and Onarinde, B.A. (2023) Food safety incidents in the red meat industry: A review of foodborne disease outbreaks linked to the consumption of red meat and its products, 1991 to 2021. *International Journal of Food Microbiology* 398, 110240. DOI: 10.1016/j.ijfoodmicro.2023.110240.

WHO (2021a) *Action Framework for Developing and Implementing Public Food Procurement and Service Policies for a Healthy Diet*. WHO: Nutrition and Food Safety, Geneva, Switzerland.

WHO (2021b) *Estimating the Burden of Food-Borne Diseases*. WHO, Geneva, Switzerland.

WHO (2022a) *Food Safety*. WHO, Geneva, Switzerland. DOI: 10.1017/S0950268821002703.

WHO (2022b) *WHO Global Strategy for Food Safety 2022–2030: Towards Stronger Food Safety Systems and Global Cooperation*. WHO, Geneva, Switzerland.

Wieck, C. and Grant, J.H. (2021) Codex in motion: Food safety standard setting and impacts on developing countries' agricultural exports. *EuroChoices* 20(1), 37–47. DOI: 10.1111/1746-692X.12293.

World Bank (2018) *Food-Borne Illnesses Cost US$ 110 Billion Per Year in Low- and Middle-Income Countries*. World Bank, Washington, DC.

Wu, J., Wood, J., Oh, K.Y., Li, Y. and Bhuyan, M.I. (2023) Impact of TBT and SPS measures on domestic value-added exports: Evidence from the United States. *Asia-Pacific Journal of Accounting and Economics* 30(5), 1150–1164.

Yamaguchi, T. and Chiang, S.-N. (2019) Food safety in a global economy: Policies and social issues. In: *Oxford Research Encyclopedia of Environmental Science*. Oxford University Press, Oxford, UK. DOI: 10.1093/acrefore/9780199389414.013.667.

Yera, H., Bory, S., Khieu, V. and Caron, Y. (2022) Human trichinellosis in Southeast Asia, 2001-2021. *Food and Waterborne Parasitology* 28, e00171. DOI: 10.1016/j.fawpar.2022.e00171.

Yu, H.H., Chin, Y.W. and Paik, H.D. (2021) Application of natural preservatives for meat and meat products against food-borne pathogens and spoilage bacteria: A review. *Foods* 10(10), 2418. DOI: 10.3390/foods10102418.
Zhou, M., Zhang, N., Zhang, M. and Ma, G. (2020) Culture, eating behavior, and infectious disease control and prevention. *Journal of Ethnic Foods* 7(1), 40. DOI: 10.1186/s42779-020-00076-y.
Zinsstag, J., Kaiser-Grolimund, A., Heitz-Tokpa, K., Sreedharan, R., Lubroth, J. *et al.* (2023) Advancing one human–animal–environment health for global health security: What does the evidence say? *The Lancet* 401(10376), 591–604. DOI: 10.1016/S0140-6736(22)01595-1.

12 Future of Food Safety: The Way Forward

Satish S. Patil*
Department of Environmental Science, Dr. Babasaheb Ambedkar Marathwada University, Aurangabad, India

Abstract
Food safety is a priority for industry operators, policymakers, and consumers. Keeping food safe has become even more important in recent years as new foodborne illnesses emerge and food chains become increasingly interconnected around the world. This article assesses the future of food security in this context; solutions that can help mitigate the impacts of climate change on the food system and outline an action plan to address current and future challenges. The present paper highlights the importance of working together with stakeholders including academic institutions and governments to develop and implement food security measures. The importance of capacity building and food security training, particularly for resource-poor developing countries, is discussed in the present paper. The requirement for continuous observation and checking to distinguish and forestall foodborne microorganisms before flare-ups start is likewise highlighted.

Keywords: Food safety, Foodborne illnesses, Climate change, Greenhouse gases, Agricultural management

12.1 Introduction

Global climate change is one of utmost international concerns. There is a perception among many that global climate change is simply global warming. In fact, global climate change is an integrated system of several atmospheric phenomena and their products. At the surface, concentrations of greenhouse or radiative gases, such as carbon dioxide, chlorofluorocarbons, methane, and nitrous oxide, have clearly increased since the onset of the Industrial Revolution. These trace gases, in addition to warming the surface air temperature by trapping some of the energy from the outgoing radiation, when transported into the stratosphere (10–15 km above the surface), destroy the beneficial ozone layer naturally present at that height. Such a loss in the thickness or thinning of the ozone layer will permit the increased penetration to the surface of deleterious wavelengths of solar radiation (ultraviolet or UV of 280–315 nm), with much concern for consequent increases in the incidence of melanoma or skin cancer.

Average global air temperature has increased by about 0.8°C above pre-industrial levels, and a 2001 report by the Intergovernmental Panel on Climate Change projected a rise from 1.4°C to 5.8°C by the year 2100. While this prediction is extremely important, there is a significant amount of spatial variability in the air temperature. The local climate in different geographic areas may

*sushshrey@rediffmail.com

DOI: 10.1079/9781800625037.0012

very well become warmer and drier, cooler and wetter, or remain unchanged. Any change in the climate is expected to manifest itself as an increase in the frequency of extreme events such as hurricanes, blizzards, heat waves, and the number of days without precipitation during the plant-growing season (drought). Increases in air temperature can accelerate crop growth and consequently shorten the growth period. In cereal crops, for example, such changes can lead to poor vernalization (e.g. hastened flowering) and reduced yield. There is also evidence that since the 1950s, in North America, China, and the former Soviet Union, night-time temperatures have increased much more than the daytime values (because of increased cloud cover due to air pollution and the formation of hygroscopic, cloud-forming aerosols in the atmosphere), with the potential to reduce reproductive development in many crop species and, consequently, their seed yield.

In developed countries, agriculture is supported by a complex system of research, education, finance, and farm supply, overlying the agricultural potentials of the arable soils. Public policy and agricultural management will attempt to develop strategies for maintaining crop production in areas with the best soils, despite shifts in climate. Of equal or greater importance than the direct effects of rising air temperatures is the indirect effect on the hydrologic cycle, leading to shifts in the dependence on irrigation where water is available. For example, projected drier summers in parts of the USA corn belt will probably result in a shift from the production of corn to grain sorghum. This has serious economic consequences. Currently, the USA ranks first in the world in corn production (41% of the global production), and the area of its cultivation is on the rise, particularly since it is the largest source of ethanol (through fermentation) used as fuel in automobiles.

As much as changes in air temperatures will have an impact on crop production, so will increase in the surface concentrations of carbon dioxide and ozone and elevated levels of ultraviolet-B radiation. Most unfortunately, much of our knowledge of these effects is based on experimental studies directed to changes in one variable at a time. Increases in the carbon dioxide concentrations are expected to increase the biomass in many crop species (particularly C_3 plants such as soybean and rice, in comparison to C_4 plants such as corn and sorghum). However, as seen in the greenhouse production of horticultural crops such as lettuce, increases in biomass may not always translate to an acceptable nutritional quality of the consumed product due to the increases in the accumulation of starch in the foliage at the expense of soluble carbohydrates and decreases in the protein content, unless nitrogen fertilization is practiced. As much as the beneficial effects of increasing carbon dioxide concentrations on crop production are frequently cited, increasing levels of surface ozone (generated from precursor pollutant emissions, primarily from fossil fuel combustion) and ultraviolet-B radiation produce deleterious effects. A rapidly growing number of studies show that the adverse effects of ozone and ultraviolet-B radiation offset the beneficial effects of carbon dioxide. Since all three variables are on the rise (accounting, however, for their spatial variability), questions remain as to the outcome of their joint effects, superimposed on other climate variables such as shifts in air temperature and soil moisture availability and biotic factors (incidence of pathogens and pests).

While plant canopies serve as a sink for carbon dioxide during photosynthesis (daytime), they are a source during the night due to respiration. Furthermore, soils are a larger source of carbon dioxide due to increased microbial respiration and litter or biomass decomposition under increasing temperatures. Equally important agricultural emissions of greenhouse gases are governed by the use of nitrogen fertilizers (increased nitrous oxide emissions from the soil as a greenhouse gas) and animal husbandry (increased ammonia emissions from manure leading to atmospheric nitrogen fertilization by deposition, for example, in precipitation). Those are major concerns in the context of the contribution of agriculture to the climate change issue (warming air temperatures due to the former or shifts in biodiversity due to the latter). In addition, rice paddies are a major source of methane (a greenhouse gas). As populations grow in Asia (by some 29% in 2030), so will the area under rice production to sustain the food supply.

Clearly, the issues associated with deteriorating air quality are a global concern. Satellite data document the spread of aerosols all the way from West Africa to Northeast India to China. India and China rank among the top countries in net human consumption of primary food production. India also ranks at the top in the withdrawal of more than 40% of its total available freshwater supply, in comparison to China (20–40%) and the USA (10–20%). By the year 2050, a majority of the global population is projected to be at a high risk of suffering from water stress. Currently, Asia supports more than half of the global population, with a projected increase of 29% by the year 2030. In comparison, the population is expected to increase in Africa by 65%, North America by 26%, and South America by 31%, and in contrast, a decrease of 6% is predicted for Europe. The increases in the populations of Asia, Africa, and South America are projected to be largely due to the birth rates, while increases in the US population would be mainly due to immigration.

It is very educative to examine the status of the present and projected global agricultural production within these overall scenarios. Currently, the ratio of rural to total populations is for the world 52%, China 65%, India 71%, and the USA 20%. Similarly, the ratio of agricultural to rural populations is for the world 80%, China 94%, India 73%, and the USA 10%. These are telling statistics of the major dependence of the two most populous countries in the world (China and India) and in addition, some others, on intense human labor for food production and sustainability, compared to the US. China, India, and the US account equally for some 40% of the total area in the world under crop production. Yet, the agricultural production in metric tons per capita of the total population is for the world 0.26, China 0.29, India 0.20, and the USA 1.4. The large difference between the statistic for the US and the others is due mainly to the use of complex and highly mechanized and managed crop production systems in the US, compared to the emphasis on manual labor elsewhere. In the US, these types of statistics have led to some adverse, shortsighted consequences. There has been a progressive decline in the number of individual farmers and a converse increase in corporate mega-farming. In addition, fluctuations in commodity prices, the so-called surplus food supply, and imports of foreign plant products have worked against the profitability of individual US farmers, thus the decline in their numbers.

The genetic base of domesticated or cultivated plants is very narrow compared to their wild relatives. Thus, many currently grown crops are considered genetically depauperate. Elite lines and hybrids derived from such germ plasmas are designed to yield well under relatively narrow and well-defined growing conditions. However, superior genotypes are used heavily in various crop-breeding programs, often resulting in considerable relatedness among cultivars grown across large geographic areas, although some cultivars may have a narrower distribution than others.

In general, the narrow genetic base and specific goals used in the breeding of virtually all modern crops make it unlikely that crop breeders will be able to accommodate large and rapid changes in the climate. On the other hand, if climate change occurs gradually, production agriculture will be able to adapt to such changes as long as farm resources are not the limiting factor. Nevertheless, if indirect selection of crops to climate change is practised, it will prove to be an inefficient process. Thus, there will be an increasing need for regions able to sustain high crop production in the future to distribute their surplus food supply to others subjected to adverse impacts, particularly to populations in dire need. That might prove to be a critical determinant of future world populations.

Agriculture is influenced by: (i) increased urbanization, (ii) increasing population migration from rural to urban communities, (iii) farming by increased mechanization and consequently decreased dependence on labor, (iv) increased area under cultivation of crops developed through the application of biotechnology and consequently better disease, pest and weed control, (v) gain of higher yields and (vi) farm consolidation through mergers (mega-corporate farms). In contrast, factors influencing food security in developing countries like India include: (i) increasing population, (ii) rapidly growing urbanization, (iii) decreasing cropland, (iv) decreasing farm resources, (v) continuing crop loss, (vi) declining crop production, and (vii) declining biodiversity.

12.1.1 Climate change and food security

12.1.1.1 Impacts on food production and availability

Climate change affects agriculture and food production in complex ways. It affects food production directly through changes in agro-ecological conditions and indirectly by affecting the growth and distribution of incomes and thus demand for agricultural produce. Impacts have been quantified in numerous studies and under various sets of assumptions. Changes in temperature and precipitation associated with continued emissions of greenhouse gases will bring changes in land suitability and crop yields. In particular, the Intergovernmental Panel on Climate Change (IPCC) considers four families of socioeconomic development and associated emission scenarios, known as the Special Report on Emissions Scenarios (SRES) A2, B2, A1, and B1. Of relevance to this review, of the SRES scenarios, A1, the "business-as-usual scenario," corresponds to the highest emissions, and B1 corresponds to the lowest. The other scenarios are intermediate between these two. Importantly for agriculture and world food supply, SRES A2 assumes the highest projected population growth of the four (United Nations high projection) and is thus associated with the highest food demand. Depending on the SRES emission scenario and climate models considered, the global mean surface temperature is projected to rise in a range from 1.8°C (with a range of 1.1–2.9°C for SRES B1) to 4.0°C (with a range from 2.4–6.4°C for A1) by 2100. In temperate latitudes, higher temperatures are expected to bring predominantly more benefits to agriculture; the areas potentially suitable for cropping will expand, the length of the growing period will increase, and crop yields may rise.

Moderate incremental warming in some humid and temperate grasslands may increase pasture productivity and reduce the need for housing and compound feed. These gains have to be set against an increased frequency of extreme events, for instance, heat waves and droughts in the Mediterranean region or increased heavy precipitation events and flooding in temperate regions, including the possibility of increased coastal storms; they also have to be set against the fact that semi-arid and arid pastures are likely to see reduced livestock productivity and increased livestock mortality.

In drier areas, climate models predict increased evapotranspiration and lower soil moisture levels. As a result, some cultivated areas may become unsuitable for cropping and some tropical grassland may become increasingly arid. Temperature rise will also expand the range of many agricultural pests and increase the ability of pest populations to survive the winter and attack spring crops.

Another important change for agriculture is the increase in atmospheric carbon dioxide (CO_2) concentrations. Depending on the SRES emission scenario, the atmospheric CO_2 concentration is projected to increase from 379 ppm today to >550 ppm by 2100 in SRES B1 to >800 ppm in SRES A1FI. Higher CO_2 concentrations will have a positive effect on many crops, enhancing biomass accumulation and final yield. However, the magnitude of this effect is less clear, with important differences depending on management type (e.g. irrigation and fertilization regimes) and crop type. Experimental yield response to elevated CO_2 shows that under optimal growth conditions, crop yields increase at 550 ppm CO_2 in the range of 10–20% for C_2 crops (such as wheat, rice, and soybean) and only 0–10% for C_1 crops (such as maize and sorghum). Yet, the nutritional quality of agricultural produce may not increase in line with higher yields. Some cereal and forage crops, for example, show lower protein concentrations under elevated CO_2 conditions.

Several recent studies have estimated the likely changes in land suitability, potential yields, and agricultural production on the current suite of crops and cultivars available today. Therefore, these estimates implicitly include adaptation using available management techniques and crops, but excluding new cultivars from breeding or biotechnology. These studies are in essence based on the FAO/International Institute for Applied Systems Analysis (IIASA) agroecological zone (AEZ) methodology. For instance, pioneering work suggested that total land and total prime land would remain virtually unchanged at the current levels of 2600 million and 2000 million ha, respectively. The same study also showed, however, more pronounced regional shifts, with a considerable increase in suitable cropland at higher latitudes (developed countries + 160 million ha) and a corresponding decline of potential cropland at lower latitudes (developing countries 110 million ha). An even more pronounced shift in the quality of cropland is predicted in developing countries. The net decline of 110 million ha is the result of a massive decline in agricultural prime land of 135 million ha, which is offset by an increase in moderately suitable land of >20 million ha.

This quality shift is also reflected in the shift in land suitable for multiple cropping. In sub-Saharan Africa alone, land for double cropping would decline by 10–20 million ha, and land suitable for triple cropping would decline by 5–10 million ha. At a regional level, similar approaches indicate that under climate change, the biggest losses in suitable cropland are likely to be in Africa, whereas the largest expansion of suitable cropland is in the Russian Federation and Central Asia.

12.1.1.2 Impacts on the stability of food supplies

Global and regional weather conditions are also expected to become more variable than at present, with increases in the frequency and severity of extreme events such as cyclones, floods, hailstorms, and droughts. By bringing greater fluctuations in crop yields and local food supplies and higher risks of landslides and erosion damage, they adversely affect the stability of food supplies and thus food security.

Neither climate change nor short-term climate variability and associated adaptation are new phenomena in agriculture, of course. Some important agricultural areas of the world, like the Midwest of the United States, the northeast of Argentina, Southern Africa, southeast Australia, have traditionally experienced higher climate variability than other regions such as Central Africa or Europe. They also show that the extent of short-term fluctuations has changed over longer periods of time. In developed countries, for instance, short-term climate variability increased from 1931 to 1960 as compared with 1901 to 1930 but decreased strongly in the period from 1961 to 1990. What is new, however, is the fact that the areas subject to high climate variability are likely to expand, whereas the extent of short-term climate variability is likely to increase across all regions. Furthermore, the rates and levels of projected warming may exceed in some regions the historical experience.

If climate fluctuations become more pronounced and more widespread, droughts and floods, the dominant causes of short-term fluctuations in food production in semi-arid and subhumid areas, will become more severe and more frequent. In semi-arid areas, droughts can dramatically reduce crop yields and livestock numbers and productivity.

Again, most of this land is in sub-Saharan Africa and parts of South Asia, meaning that the poorest regions with the highest level of chronic undernourishment will also be exposed to the highest degree of instability in food production. How strongly these impacts will be felt will crucially depend on whether such fluctuations can be countered by investments in irrigation, better storage facilities, or higher food imports. In addition, a policy environment that fosters freer trade and promotes investments in transportation, communications, and irrigation infrastructure can help address these challenges early on.

12.1.1.3 Impacts of climate change on food utilization

Climate change will also affect the ability of individuals to use food effectively by altering the conditions for food safety and changing the disease pressure from vector-, water- and foodborne diseases. The IPPC Working Group II provides a detailed account of the health impacts of climate change in its fourth assessment report. It examines how the various forms of diseases, including vectorborne diseases such as malaria, are likely to spread or recede with climate change. This article focuses on a narrow selection of diseases that affect food safety directly (food and waterborne diseases).

The main concern about climate change and food security is that changing climatic conditions can initiate a vicious circle where infectious disease causes or compounds hunger, which in turn makes the affected populations more susceptible to infectious disease. The result can be a substantial decline in labor productivity and an increase in poverty and even mortality. Essentially, all manifestations of climate change, be they drought, higher temperatures, heavy rainfalls, have an impact on disease pressure, and there is growing evidence that these changes affect food safety and food security. The recent IPCC report also emphasizes that increases in daily temperatures will increase the frequency of food poisoning, particularly in temperate regions. Warmer seas may contribute to increased cases of human shellfish and reef-fish poisoning (ciguatera) in tropical regions and a poleward expansion of the disease. However, there is little new evidence that climate change significantly alters the prevalence of these diseases. Several studies have confirmed and quantified the effects of temperature on common forms of food poisoning, such as salmonellosis. These studies show an approximately linear increase in reported cases for each degree increase in weekly temperature. Moreover, there is evidence that temperature variability affects the incidence of diarrheal disease. A number of studies found that rising temperatures were strongly associated with increased episodes of diarrheal disease in adults and children. These findings have been corroborated by analyses based on monthly temperature observations. Several studies report a strong correlation between monthly temperature and diarrheal episodes in the Pacific Islands, Australia, Israel, and India.

Extreme rainfall events can increase the risk of outbreaks of waterborne diseases, particularly where traditional water management systems are insufficient to handle the new extremes. Likewise, the impacts of flooding will be felt most strongly in environmentally degraded areas and where basic public infrastructure, including sanitation and hygiene, is lacking. This will raise the number of people exposed to waterborne diseases (e.g. cholera) and thus lower their capacity to effectively use food.

12.1.1.4 Impacts of climate change on access to food

Access to food refers to the ability of individuals, communities, and countries to purchase sufficient quantities and qualities of food. Over the last 30 years, falling real prices for food and rising real incomes have led to substantial improvements in access to food in many developing countries. Increased purchasing power has allowed a growing number of people to purchase not only more food but also more nutritious food with more protein, micronutrients, and vitamins.

East Asia and to a lesser extent the Near-East/North African region have particularly benefited from a combination of lower real food prices and robust income growth. From 1970 to 2001, the prevalence of hunger in these regions, as measured by FAO's indicator of undernourishment, has declined from 24% to 10.1% and from 44% to 10.2% respectively. In East Asia, it was endogenous income growth that provided the basis for the boost in demand for food, which was largely produced in the region; in the Near-East North African region, demand was spurred by exogenous revenues from oil and gas exports, and additional food supply came largely from imports. But in both regions, improvements in access to food have been crucial in reducing hunger and malnutrition.

FAO's longer-term outlook to 2050 suggests that the importance of improved demand-side conditions will even become more important over the next 50 years. The regions that will see the strongest reductions in the prevalence of undernourishment are those that are expected to see the highest rates of income growth. South Asia stands to benefit the most. Spurred by high-income growth, the region is expected to reduce the prevalence of undernourishment from >22% currently to 12% by 2025 and just 4% by 2050. Progress is also expected for sub-Saharan Africa, but improvements will be less pronounced and are expected to set in later. Over the next 15 years, for instance, the prevalence of undernourishment will decline less than in other regions, from 33% to a still worrisome 21%, as significant constraints (soil nutrients, water, infrastructure, etc.) will limit the ability to further increase food production locally, while continuing low levels of income rule out the option of importing food. In the long run, however, sub-Saharan Africa is expected to see a more substantial decline in hunger; by 2050, <6% of its total population is expected to suffer from chronic hunger. It is important to note that these FAO projections do not take into account the effects of climate change.

By coupling agroecologic and economic models, others have gauged the impact of climate change on agricultural gross domestic product (GDP) and prices. At the global level, the impacts of climate change are likely to be very small; under a range of SRES and associated climate change scenarios, the estimates range from a decline of 1.5% to an increase of 2.6% by 2080. At the regional level, the importance of agriculture as a source of income can be much more important. In these regions, the economic output from agriculture itself (over and above subsistence food production) will be an important contributor to food security. The strongest impact of climate change on the economic output of agriculture is expected for sub-Saharan Africa, which means that the poorest and already most food-insecure region is also expected to suffer the largest contraction of agricultural incomes. For the region, the losses in agricultural GDP, compared with no climate change, range from 2% to 8% for the Hadley Centre Coupled Model (version 3), and 7–9% for the Commonwealth Scientific and Industrial Research Organisation projections.

12.1.1.5 Impacts on food prices

Essentially all SRES development paths describe a world of robust economic growth and a rapidly shrinking importance of agriculture in the long run, and thus, a continuation of a trend that has been underway for decades in many developing regions. SRES scenarios describe a world where income growth will allow the largest part of the world's population to address possible local production shortfalls through imports and, at the same time, find ways to cope with safety and stability issues of food supplies. It is also a world where real incomes rise more rapidly than real food prices, which suggests that the share of income spent on food should decline and that even high food prices are unlikely to create a major dent in the food expenditures of the poor. However, not all parts of the world perform equally well in the various development paths, and not all development paths are equally benign for growth. Where income levels are low and shares of food expenditures are high, higher prices for food may still create or exacerbate a possible food security problem.

Several studies have ventured to measure the likely impacts of climate change on food prices. The basic conclusions that emerge from these studies are that first, on average, food prices are expected to rise moderately in line with moderate increases in temperature (until 2050); some studies even foresee a mild decline in real prices until 2050. Second, after 2050, and with further increases in temperatures, prices are expected to increase more substantially. In some studies and for some commodities (rice and sugar), prices are forecast to increase by as much as 80% above their reference levels without climate change. Third, price changes expected from the effects of global warming are, on average, much smaller than price changes from socioeconomic development paths. For instance, the SRES A2 scenario would imply a price increase in real cereal prices by 170%. The (additional) price increase caused by climate change (in the Hadley Centre Coupled Model, version 3, climate change case) would only be 14.4%. Overall, this appears to be the sharpest price increase reported, and it is not surprising that this scenario would imply a stubbornly high number of undernourished people until 2080. However, it is also needless to say that a constant absolute number of undernourished people would still imply a sharp decline in the prevalence of hunger, and, given the high population assumptions in the SRES A2 world (>13.6 billion people globally and >11.6 billion in the developing world), this would imply a particularly sharp drop in the prevalence from currently 17% to 7% by 2080.

12.1.2 Risks to food security

12.1.2.1 Fossil fuel dependence

While agricultural output increased as a result of the Green Revolution, the energy input into the process (that is, the energy that must be expended to produce a crop) has also increased at a greater rate, so that the ratio of crops produced to energy input has decreased over time. Green Revolution techniques also heavily rely on chemical fertilizers, pesticides, and herbicides, some of which must be developed from fossil fuels, making agriculture increasingly reliant on petroleum products. Between 1950 and 1984, as the Green Revolution transformed agriculture around the globe, world grain production increased by 250%. The energy for the Green Revolution was provided by fossil fuels in the form of fertilizers (natural gas), pesticides (oil), and hydrocarbon-fueled irrigation.

David Pimentel, professor of ecology and agriculture at Cornell University, and Mario Giampietro, senior researcher at the National Research Institute on Food and Nutrition (INRAN), place in their study Food, Land, Population, and the US Economy the maximum US population for a sustainable economy at 200 million. To achieve a sustainable economy and avert disaster, the United States must reduce its population by at least one-third, and the world population will have to be reduced by two-thirds, says the study. The authors of this study believe that the mentioned agricultural crisis will only begin to impact us after 2025 and will not become critical until 2050. The oncoming peaking of global oil production (and subsequent decline of production), along with the peak of North American natural gas production, will very likely precipitate this agricultural crisis much sooner than expected. Geologist Dale Allen Pfeiffer claims that coming decades could see spiraling food prices without relief and massive starvation on a global level such as never experienced before.

However, one should take note that (numbers taken from the CIA World Factbook) the country of Bangladesh achieved food self-sufficiency in 2002 with both a far higher population density than the USA (~1000 inhabitants per square kilometer in comparison to just 30 for the USA, so this is more than 30 times as many), and at only a tiny fraction of the USA's usage of oil, gas, and electricity. Also, pre-industrial Chinese mini-farmers/gardeners developed techniques to feed a population of more than 1000 people per square kilometer. Hence, the dominant problem is not energy availability but the need to stop and revert soil degradation.

In agriculture and animal husbandry, the Green Revolution popularized the use of conventional hybridization to increase yield by creating high-yielding varieties. Often, the handfuls of hybridized breeds originated in developed countries and were further hybridized with local varieties in the rest of the developing world to create high-yield strains resistant to local climate and diseases. Local governments and industry have been pushing hybridization, which has resulted in several of the indigenous breeds becoming extinct or threatened. Through disuse because of unprofitability and uncontrolled intentional and unintentional cross-pollination and crossbreeding (genetic pollution), formerly huge gene pools of various wild and indigenous breeds have collapsed, causing widespread genetic erosion and genetic pollution. This has resulted in a loss of genetic diversity and biodiversity as a whole. A genetically modified organism (GMO) is an organism whose genetic material has been altered using the genetic engineering techniques generally known as recombinant DNA technology. Genetically Modified (GM) crops today have become a common source of genetic pollution, not only of wild varieties but also of other domesticated varieties derived from relatively natural hybridization.

Genetic erosion coupled with genetic pollution may be destroying unique genotypes, thereby creating a hidden crisis that could result in a severe threat to our food security. Diverse genetic material could cease to exist, which would impact our ability to further hybridize food crops and livestock against more resistant diseases and climatic changes.

12.1.2.2 Genetic erosion in agricultural and livestock biodiversity

Genetic erosion in agricultural and livestock biodiversity is the loss of genetic diversity, including the loss of individual genes, and the loss of particular combinants of genes (or gene complexes) such as those manifested in locally adapted landraces of domesticated animals or plants adapted to the natural environment in which they originated. The term "genetic erosion" is sometimes used in a narrow sense, such as for the loss of alleles or genes, as well as more broadly, referring to the loss of varieties or even species. The major driving forces behind genetic erosion in crops are variety replacement, land clearing, overexploitation of species, population pressure, environmental degradation, overgrazing, policy, and changing agricultural systems. The main factor, however, is the replacement of local varieties of domestic plants and animals by high-yielding or exotic varieties or species. A large number of varieties can also often be dramatically reduced when commercial varieties (including GMOs) are introduced into traditional farming systems. Many researchers believe that the main problem related to agroecosystem management is the general tendency toward genetic and ecological uniformity imposed by the development of modern agriculture.

12.1.2.3 *Intellectual property rights*

There is much debate on whether Intellectual Property Rights (IPRs) hurt or harm independent development in terms of agriculture and food production. Hartmut Meyer and Annette von Lossau describe both sides of the issue, while saying "Among scholars, the thesis that the impetus to self-determined development and the protection of intellectual property go hand in hand is disputed—to put it mildly" (Parvathamma, 2015). Many studies have concluded that there is virtually no positive correlation between establishing self-sustained economic growth and ensuring protection of intellectual property rights.

12.1.2.4 *Price setting*

On April 30, 2008, Thailand announced the project of the creation of the Organisation of Rice Exporting Countries with the potential to develop into a price-fixing cartel for rice, treating food the same as other internationally traded commodities. On October 23, 2008, Associated Press reported the following:

Former President Clinton told a UN gathering Thursday (October 16, 2008) that the global food crisis shows "we all blew it, including me" by treating food crops "like color TVs" instead of as a vital commodity for the world's poor. Clinton criticized decades of policymaking by the World Bank, the International Monetary Fund, and others, encouraged by the US, that pressured Africans in particular into dropping government subsidies for fertilizer, improved seed, and other farm inputs as a requirement to get aid. Africa's food self-sufficiency declined, and food imports rose. Now, skyrocketing prices in the international grain trade on average more than doubling between 2006 and early 2008 have pushed many in poor countries deeper into poverty.

12.1.3 Quantifying the impacts on food security

Several studies have recently quantified the impacts of climate change on food security. In terms of quantifying agronomic yield change projections, these studies are either based on the AEZ tools developed by the IIASA analysis or the Decision Support System for Agrotechnology Transfer suite of crop models; all use the IIASA Basic Linked System (BLS) economic model for assessing economic impacts. These tools, with some modifications relating to how crop-yield changes are simulated, have also been used by others to undertake similar assessments and provide sensitivity analyses across a range of SRES and general circulation model (GCM) projections. Many other simulations have also examined the effects of climate change with and without adaptation (induced technological progress, domestic policy change, international trade liberalization, etc.), with and without mitigation (e.g. CO_2 stabilization, variants for temperature, rainfall change, and distribution), or provide impact assessments for different speeds of climate change. This section focuses on the quantitative results for food security, trying to illuminate some of the differences and extract the main messages that emerge from the various studies. Unless indicated, all simulation results discussed below include the combined effects of climate change and elevated CO_2 on crops. The key messages can be summarized as follows:

First, it is very likely that climate change will increase the number of people at risk of hunger compared with reference scenarios with no climate change; the exact impacts will, however, strongly depend on the projected socioeconomic developments. For instance, it is estimated that climate change would increase the number of undernourished in 2080 by 5–26%, compared with no climate change, or by 5–10 million (B1 SRES) and 120–170 million people (A2 SRES), with within-SRES ranges depending on GCM climate projections. Using only one GCM scenario, others projected small reductions by 2080: −5%, or −10 million (B1) to −30 million (A2) people, and slight increases of +13–26%, or 10 million (B2) to 30 million (A1) people.

Second, it is likely that the magnitude of these climate impacts will be small compared with the impact of socioeconomic development. The limitations of socioeconomic forecasts, these studies suggest that robust economic growth and a decline in population growth projected for the 21st century will, in all but one scenario (SRES A2), significantly reduce the number of people at risk of hunger in 2080. At any rate, the prevalence of undernourishment will decline as all scenarios assume that the world population will continue to grow to 2080, albeit at lower rates. Compared with FAO estimates of 820 million undernourished in developing countries today, several studies estimate reductions of >75% by 2080, or 560 million to 700 million people, projecting 100–240 million undernourished by 2080 (A1, B1, and B2). As mentioned, the only exception is scenario A2, where the number of the hungry is forecast to decrease only slightly by 2080; but the higher population growth rates in A2 compared with other scenarios mean that the prevalence of undernourishment will decline drastically. However, these analyses also confirm that progress will be unevenly distributed over the developing world, and more importantly, progress will be slow during the first decades of the outlook. With or without climate change, the millennium development goal of halving the prevalence of hunger by 2015 is unlikely to be realized before 2020–2030.

In addition to socioeconomic pressures considered within the IPCC SRES scenarios, food production may increasingly compete with bioenergy in coming decades; studies addressing possible consequences for world food supply have only started to surface, providing both positive and negative views. Importantly, none of the major world food models discussed herein have yet considered such competition.

Third, sub-Saharan Africa is likely to surpass Asia as the most food-insecure region. However, this is largely independent of climate change and is mostly the result of the socioeconomic development paths assumed for the different developing regions in the SRES scenarios. Throughout most SRES and climate change scenarios, sub-Saharan Africa accounts for 40–50% of global hunger by 2080, compared with 24% today; in some simulations, sub-Saharan Africa even accounts for 70–75% of global undernourishment by 2080. These high estimates have emerged from slower growth variants of the A2 and B2 scenarios; also, an A2 variant with slower population growth yields a sharper concentration of hunger in sub-Saharan Africa. For regions other than sub-Saharan Africa, results largely depend on GCM scenarios and therefore are highly uncertain.

Fourth, although there is significant uncertainty on the effects of elevated CO_2 on crop yields, this uncertainty is carried to a much lesser extent on food security. This result emerges from a comparison of climate change simulations with and without CO_2 fertilization effects on crop yields. As can be seen, higher CO_2 fertilization would not greatly affect global projections of hunger. In view of the fact that essentially all SRES worlds are characterized by much higher real incomes, much-improved transportation, and communication options, and still sufficient global food production, the somewhat smaller supplies will not be able to make a dent in global food security outcomes. Many studies find that climate change without CO_2 fertilization would reduce the number of undernourished people by 2080 only by some 20–140 million (120–380 million for SRES A1, B1, and B2 without, compared with 100–240 million with CO_2 fertilization effect). The exception again in these studies is SRES A2, under which the assumption of no CO_2 fertilization results in a projected range of 950–1300 million people undernourished in 2080, compared with 740–850 million projected under climate change but with CO_2 effects on crops.

Finally, recent research suggests large positive effects of climate stabilization for the agricultural sector. However, as the stabilizing effects of mitigation measures can take several decades to be realized from the moment of implementation, the benefits for crop production may be realized only in the second half of this century. Importantly, even in the presence of robust global long-term benefits, regional and temporal patterns of winners and losers that can be projected with current tools are highly uncertain and depend critically on the underlying GCM projections.

12.1.4 Uncertainties and limitations

The finding that socioeconomic development paths have an important bearing on future food security and that they are likely to top the effects of climate change should not or at least not only be interpreted as a probability-based forecast. This is because SRES scenarios offer a range of possible outcomes "without any sense of likelihood." Yet, SRES scenarios, like all scenarios, do not overcome the inability to accurately project future changes in economic activity, emissions, and climate.

Second, the existing global assessments of climate change and food security have only been able to focus on the impacts on food availability and access to food, without quantification of the likely important climate change effects on food safety and vulnerability (stability). This means that these assessments neither include potential problems arising from additional impacts due to extreme events such as drought and floods, nor do they quantify the potential impacts of changes in the prevalence of foodborne diseases (positive as well as negative) or the interaction of nutrition and health effects through changes in the proliferation of vectorborne diseases such as malaria. On the food availability side, they also exclude the impacts of a possible sea level rise on agricultural production or those that are associated with possible reductions of marine or freshwater fish production.

Third, it is important to note that even in terms of food availability, all current assessments of the world food supply have focused only on the impacts of mean climate change; they have not considered the possibility of significant shifts in the frequency of extreme events on regional production potential, nor have they considered scenarios of abrupt climate or socioeconomic change. Any of these scenario variants is likely to increase the already negative projected impacts of climate change on the world food supply. Models that take into account the specific biophysical, technological, and market responses necessary to simulate realistic adaptation to such events are not yet available.

Fourth, this review finds that recent global assessments of climate change and food security rest essentially on a single modeling framework, the IIASA system, which combines the FAO/IIASA AEZ model with various GCM models and the IIASA BLS system, or close variants of the IIASA system. This has important implications for uncertainty, given that the robustness of all these assessments strongly depends on the performance of the underlying models. There is, therefore, a clear need for continued and enhanced validation efforts of both the agroclimatological and food trade tools developed at IIASA and widely used in the literature.

Finally, it is to be noted that assessments that do not only provide scenarios but also attach probabilities for particular outcomes to come true could provide an important element for improved or at least better-informed policy decisions. A number of possibilities are offered to address the related modeling challenges. One option would be to produce such estimates with probability-based estimates of the (key) model parameters. Alternatively, the various scenarios could be constructed so that they reflect expert judgments on a particular issue. It would be desirable to attach probabilities to existing scenarios. Information on how likely the suggested outcomes are would contribute greatly to their usefulness for policymakers and help justify (or otherwise) policy measures to adapt to or mitigate the impacts of climate change on food security.

Climate change will affect all four dimensions of food security, namely food availability (i.e. production and trade), access to food, stability of food supplies, and food utilization. The importance of the various dimensions and the overall impact of climate change on food security will differ across regions and over time and, most importantly, will depend on the overall socioeconomic status that a country has accomplished as the effects of climate change set in.

Essentially, all quantitative assessments show that climate change will adversely affect food security. Climate change will increase the dependency of developing countries on imports and accentuate the existing focus on food insecurity in sub-Saharan Africa and to a lesser extent in South Asia. Within the developing world, the adverse impacts of climate change will fall disproportionately on the poor. Many quantitative assessments also show that the socioeconomic environment in which climate change is likely to evolve is more important than the impacts that can be expected from the biophysical changes of climate change.

Less is known about the role of climate change on food stability and utilization, at least in quantitative terms. However, it is likely that differences in socioeconomic development paths will also be the crucial determinant for food utilization in the long run and that they will be decisive for the ability to cope with problems of food instability, be they climate-related or caused by other factors.

Finally, all quantitative assessments we reviewed show that the first decades of the 21st century are expected to see low impacts of climate change but also lower overall incomes and still a higher dependence on agriculture. During these first decades, the biophysical changes as such will be less pronounced, but climate change will affect those particularly adversely who are still more dependent on agriculture and have lower overall incomes to cope with the impacts of climate change. By contrast, the second half of the century is expected to bring more severe biophysical impacts but also a greater ability to cope with them. The underlying assumption is that the general transition in income formation away from agriculture toward non-agriculture will be successful.

How strongly the impacts of climate change will be felt over all decades will crucially depend on the future policy environment for the poor. Freer trade can help to improve access to international supplies; investments in transportation and communication infrastructure will help provide secure and timely local deliveries; irrigation, promotion of sustainable agricultural practices, and continued technological progress can play a crucial role in providing steady local and international supplies under climate change.

12.1.5 Food security, food systems, and the link to climate

There are several definitions of what constitutes food systems each formulated in relation to a specific range of issues. For GECAFSS purposes, food systems are defined as a set of dynamic interactions between and within the biogeophysical and human environments which result in the production, processing, distribution, preparation, and consumption of food. They encompass components of: (i) food availability (with elements related to production, distribution, and exchange), (ii) food access (with elements related to affordability, allocation, and preference), and (iii) food utilization. Food systems, then, involve much broader considerations than productivity and production alone. They underpin food security, which is the state achieved when food systems operate such that "all people, at all times, have physical and economic access to sufficient, safe and nutritious food to meet their dietary needs and food preferences for an active and healthy life" (FAO, 1996). Food security is diminished when food systems are stressed. This can be caused by a range of factors in addition to climate and other environmental changes (e.g. conflict, changes in international trade agreements and policies, HIV/AIDS) and may be particularly severe when these factors act in combination. Access to culturally acceptable food by individuals and communities, or means for its procurement, is increasingly being elaborated into human rights legislation.

Food systems may be simple, as in the case of a subsistence farmer who produces, processes, and consumes food on a farm. However, there are comparatively few individuals or households in the world that are completely self-reliant for food throughout the year, and in almost all cases, there is an element of bartering, exchange, or the cash economy to bring food into the household.

In many places, the food system has changed radically in the last century and continues to become increasingly complex. The intensification of agricultural production since the 1940s has been accompanied by profound changes in the organization of food systems around the world, including changes in distribution, marketing, affordability, and preferences for particular food items. These changes are especially obvious in the USA and Europe, where market globalization has occurred with global sourcing of products by retailers directly from producers in the case of fresh fruit and vegetables and from a few large manufacturers of other food products. Such changes have also shifted economic and political power from farmers to retailers, from national legislative bodies to regional and global organizations, and from the state to multinational corporations. For example, in a study of selected OECD countries in Europe, Grievink (2003) determined that the food chain has some 160 million consumers of whom about 3.2 million are also farmers or food producers, but the link between these groups is increasingly determined by the small number of food processors/manufacturers (about 90,000) and the even smaller number of buyers (about 100) for the supermarket chains. Components of the food chain in selected OECD countries of Europe indicate the two inverted pyramid structures that relate farmers to consumers (Grievink, 2003).

Food systems around the world are changing very rapidly as urbanization and globalization proceeds apace. The urbanization of many predominantly rural countries in the last three decades has been accompanied by the rapid growth of supermarkets in many, often accompanied by foreign investment by global retail chains. However, even in poor countries, such as Kenya, where per capita gross domestic product (GDP) was less than US$400 per year in 2002, supermarkets have grown from a tiny niche market in 1997 to be greater than 20% of urban food retailing today. This growth has been a consequence of three major factors. First, rapid urbanization has seen the proportion of the population based in urban areas grow from 0.13 in 1975 to 0.36 in 2000, with an expectation of greater than 0.5 in 2013; second, trade and domestic market liberalization since 1993 has seen the removal of import controls and the deregulation of prices; and third, movement toward price-based competition between the indigenous chains. Success within Kenya is now spreading to other East African countries, with important effects on the market conditions faced by farmers, including the decline of traditional wholesalers (and the smallholder producers from whom they buy) and the increase in direct purchases from larger farms.

These marked changes in access and utilization of food around the world provide a context in which to evaluate the likely effects of climate and other environmental changes on crop production and food security. Not all food systems or parts of food systems are equally vulnerable to environmental changes, because the capacity to cope with existing variability in biophysical and socioeconomic systems, and the ability of humans to perceive environmental changes and to adapt food systems, differs. Human vulnerability includes both the likelihood of exposure to stresses and the capacity to cope with such stresses (i.e. sensitivity). Vulnerability and poverty are often interrelated because both the likelihood of exposure to stresses is greater among the poor and because a large proportion of their resources are spent either purchasing or producing food, thereby reducing their capacity to cope with perturbations (Hume *et al.*, 2001). Moreover, pursuit of food security frequently involves trade-offs with expenditure on health and still further reducing the ability to improve longer-term living conditions or resilience to stress and shock. Food insecurity is experienced at a range of spatial scales from individual households to regions, as well as a range of timescales, and reductions of vulnerability at one scale do not automatically flow to the next scale. For example, regional policy decisions do not always convert to successful local implementation, especially if distribution services, are inadequate or food preferences are ignored (Ellis, 2003).

Climate variation is one of several interacting factors that affect food security. For example, in studies of household food security in Southern Africa, climate/environment was only one of some 33 drivers mentioned as important by householders (Misselhorn, 2005). The mix of drivers varied across the region, but in all communities, many interacting factors resulted in vulnerability to food shortages. Overall, however, climate/environment was one of the seven factors influencing food security that were frequently cited, because of its role both as an ongoing issue (57% of cases, where it was mentioned) and as a "shock" (43%). The impacts of sudden shocks such as drought are felt, then, on top of ongoing long-term stresses, and the low ability to cope with such shocks and to mitigate long-term stresses means that the employment of coping strategies that might be available to others is at a too high cost or, simply, unavailable.

Typically, reliance on purchased food increases in drought years due to losses in food production leading to an increase in poverty due to the synergistic action of other drivers such as rising food prices and unemployment. Biggs *et al.* (2004) record that the food security crisis in Southern Africa in 2002–2003 was not simply a result of drought alone and, indeed, climatic stress was not as severe as in previous crises. Maize production during the preceding growing season was only 5.5% less than the previous 5-year average, so food stocks at the start of the climatic shock were not unusually low. It was concluded that the crisis was indicative of entrenched vulnerability resulting from a range of regional and global political and economic factors (Vogel and Smith, 2002) including high food prices, legacies of structural adjustment, government policies, conflict and war, policies on genetically modified foods, and poor responses to the HIV/AIDS pandemic. The key here was that conditions that weakened food systems in the region were already in place. The moderate climatic shock intensified food insecurity and the long-term vulnerability of the region.

A cycle of deep-rooted socioeconomic instability that includes unemployment, exclusion from markets, low levels of education, poverty, and vulnerability cause food insecurity, although the drivers and their effects differ throughout the region. Vulnerability is a result of numerous interrelated factors in every community. The two most significant contributing factors to food shortages are poverty and climate. The effects of unexpected shocks, like drought or a change in access to resources, are felt on top of them.

12.1.6 Food security and the risks of a changing climate

The growth in agriculture this century is unprecedented due to the development of new technologies to increase crop yields. However, climate variability causes considerable fluctuations in crop yields and productivity. On top of that, day-to-day variations in weather that manifest themselves in the form of hurricanes and typhoons, floods, and dry spells often lead to mass displacement of populations and cause damage to food production systems, resulting in food shortages and famine. Beyond regional, seasonal, and year-to-year climate variability, global warming is predicted to strongly affect agricultural production in some areas, especially in the tropics and subtropics.

There are potentially serious consequences on a regional scale for food security because of climate variability and change. Many models also predict a considerable increase in extreme weather conditions as a consequence of climate change. There may be an increased risk of hunger and famine in some locations. Many of the world's poorest people are most at risk, especially those living in subtropical and tropical areas, and dependent on food production systems in semi-arid and arid regions. These areas are found in sub-Saharan Africa; south, east, and south-east Asia; tropical parts of Latin America; and some Pacific Island nations. Atmospheric and oceanic conditions determine the breeding grounds for fish generally and fish production overall. Globally, climate change is unlikely to affect marine fish production but will probably produce changes at the regional level. Principal changes will be longer growing seasons and lower winter fish mortality. Oceanic changes could lead to migrations in species and affect migration routes, with subsequent changes in fish stocks in the economic zones of some nations. The threat of climate change over decades to centuries caused by human-induced (anthropogenic) increases in greenhouse gases in the atmosphere is a matter of global concern. This concern is being addressed by the UN Framework Convention on Climate Change through which nations have committed themselves to achieve stabilization of greenhouse gas concentrations in the atmosphere "at a level that will prevent dangerous anthropogenic interference with the climate system."

12.1.7 The impact of climate variability on food production

Climate variability has always had a significant impact on food production. Reliable climate records have been kept for many regions of the globe only over the last 100 years. Around the mean global surface temperature increase of between 0.3°C and 0.6°C since the late 19th century, considerable variability occurs. About 20–30% of the temperature variability from the global trend, especially in the Southern Hemisphere, is linked to the Southern Oscillation, an irregular variation in the tropical oceans and atmosphere between the Pacific and Indo-Australian areas. Rainfall in the Sahel from the late 1960s to 1993 was well below amounts received earlier in the century, about half that of the wet 1950s.

Food production is well-adjusted to the mean climatic conditions of the region and can cope with moderate variations. However, climatic variability creates hazards to which agricultural ecosystems are not well adapted. High temperatures exacerbate the effects of drought, damage crops, and their establishment, and reduce yields. Low temperatures, frost, and heavy snowfall curtail yields and destroy crops. Heavy rainfall and flooding cause inundation and waterlogging of productive land. Drought directly impacts agricultural yields by reducing plant growth and animal production. High winds cause mechanical damage to crops, soil erosion, and land degradation. A phenomenon that is related to the occasional warming of the eastern part of the Pacific Ocean is the El Niño/ Southern Oscillation. El Niño events affect the global climate, particularly in the Southern Hemisphere.

12.1.7.1 The toll of chronically poor and erratic rainfall

Although catastrophic weather phenomena receive wide media coverage, it should be noted that far more agricultural production is lost in pockets of chronically poor rainfall or other adverse weather during otherwise average years. In addition, climate is characterized by persistence—a tendency for good or bad years to occur in runs, which highlights the need for storage of food and particularly affects livestock. Finally, due to population growth, agriculture had to expand more and more into marginal land, including those with soils with poor water storage capacity. Such areas are more vulnerable to irregular rainfall, resulting in more frequent drought even under relatively stable climatic conditions. These even further increase vulnerability to climate change.

12.1.7.2 Adaptation strategies

The impacts of climate variability and change require a number of strategies to make food security more robust:

- New varieties
- Adaptation of new crops
- Efficient land management
- Minimum and reduced tillage technologies
- Control and prevention of land degradation and desertification
- Mitigation of the effects of drought
- Improved seasonal climate prediction
- Fisheries stock assessment and management
- Preservation and restoration of wetlands and estuaries.

12.1.8 Global warming and future climate

During the past century, human activities have changed the atmospheric concentrations and distributions of greenhouse gases and aerosols. Concentrations of carbon dioxide, methane, and nitrous oxide have grown by about 30%, 145%, and 15% respectively. These increases have been from the burning of fossil fuels, land-use changes, and agriculture. Greenhouse gases have a warming effect on surface temperatures. In contrast, tropospheric aerosols, resulting from fossil fuel combustion and burning of biomass, have a cooling effect on climate. The balance of evidence now suggests a discernible human influence on global climate, and that the warming effect of greenhouse gases outweighs the cooling effect of aerosols.

It is now believed that global temperatures could increase at a faster rate than they have up to now. Results of advanced computer models project a best estimate increase of global mean surface air temperatures by about 2°C, relative to 1990, by 2100. General warming is expected to lead to an increase in high-temperature extremes, with decreases in extremes of cold with repercussions on the hydrological cycle with changes in the occurrence of severe droughts and floods. Most climate models indicate an increase in rainfall intensity leading to more extreme rainfall events.

12.2 Conclusions

In conclusion, for the future of food safety, a multidisciplinary strategy integrating research, technology development, government regulations, and consumer education is essential. In order to keep food safe and quality, there is a need for continuous innovation and adaptation against the ever-increasing complexity of global food supply chains as well as the introduction of new pathogens. To identify and decrease food handling hazards, research in microbial science, genetic sequencing, and food handling advances is required. Moreover, to consider food makers and providers responsible for arriving at more prominent food handling guidelines, creative administrative systems and consistence strategies should be laid out. By collaborating among all entertainers in the pecking order, food management can move toward a future where food management is seen as an essential right and need of all consumers. Intelligent endeavors to advance development, emphasize guidelines and raise consumer mindfulness will eventually be gainful for the security of food. Later on, we can guarantee a more secure and more reasonable food supply through this practice.

References

Biggs, R., Scholes, R.J. and Reyers, B. (2004) *Assessing Biodiversity Intactness at Multiple Scales*. Bridging Scales and Epistemologies Conference, Alexandria, Egypt.

Ellis, R. (2003) *Task-Based Language Learning and Teaching*. Oxford University Press, Oxford, UK.

FAO (1996) *The State of Food and Agriculture*. Food and Agriculture Organization of the United Nations, Rome, Italy.

Grievink, J.W. (2003) The changing face of the global food industry. In: *Conference Presentation at Changing Dimensions of the Food Economy: Exploring the Policy Issues Edition*, OECD Conference, The Hague, the Netherlands.

Hume, E., Norval, S. and Jeroen, W. (2001) Surface syllable structure and segment sequencing. *HIL Occasional Papers* 4(1), 1–25.

Misselhorn, A.A. (2005) What drives food insecurity in Southern Africa? A meta-analysis of household economy studies. *Global Environmental Change* 15(1), 33–43. DOI: 10.1016/j.gloenvcha.2004.11.003.

Parvathamma, G.L. (2015) An assessment on food security in developing economies problems and policy initiatives. *OSR Journal of Economics and Finance* 6(1), 01–10.

Vogel, C. and Smith, J. (2002) The politics of scarcity: Conceptualising the current food security crisis in Southern Africa. *South African Journal of Science* 98(7–8), 315–317.

Index

Note: Page numbers in **bold** type refer to **figures.**
Page numbers in *italic* type refer to *tables.*

Printed and bound by CPI Group (UK) Ltd, Croydon, CR0 4YY
12/12/2024
14611049-0001